Springer Complexity

Springer Complexity is an interdisciplinary program publishing the best research and academic-level teaching on both fundamental and applied aspects of complex systems – cutting across all traditional disciplines of the natural and life sciences, engineering, economics, medicine, neuroscience, social and computer science.

Complex Systems are systems that comprise many interacting parts with the ability to generate a new quality of macroscopic collective behavior the manifestations of which are the spontaneous formation of distinctive temporal, spatial or functional structures. Models of such systems can be successfully mapped onto quite diverse "real-life" situations like the climate, the coherent emission of light from lasers, chemical reaction-diffusion systems, biological cellular networks, the dynamics of stock markets and of the internet, earthquake statistics and prediction, freeway traffic, the human brain, or the formation of opinions in social systems, to name just some of the popular applications.

Although their scope and methodologies overlap somewhat, one can distinguish the following main concepts and tools: self-organization, nonlinear dynamics, synergetics, turbulence, dynamical systems, catastrophes, instabilities, stochastic processes, chaos, graphs and networks, cellular automata, adaptive systems, genetic algorithms and computational intelligence.

The two major book publication platforms of the Springer Complexity program are the monograph series "Understanding Complex Systems" focusing on the various applications of complexity, and the "Springer Series in Synergetics", which is devoted to the quantitative theoretical and methodological foundations. In addition to the books in these two core series, the program also incorporates individual titles ranging from textbooks to major reference works.

Understanding Complex Systems

Founding Editor: J.A. Scott Kelso

Future scientific and technological developments in many fields will necessarily depend upon coming to grips with complex systems. Such systems are complex in both their composition – typically many different kinds of components interacting simultaneously and nonlinearly with each other and their environments on multiple levels – and in the rich diversity of behavior of which they are capable.

The Springer Series in Understanding Complex Systems series (UCS) promotes new strategies and paradigms for understanding and realizing applications of complex systems research in a wide variety of fields and endeavors. UCS is explicitly transdisciplinary. It has three main goals: First, to elaborate the concepts, methods and tools of complex systems at all levels of description and in all scientific fields, especially newly emerging areas within the life, social, behavioral, economic, neuro- and cognitive sciences (and derivatives thereof); second, to encourage novel applications of these ideas in various fields of engineering and computation such as robotics, nano-technology and informatics; third, to provide a single forum within which commonalities and differences in the workings of complex systems may be discerned, hence leading to deeper insight and understanding.

UCS will publish monographs, lecture notes and selected edited contributions aimed at communicating new findings to a large multidisciplinary audience.

Leonid I. Perlovsky · Robert Kozma
Editors

Neurodynamics of Cognition and Consciousness

With 139 Figures and 17 Tables

Editors
Leonid I. Perlovsky
School of Engineering and Applied Sciences
Harvard University
33 Oxford St.
Cambridge MA 02138, USA
leonid@hrl.harvard.edu

Robert Kozma
Computational Neurodynamics Laboratory
373 Dunn Hall
Department of Computer Science
University of Memphis
Memphis, TN 38152, USA
rkozma@memphis.edu

Library of Congress Control Number: 2007932441

ISSN 1860-0832
ISBN 978-3-540-73266-2 Springer Berlin Heidelberg New York

Springer is a part of Springer Science+Business Media
springer.com

Typesetting: Integra Software Services Pvt. Ltd., India

Cover design: WMX, Heidelberg

Printed on acid-free paper SPIN: 11820758 89/3180/Integra 5 4 3 2 1 0

Preface

Higher cognition and consciousness became the objective of intensive scientific studies and often spirited debates in the past decades. The brain has immensely complex structure and functionality, while our understanding of its operation is yet in its infancy. Some investigators declare that higher-level cognition and consciousness are ultimately beyond the realms of human scientific understanding and reasoning, and research into these areas is a futile exercise. Yet, the knowledge instinct deeply rooted in human activity drives researchers toward exploring the mysteries of human thought. These efforts lead to the development of increasingly sophisticated scientific methods of understanding the operation of brains and the very nature of consciousness. The studies rely on the immense amount of experimental data becoming available as the result of advanced brain monitoring techniques. The experimental techniques require, in turn, the development of novel methods of theoretical analysis and interpretation tools. Our work is part of these endeavors.

This volume presents an in-depth overview of theoretical and experimental approaches to cognitive dynamics and applications in designing intelligent devices. The research efforts leading to this book have started about 2 years ago when we embarked on organizing a series of special and invited sessions which took place in July 2006 at the *World Congress of Computational Intelligence WCCI2006,* in beautiful Vancouver, Canada. The talks covered various aspects of neurodynamics of higher-level cognitive behavior, consciousness, and modeling emotions.

The main theme of this volume is the dynamics of higher cognitive functions. The authors of this volume provide a line of arguments explaining why dynamics plays central role in intelligence in biological brains and in man made artifacts? Researches too often make the mistake of identifying intelligence with the projection of intelligence into a brilliant symbolic form, whereas intelligence is the unity of the path (or dynamic trajectory) leading to the observed formalisms and the intermitted appearance of the formalism itself. Intelligence can be understood only in the context of the complementary nature as it is described by Kelso and collaborators. Neurodynamics has a peculiar property described as the "edge of stability" or "metastability." Accordingly, the brain as a complex dynamic system is in perpetual movement from one state to another. When the brain reaches a dominant state, it does not rests there, rather it immediately moves on and decays into an unordered state, only to emerge a moment later to another prominent state. Freeman has identified neurophysiologic correlates of this metastable wandering along the landscape of brain dynamics in the form of spatio-temporal patterns of oscillations, sudden jumps or phase transitions of local field potentials.

This book explores various aspects of the neurodynamics of metastable cognitive states. It covers a wide range of research areas related to dynamics of cognition, including experimental studies, dynamical modeling and interpretation of cognitive experiments, and theoretical approaches. Spatio-temporal structures of neural activity and synchronization are manifested as propagating phase cones or phase boundaries over the cortex. Methods to detect, identify and characterize such transient structures are described. Identification of transients is a very hard pattern recognition problem as the simultaneous overlapping dynamical processes provide a noisy and cluttered domain. Advanced techniques of dynamical logic progress from vague concepts to increasingly crisp and articulate forms, which is a promising approach to detect the required complex spatio-temporal correlates of cognitive functions. Significant part of the volume is devoted to the description of various components of the action-perception cycle and sensory processing domains, from cellular to system levels, and applications in intelligent designs.

This volume is of great interest for researchers and graduate students working on practical and modeling aspects of cognitive dynamics. It provides a comprehensive introduction to the field, which can be used as a supplementary textbook for cognitive science and computer science and engineering graduate courses covering intelligent behavior in biological and artificial systems. Our sincere hope is that this volume provides a overview of major aspects of the field and can support future explosive development of this exciting research area.

We were fortunate to be able to rely on the help and cooperation of a large number of people in creating this book. Gary Yen and Lipo Wang were General and Program Chairs of the 2006 World Congress of Computational Intelligence, where the ideas presented in this book has been first introduced. Their continued support to this endeavor is highly appreciated. The support of Springer's senior editor Thomas Ditzinger was instrumental in having this book completed. Our book appears in the Springer series "Understanding Complex Systems," which has been founded by J.A. Scott Kelso, and which has published a lot of exciting titles on various aspects of complexity since its inception 3 years ago. The encouragement of Janusz Kaczprzyk has been very useful at the start of our project, as well as the technical help of Heather King and Hema Latha during the production process. Without the high quality editorial work by Dr. Dinh Dzung Luong we would not be able to present this book within this relatively short time period to the broad research community.

Lexington-Cambridge, MA, Leonid Perlovsky
Summer 2007 Robert Kozma

Contents

List of Contributors

Peter Andras
School of Computing Science
University of Newcastle
Newcastle upon Tyne, NE1 7RU, UK
peter.andras@ncl.ac.uk

Theodore W. Berger
Department of Biomedical Engineering
University of Southern California
Los Angeles, CA 90089, USA
berger@bmsr.usc.edu

Hugues Bersini
IRIDIA
Universite Libre de Bruxelles
Avenue Franklin Roosevelt 50 B-1050
Brussels - Belgium
nospambersini@ulb.ac.be

Steven L. Bressler
The Human Brain and Behavior Laboratory
Center for Complex Systems and Brain Sciences
Florida Atlantic University
777 Glades Road
Boca Raton, FL 33431, USA
bressler@fau.edu

Bruce G Charlton
School of Biology and Psychology
University of Newcastle
Newcastle upon Tyne, NE1 7RU, UK
bruce.charlton@ncl.ac.uk

Siu-Yeung Cho
Division of Computing Systems
School of Computer Engineering
Nanyang Technological University
Singapore 639798
assycho@ntu.edu.sg

Yoonsuck Choe
Department of Computer Science
Texas A&M University
College Station, TX 77843-3112, USA
choe@tamu.edu

Emilio Del-Moral-Hernandez
Polytechnic School
University of Sao Paulo
Department of Electronic Systems Engineering
Av. Prof. Luciano Gualberto,
Trav. 3, n. 158,
05508-970 Sao Paulo, Brazil
Emilio.del.moral@ieee.org

Walter J Freeman
Division of Neurobiology, Donner 101
Department of Molecular & Cell Biology
University of California at Berkeley
Berkeley, CA, 94720-3206, USA
dfreeman@berkeley.edu

Khan M. Iftekharuddin
Department of Electrical and Computer Engineering

University of Memphis
Memphis, TN 38152, USA
iftekhar@memphis.edu

J. A. Scott Kelso
The Human Brain and Behavior Laboratory
Center for Complex Systems and Brain Sciences
Florida Atlantic University
777 Glades Road
Boca Raton, FL 33431, USA
kelso@ccs.fau.edu

Robert Kozma
Computer Science, 373 Dunn Hall
University of Memphis
Memphis, TN 38152, USA
rkozma@memphis.edu and
US Air Force Research Laboratory
80 Scott Drive, Sensory Directorate
Hanscom AFB, MA 01731-2909 USA

Daniel S. Levine
Department of Psychology
University of Texas at Arlington
Arlington, TX 76019-0528 USA
levine@uta.edu

Guang Li
National Laboratory of Industrial Control Technology
Zhejiang University
Hangzhou 310027, China
guangli@cbeis.zju.edu.cn

Yaqin Li
Department of Electrical and Computer Engineering
University of Memphis
Memphis, TN 38152, USA
yaqinli@memphis.edu

Bing Lu
Department of Biomedical Engineering
University of Southern California
Los Angeles, CA 90089, USA
blu@usc.edu

Navendu Misra
Department of Computer Science
Texas A&M University
College Station, TX 77843-3112, USA
navendu@tamu.edu

Colin Molter
Laboratory for Dynamics of Emergent Intelligence
RIKEN BSI, 2-1 Hirosawa, Wako
Saitama 351-0198, Japan
cmolter@brain.riken.jp

Leonid I. Perlovsky
School of Engineering and Applied Sciences
Harvard University, 336 Maxwell Dworkin
33 Oxford St, Cambridge MA 02138
USA

leonid@hrl.harvard.edu and
US Air Force Research Laboratory
80 Scott Drive, Hanscom AFB, MA 01731-2909 USA
Leonid.Perlovsky@hanscom.af.mil

Utku Salihoglu
IRIDIA
Universite Libre de Bruxelles
Avenue Franklin Roosevelt 50
B-1050 Brussels, Belgium
usalihog@iridia.ulb.ac.be

Faraz Siddiqui
Department of Electrical and Computer Engineering
University of Memphis
Memphis, TN 38152, USA
fsiddiqu@memphis.edu

Emmanuelle Tognoli
The Human Brain and Behavior Laboratory
Center for Complex Systems and Brain Sciences

Florida Atlantic University
777 Glades Road
Boca Raton, FL 33431, USA
tognoli@ccs.fau.edu

Paul J. Werbos
Room 675
National Science Foundation
Arlington, VA 22203, USA
pwerbos@nsf.gov

Jia-Jun Wong
Division of Computing Systems
School of Computer Engineering
Nanyang Technological University
Singapore 639798
jjwong@ntu.edu.sg

Walter M. Yamada
Department of Biomedical Engineering
University of Southern California
Los Angeles, CA 90089, USA
yamada@usc.edu

Jin Zhang
Department of Biomedical Engineering
Zhejiang University
Hangzhou, 310027, China
zhangjin@163.com

Neurodynamics of Cognition and Consciousness

Leonid I. Perlovsky and Robert Kozma

Abstract Dynamic aspects of higher cognitive functions are addressed. Dynamical neural networks with encoding in limit cycle and non-convergent attractors have gained increasing popularity in the past decade. Experimental evidence in humans and other mammalians indicates that complex neurodynamics is crucial for the emergence of higher-level intelligence and consciousness. We give an overview of research activities in the field, including dynamic models of consciousness, experiments to identify neurodynamic correlates of cognitive functions, interpretation of experimental findings, development of dynamical neural memories, and applications of dynamical approaches to intelligent system.

1 From Unconscious Chaos to Less Chaos and More Consciousness

A ubiquitous property of neurodynamics of consciousness is evolution from vague, fuzzy, and unconscious states to more concrete, and conscious. These neurodynamical processes are the essence of perception, cognition, and behavioral decision-making. More specific and conscious states correspond to recognized patterns and executed decisions. Neurodynamics proceeds from less knowledge to more knowledge and from less consciousness to more consciousness. Records of brain activity using EEG arrays and single-neuron evoked potential measurements indicate that brain states exhibit dynamic features of chaotic attractors [1, 2]. Vague and less conscious states are characterized by high-dimensional chaotic attractors. More concrete and conscious states are characterized by lower-dimensional "less chaotic" attractors.

Transitions from high-dimensional chaotic states to lower-dimensional and "less chaotic" states form a sequence of increasingly structured dynamics. Ultimately the dynamics leads to conditions facilitating conscious decision making and deliberate action by the individual [2, 3]. Following the action, a new situation is generated with modified internal states and goals and with changed environmental conditions. In this novel situation the neurodynamics starts again from high dimensional chaotic states and proceeds to lower-dimensional "less chaotic" states; this is called the intentional action-perception cycle.

L. I. Perlovsky and R. Kozma, *Neurodynamics of Cognition and Consciousness.*

In a cognitive cycle, neurodynamics evolves from less conscious to more conscious states, from vague and uncertain to more concrete knowledge, which is described at an abstract level of dynamic logic [4]. According to dynamic logic, brain states reflect the surrounding world and they are characterized by representations-models and by measures of similarity between the models and input signals. Vague, less conscious states are described by uncertain models with low similarity values. More conscious states correspond to concrete perceptions, cognitions, and decisions; they are described by concrete models and high similarity values. These more conscious models are better adapted-matched to input signals. Adaptation of models is driven by maximization of similarity. This drive is a mathematical representation of a fundamental instinct for more knowledge and more consciousness [5].

To summarize, chaotic neurodynamics and dynamic logic are equivalent descriptions of the dynamics of brain states. Dynamic logic-based modeling field theory provides the modeling framework, which evolves through the action-perception cycle. Dynamic logic can provide a cognitively-motivated model-based approach to describe the emergence of models of increasing clarity as the cognitive cycle progresses. Initial states are vague, uncertain, and less conscious. They are described by highly chaotic neurodynamics. They evolve into more concrete, certain, conscious states, described by less chaotic neurodynamics. Transitions from more chaotic states to less chaotic ones correspond to processes of perception, cognition, and decision making.

2 Neurocognition

Recordings of EEG activity in animals and humans demonstrate the spatial coherence of oscillations carrying spatiotemporal patterns, as described by Freeman in Chap. 2. The recordings reveal a shared oscillation in cortical potential over extended cortical areas in the beta band (12 Hz to 30 Hz) and gamma band (30 Hz to 80 Hz). These fluctuations serve as a carrier wave for perceptual information by means of spatial patterns with amplitude modulation (AM). The spatiotemporal patterns of the frequency of neural firing are correlated with the AM patterns. The shared carrier waveform is usually aperiodic and unpredictable, reflecting the chaotic dynamics of sensory cortices. These observations have been interpreted in terms of dynamic system theory [1, 6, 7, 8, 9]. A complex brain state is characterized by a trajectory over a chaotic attractor landscape. The system dynamics may reside for a brief time period in a localized attractor basin, before it transits to another basin. The presence of a given sensory stimulus may constrain the trajectory to a lower dimensional attractor basin. Once the sensory stimulus is removed, the trajectory switches to a higher-dimensional dynamics (less concrete, less conscious state) until the next sensory stimulus constrains the dynamics again to a specific perceptual state.

Experimental evidence indicates metastability of the dynamics, as described by Kelso and Tognoli in Chap. 3. Metastability is an inherent property of cortical dynamics, which is also manifested in coordination patterns of behaviors [10].

Metastability is the consequence of the competition of complementary tendencies of integration and fragmentation between cortical areas [11]. Intermittent oscillations, similar to those observed experimentally in metastable cortical states, have been mathematically described by chaotic itinerancy [12]. In chaotic itinerancy, the attractor landscape is characterized by attractor ruins. Accordingly, the trajectory is constrained intermittently to the neighborhood of the attractor ruin, but it never settles to the attractor. Rather it jumps from one attractor ruin to another and it generates a metastable dynamic pattern of spatio-temporal oscillations. Neuromodulatory effects through the hypothalamus and brain stem play role in the observed transitions (Chap. 2 by Freeman).

Experimental evidence of the formation of a global cortical neurocognitive state is provided by Bressler (Chap. 4). The cortex consists of a large number of areas profusely interconnected by long-range pathways in a complex topological structure. An important aspect of cortical connectivity is that each cortical area has a specialized topological position within the cortex, i.e. a unique pattern of interconnectivity with other cortical areas. To a large degree, the function of every cortical area is determined by its unique patterning of long-range connectivity. At the same time, the short-range interconnectivity of local circuits within cortical areas is generally similar throughout the cortex, implying that no area has a specialized monitoring function by virtue of its internal organization. These considerations suggest that cortical monitoring and integrative functions are the result of cooperative interaction among many distributed areas, and not the sole property of any one area or small group of areas [13]. Bressler shows that the cortex dynamically generates global neurocognitive states from interactions among its areas using short- and long-range patterning of interconnectivity within the cortex.

Various models address questions related to the hierarchies in neural systems. Dynamic models span from the cellular level to populations, including massively recurrent architectures with complex dynamics [14, 15, 16]. Biologically-motivated models of sensory processing have been successfully implemented in a wide range of areas [17, 18]. Clearly, dynamical approaches to neural modeling provide powerful and robust tools of solving difficult real life problems.

3 Cognition, Emotions, and Brain

Emotions were considered opposite to intelligence since Plato and Aristotle. Similar were attitudes to emotions during initial attempts to construct artificial intelligence; these attempts were based on logical-conceptual mechanisms. Recently, however, fundamental role of emotions in cognition have being recognized. What are the brain modules involved in interaction between emotions and cognition, and how these interactions can be modeled mathematically? In Chap. 5, Perlovsky analyses the interplay of conceptual and emotional mechanisms [5]. It is related to the role of prefrontal cortex, hippocampus, and amygdala in Chap. 8 by Levine [19]. The brain uses emotional memories to switch between appropriate sets of behavioral rules in various circumstances.

Special emotions accompany knowledge creation and improvement. Perlovsky (Chap. 5) describes our drive to learning, to improving understanding of the surrounding world, due to the knowledge instinct [4]. Satisfaction of this instinct is experienced as positive aesthetic emotions. These emotions are the foundation for all our higher cognitive abilities. Mathematical technique describing these instinctual-emotional mechanisms of learning overcome difficulties encountered by previous artificial intelligence and neural network approaches, difficulties known as "curse of dimensionality," "exponential explosion," and "combinatorial complexity."

The relationship between emotional and rule-based aspects of decision making is analyzed by Levine (Chap. 8). The amygdala is closely connected with hypothalamic and midbrain motivational areas and it appears to be the prime region for attaching positive or negative emotional valence to specific sensory events. The amygdala is involved in emotional responses, from the most primitive to the most cognitively driven. The complex interplay of attention and emotion has been captured in various network models involving amygdala as well as various parts of the prefrontal cortex [19]. Emotional states of humans have been successfully identified using facial imaging based on neural network models [20].

In the dynamic brain model KIV introduced by Kozma (Chap. 7), amygdala plays an important role in its interaction with the cortico-hippocampal system. It gives an emotional bias to mechanisms progressing from sensory cortical areas toward decision making and motor system [3, 21]. In a more general context, the role of the internal state of the sensory-motor agent is analyzed by Misra and Choe (Chap. 9). Instead of using a pure reactive agent, they develop an approach in which the sensory-motor agent has a long-term memory of its previous experiences (visual and skill memories) [22]. It has been demonstrated that especially skill memory leads to improved performance in the recognition task.

4 Hierarchy

The mind is not a strict hierarchy. Significant feedback exists between higher and lower levels. This architecture is sometimes called heterarchy. For simplicity, we will sometimes call it hierarchy. At lower levels of the hierarchy there are mechanisms of recognition and understanding of simple features and objects. Higher up are situations, relationships, abstract notions.

A hierarchy of structures, functions, and dynamics over spatial and temporal brain scales are described by the K models by Kozma (Chap. 8). K sets are multi-scale models, describing increasing complexity of structure and dynamical behavior [3, 23]. K sets are mesoscopic (intermediate-scale) models introduced by Freeman in the 70's, and they represent an intermediate-level of hierarchy between microscopic neurons and macroscopic brain structures. The basic building block is the K0 set which describes the dynamics of a cortical micro-column with about 10 thousand neurons. K-sets are topological specifications of the hierarchy of connectivity in neuron populations in the 6-layer cortex. A KI set contains K0 sets from a given layer with specific properties. KII includes KI units from different populations, i.e., excitatory and inhibitory ones. KIII has several KII sets modeling various cortical areas. KIV covers cortical areas across the hemisphere. KV is the highest level of hierarchy describing neocortex. The dynamics of K sets has the following hierarchy: K0 has zero fixed point attractor; KI has non-zero fixed point attractor; KII has limit cycle oscillations; KIII exhibits chaos; and KIV shows intermittent spatio-temporal chaos. The function of KIII sets can be sensory processing and classification using a single channel; KIII may correspond to visual sensory system, olfactory system, hippocampus, midline forebrain, etc. KIV performs multisensory fusion and decision making. KV has components of higher cognition and conscious functions.

Werbos (Chap. 6) gives a careful, qualified endorsement to hierarchical brain models. The spatio-temporal complexity of cognitive processing in brains would benefit from a hierarchical partitioning and nature certainly exploits this opportunity [24, 25, 26]. He cites evidence for such temporal hierarchy in basal ganglia. The hierarchy in time, however, is not crisp, but apparently fuzzy. Recent research indicates that the connection from dorsolateral cortex to the basal ganglia proposes the "verb" or "choice of discrete decision block type" to the basal ganglia. It also suggests that the "hierarchy'' is implicit and fuzzy, based on how one decision may engage others – but may in fact be forgotten at times because of limited computational resources in the brain.

Dynamic logic and neural modeling fields describe operation of the knowledge instinct within the mind hierarchy [4, 5]. Aesthetic emotions related to improvement of knowledge, operate at every hierarchical level. At lower levels, where knowledge is related to everyday objects, aesthetic emotions are barely noticeable. At the top of the hierarchy, knowledge is related to purposes and meanings of life; improvement of this knowledge is felt as emotion of the beautiful. Operation of the knowledge instinct in the hierarchy is manifested as differentiation and synthesis. Differentiation refers to creation of diverse knowledge; it occurs at every level of the hierarchy. As more diverse knowledge is created, less emotional value is vested in every element of knowledge. Synthesis refers to connecting this diverse knowledge to more general concepts at higher levels. In these connections diverse knowledge acquires emotional value. This emotional value of knowledge is necessary for creating more knowledge, for differentiation. Thus, synthesis and differentiation are in complex relationships, at once symbiotic and oppositional.

5 Cultural Dynamics

Differentiation and synthesis, created in the individual minds, determine consciousness in individuals and also drive collective consciousness and evolution of entire cultures. Chapter 5 gives a mathematical description of this evolution. When synthesis predominates in collective consciousness, knowledge is emotionally valued, and more knowledge is created. However, more knowledge leads to reduced emotional value of every piece of knowledge; thus differentiation destroys synthesis, which is the condition of differentiation. This interaction describes cultural dynamics with alternating periods of cultural flourishing and destruction, which might characterize

evolution of the Western culture
indexCultureindexWestern culture during the last 4,000 years. An opposite dynamics is characterized by permanent predominance of synthesis. When every conceptual piece of knowledge is vested with high emotional value, differentiation is stifled, and limited knowledge is vested with more and more emotional value, while culture stagnates. Within each society levels of differentiation and synthesis can be measured and used to predict cultural evolution around the globe, including our own.

6 Overview of the Book

An important aspect of cognitive processing is the intermittent character of brain waves oscillating between high-dimensional spatio-temporal chaos and lower-dimensional more ordered states. Such an oscillatory dynamics is observed at the theta range, i.e., at a rate of approximately 5 cycles per second, as described in the cinematographic model of cognition by Freeman (Chap. 2). This process is characterized as metastability in the context of coordination dynamics by Kelso and Tognoli (Chap. 3). The formation of global neurocognitive states between mammalian cortical areas is documented and analyzed by Bressler (Chap. 4). A key property of neurodynamics is the evolution from less conscious to more conscious states, from vague and uncertain to more concrete knowledge, which is described at an abstract level of dynamic logic by Perlovsky (Chap. 5).

Cortical circuitries describing various aspects of the dynamics of cognitive processing have been analyzed by several authors. The capabilities of the mammalian brain are studied in the context of adaptive dynamic programming features by Werbos (Chap. 6). In particular, principles of adaptive critic systems are elaborated, and a roadmap is outlined toward artificially intelligent designs based on the thalamo-cortical circuitry. The role of the cortico-hippocampal system in intentional behavior generation and decision making is studied by Kozma (Chap. 7). This system also incorporates an internal motivation unit and amygdala as the emotional subsystem of the autonomous agent. The interplay of the hippocampus and amygdala in forming a context-dependent behavior balancing between goal-orientedness and emotional states is described by Levine (Chap. 8).

Dynamical models of cognitive functions, including categorization and pattern recognition gain popularity in a wide range of applications due to their robustness and biological relevance. The close link between visual recognition and motor representations is analyzed by Misra and Choe (Chap. 9). A computational model of the visual sensory system is used for object recognition based on dynamic programming principles by Iftekharuddin et al. (Chap. 10). Emotional states based on facial expression are successfully identified by Wong and Cho (Chap. 11). The dynamic KIII model, motivated by the olfactory system, is used successfully to solve difficult image recognition, voice recognition, and other classification problems by Li et al. (Chap. 12).

Chaotic oscillations identified in brains strongly motivate research in dynamic memories. Hernandez in Chap. 13 studies the role of recursive connectivity in the neural circuitry for creating rich dynamics of cognitive functions. A rigorous study of cyclic attractors and their potential role as robust memory devices is given by Molter, Salihoglu, and Bersini (Chap. 14). Long-term memories modeled as self-reproducing communication networks by Charlton and Andras (Chap. 15). A novel dynamical model of the synaptic interaction in multiple vesicle pools is developed by Lu, Yamada, and Berger (Chap. 16).

7 Conclusions

This chapter provides an introduction to the volume on the neurodynamics of cognition and consciousness. Contributions from leading experts of the field provide a cutting-edge review of this challenging frontier of neuroscience and intelligent systems research. We hope it will help interested researchers to get familiar with research achievements and open new directions.

References

1. W. J. Freeman. Neurodynamics. An Exploration of Mesoscopic Brain Dynamics. London: Springer, 2001.
2. Nunez, R.E., Freeman, W.J. (1999) "Restoring to cognition the forgotten primacy of action, intention, and emotion," J. Consciousness Studies, 6 (11–12), ix-xx.
3. Kozma, R., and Freeman, W.J. Basic Principles of the KIV Model and its application to the Navigation Problem, J. Integrative Neurosci., 2, 125–140, 2003.
4. Perlovsky, L.I. Neural Networks and Intellect. Oxford Univ. Press, New York, NY, 2001.
5. L. I. Perlovsky. Toward physics of the mind: Concepts, emotions, consciousness, and symbols. Physics of Life Reviews, 3:23–55, 2006.
6. W. J. Freeman. Origin, structure, and role of background EEG activity. Part 1. Phase. Clinical. Neurophysiology 115: 2077–2088, 2006.
7. W. J. Freeman. Origin, structure, and role of background EEG activity. Part 2. Amplitude. Clinical. Neurophysiology 115: 2089–2107, 2006.
8. W. J. Freeman. Origin, structure, and role of background EEG activity. Part 3. Neural frame classification. Clinical. Neurophysiology 116 (5): 1118–1129, 2005.
9. W. J. Freeman. Origin, structure, and role of background EEG activity. Part 4. Neural frame simulation. Clinical. Neurophysiology 117/3: 572–589, 2006.
10. J. A. S. Kelso. Dynamic Patterns: The Self-Organization of Brain and Behavior. Cambridge: MIT Press, 1995.
11. Kelso, J.A.S., Engstrøm, D.: The Complementary Nature. MIT Press, Cambridge, 2006.
12. Tsuda. I. Towards an interpretation of dynamic neural activity in terms of chaotic dynamical systems. Behavioral and Brain Sciences 24:793–810, 2001.
13. S. L. Bressler and J. A. S. Kelso. Cortical coordination dynamics and cognition. Trends in Cognitive Science 5:26–36, 2001.
14. J.S. Liaw and T.W. Berger. Dynamic synapse: a new concept of neural representation and computation. Hippocampus, 6:591–600, 1996.
15. Del-Moral-Hernandez, E.: Non-homogenous Neural Networks with Chaotic Recursive Nodes: Connectivity and Multi-assemblies Structures in Recursive Processing Elements Architectures, Neural Networks, 18, 532–540, 2005.

16. C. Molter and U. Salihoglu and H. Bersini, The road to chaos by time asymmetric Hebbian learning in recurrent neural networks, Neural Computation, 19(1), 100, 2007.
17. Li, X., Li, G., Wang, L., Freeman, W.J.: A study on a Bionic Pattern Classifier Based on Olfactory Neural System, Int. J. Bifurcation Chaos., 16, 2425–2434, 2006.
18. Iftekharuddin, K.M., Power, G.: A biological model for distortion invariant target recognition. In: Proc. of IJCNN, Washington DC, U.S.A., IEEE Press, pp. 559–565, 2001.
19. D. S. Levine. Angels, devils, and censors in the brain. ComPlexus, 2:35–59, 2005.
20. Wong, J.-J. and S.-Y. Cho, A Brain-Inspired Framework for Emotion Recognition. Neural Information Processing, 10(7), pp. 169–179, 2006.
21. R. Kozma and W. J. Freeman. Chaotic resonance: Methods and applications for robust classification of noisy and variable patterns. International Journal of Bifurcation and Chaos 10: 2307–2322, 2001.
22. Choe, Y., Bhamidipati, S.K.: Autonomous acquisition of the meaning of sensory states through sensory-invariance driven action. In Ijspeert, A.J., Murata, M., Wakamiya, N., eds.: Biologically Inspired Approaches to Advanced Information Technology. Lecture Notes in Computer Science 3141, Berlin, Springer, 176–188, 2004.
23. Kozma, R., Freeman, W.J., Erdi, P. The KIV Model - Nonlinear Spatio-temporal Dynamics of the Primordial Vertebrate Forebrain, Neurocomputing, 52–54, 819–825.
24. P.Werbos, Brain-Like Design To Learn Optimal Decision Strategies in Complex Environments, in M.Karny et al eds, Dealing with Complexity: A Neural Networks Approach. Springer, London, 1998.
25. J.Albus, Outline of Intelligence, IEEE Trans. Systems, Man and Cybernetics, 21(2), 1991.
26. P.Werbos, What do neural nets and quantum theory tell us about mind and reality? In K. Yasue, M. Jibu & T. Della Senta, eds, No Matter, Never Mind : Proc. of Toward a Science of Consciousness. John Benjamins, 2002.

Part I
Neurocognition and Human Consciousness

Proposed Cortical "Shutter" Mechanism in Cinematographic Perception

Walter J. Freeman

Abstract Brains are open thermodynamic systems, continually dissipating metabolic energy in forming cinematographic spatiotemporal patterns of neural activity. In this report patterns of cortical oscillations are described as 'dissipative structures' formed near an operating point at criticality far from equilibrium. Around that point exists a small-signal, near-linear range in which pairs of impulse responses superpose. Piece-wise linearization extends analysis into nonlinear ranges. Resulting root loci are interpreted as projections from a phase plane, in which the three phase boundaries are graphed in the coordinates of rate of change in a dynamic order parameter (negentropy) on the ordinate analogous to static pressure *vs.* rate of energy dissipation (power) analogous to static temperature on the abscissa. The graph displays the neural mechanism that implements phase transitions and enables the limbic system to repeat the action-perception cycle at 3–7 Hz. The mechanism is null spikes ('vortices') in Rayleigh noise in background electrocorticogram (ECoG) that serve as a shutter by triggering phase transitions.

1 Introduction

Measurements from depth electrodes of microscopic axonal action potentials [1, 2], pial surface electrodes of mesoscopic dendritic potentials giving the electrocorticogram (ECoG [3, 4, 5, 6]), and scalp electrodes giving the macroscopic electroencephalogram (EEG) show that brains, as chaotic systems [7], don't merely *filter* and *process* sensory *information*. Brains import raw sense data that is represented by *microscopic* stimulus-driven spike activity. They replace it by constructing *mesoscopic* percepts that are manifested in spatiotemporal patterns of wave packets [8, 9, 10]. These local patterns combine into *macroscopic* states that involve much or even all of each cerebral hemisphere [11, 12]. They do this several times a second in the sensory system in sequences of wave packets [10, 12] both asynchronously with local time-varying frequencies [13] and globally synchronized at the same frequency [14].

This Chapter draws on diverse experimental evidence to outline a neural mechanism for the repetitive state transitions that initiate construction of cinematographic sequences, for which clinical evidence has been cited [15]. Section 2 describes the

L. I. Perlovsky, R. Kozma, *Neurodynamics of Cognition and Consciousness.*

spatiotemporal patterns of electrocorticograms (ECoG) that are recorded from the olfactory system during arousal and reinforcement learning. The same basic patterns occur in neocortical sensory systems. Section 3 introduces mutual excitation and the positive feedback by which cortical background activity is created and stabilized. Section 4 introduces inhibition and the negative feedback by which the carrier oscillations of wave packets are generated and stabilized. The concept is developed of bistability through input-dependent changes in nonlinear feedback gain that switch sensory cortices between receiving and transmitting states. Section 5 summarizes evidence for conditional stabilization at self-organized criticality, by which state transitions between these two states of cortex can be described as phase transitions in metastability near pseudo-equilibrium. Section 6 derives a diagram of thermodynamic phase space and phase transitions for cortex far from equilibrium that summarizes cortical operations in the action-perception cycle [16]. Section 7 describes the 'shutter', discusses its significance in perception, and summarizes.

2 Evoked *versus* Induced Activity in the Olfactory and Neocortical Sensory Systems

The creative property of nonlinear dynamics is not readily apparent in the homeostatic feedback mechanisms of brain reflexes, which insure the stability of brain function by keeping the internal environment (temperature, pressure, volume, and chemical constitution) of the brain near optimal levels despite environmental vicissitudes. It becomes clear in modeling perception, which requires creative interaction with the external environment for achieving life goals. A relatively simple example comes from study of neural activity in the olfactory bulb. The olfactory system is a semi-autonomous module that interacts with other parts of the forebrain by exchanging neural information. It receives sensory input and also centrifugal controls through release of neuromodulators from brain stem nuclei. An orchestrated mix of neuroamines and neuropeptides [50] modulates the receptivity of the bulb and olfactory cortex in arousal, search, and learning (Fig. 1).

Endogenous oscillatory activity persists after the olfactory system has been surgically isolated from the rest of the brain, which shows that its basic functions are self-organizing. However, its aperiodic chaotic activity disappears when its parts have been surgically disconnected [14], showing that its aperiodic activity is a global property that is not due to the entrainment of single neurons acting as chaotic generators. This is important, because the stimuli that are recognized in perception are spatiotemporal patterns, such as a facial expression, a musical phrase, the fit of a jacket on the torso, etc. The sensory receptor activity they excite or inhibit are characterized by spatial relationships between each part and every other part of a pattern, so that the determinants of perception in sensory cortical function must also be global, not local as in *feature detector* neurons.

Simultaneous recordings of ECoG activity from arrays of 64 electrodes placed on the olfactory bulb and cortex of rabbits demonstrate the spatial coherence of ECoG oscillations carrying spatiotemporal patterns. In every subject the recordings

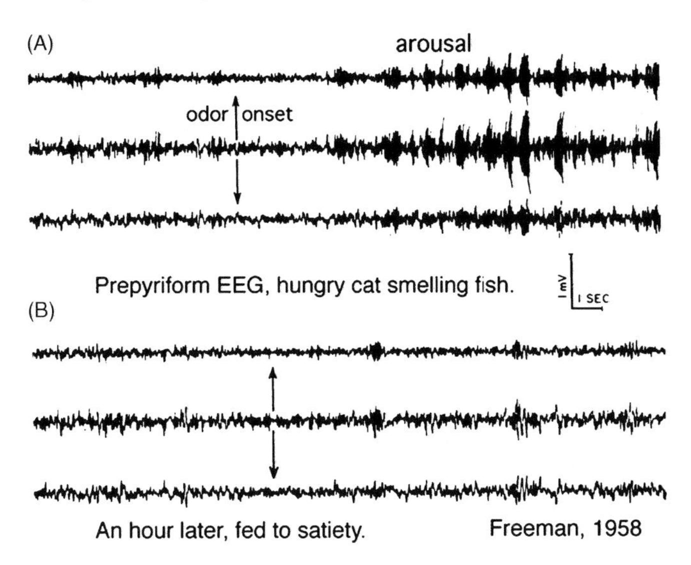

Fig. 1 (**A**) a food-deprived cat at rest is aroused by an odor of fish and searches for it by sniffing, (**B**) after feeding to satiety there is no arousal [14], Fig. 7.17, p. 442

reveal a shared oscillation in cortical potential over the entire array at the same instantaneous frequency [29, 32, 53]. The fluctuation serves as a carrier wave for perception by means of spatial patterns of amplitude modulation (AM) (Fig. 2, left). The spatiotemporal patterns of the frequency of firing of bulbar and cortical neurons are correlated with the AM patterns. The shared carrier waveform is usually aperiodic and unpredictable, reflecting the chaotic dynamics of sensory cortices.

The spatial AM patterns in the sensory cortices cannot literally represent the stimuli that are transmitted to them over their sensory input pathways, because they lack invariance with respect to unchanging stimuli [5, 18, 29, 32, 48, 50, 53, 57]. They change instead with changes in reinforcements, the learning of other new conditioned stimuli (CS) in serial conditioning, and other contextual changes that are associated with stimuli during periods of training to respond to them, in brief, the meanings of the CS, which are as unique for each of the subjects as are the AM patterns. These properties have also been found in ECoG of visual, auditory and somatic neocortices, and they hold in both the beta and gamma ranges. Owing to the fact that the waveform is everywhere similar, the AM pattern of each frame can be represented by a contour plot (Fig. 2, left) of the 64 root mean square amplitudes of the beta (12–30 Hz) or gamma (30–80 Hz) oscillation in each frame. The 64 values specify a vector and a point in 64-space. Similar patterns form clusters of points in 64-space, which can be visualized by use of discriminant analysis to project them into 2-space (Fig. 2, right). In discriminative conditioning with 2 or more CS, the classification of frames is done by calculating the Euclidean distances of the data points from each point/frame to the centers of gravity of the clusters and finding the shortest distance.

When ECoG is observed over time spans of minutes to hours or years, the olfactory dynamic mechanism appears robustly stable in a wide range of amplitudes.

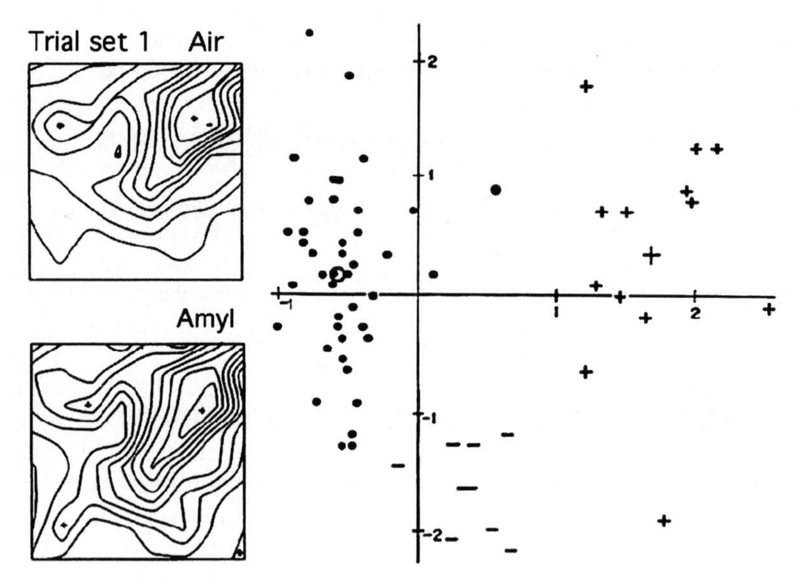

Fig. 2 Left: examples of contours of root mean square amplitude of 64 ECoG segments from and 8×8 (4×4 mm) array on the olfactory bulb. From [32]. **Right**: clusters of points projected from 64-space by discriminant analysis showing classification of patterns by Euclidean distances of points from the three centers of gravity for control (air •), a reinforced odorant (amyl acetate +), and an unreinforced odorant (butyl alcohol -). From [53]

The stability in the olfactory system and in neocortex has been explored in detail in terms of *chaotic attractor landscapes* [42, 57], *metastability* based in "*coordination dynamics*" [11, 40], *chaotic itinerancy* [59], and *explicit vs. spontaneous symmetry breaking* among multiple ground states in dissipative many-body physics [33]. In the course of normal behavior the states are changed by neuromodulatory inputs from the hypothalamus and brain stem that induce transitions such as those between stable states of waking and sleep. Abnormally, the waking state is destabilized by intense excitation leading to transmitter depletion, causing the brain to transit to an alternative state that is characterized by a form of epilepsy [16]: complex partial seizures that include behavioral *absence* (loss of consciousness with failure to attend, perceive or learn). An important normal form of destabilization occurs during learning to identify a new odorant, which requires the actions of neuromodulatory nuclei in the brain stem that release neuroamines and neuropeptides under limbic control.

These properties indicate the need to distinguish among three types of change in both olfactory and neocortical dynamics. First, a physiological stimulus to receptors causes a pattern of action potentials that through relays from receptors injects a pattern of spikes into the bulb that represents the stimulus. The driven cortical response is an evoked potential. By itself the input does not force a state transition, but it perturbs the dynamics and, when ended, allows the system to relax to its prestimulus state without changing the cortical dynamics. Second, learning establishes an attractor landscape in the cortex. Each attractor is surrounded by its basin of attraction that corresponds to the generalization gradient for a category of stimulus

[48]. Formation of a new basin and its attractor is an irreversible structural bifurcation [41, 42]. Third, an act of perception is triggered by the representation of a stimulus when the relevant action potentials select an attractor in the landscape. The percept is dependent on a spatiotemporal activity pattern that is shaped by synaptic weights that e shaped by prior Hebbian and non-Hebbian learning [13]. These modified synapses that store diverse forms of experience form nerve cell assemblies that are intermingled and overlapping. At each moment of engagement by a subject with the environment, a selection must be made in the primary sensory cortices by priming the relevant cell assemblies, that is, by enhancing their excitability and sensitivity to anticipated input. That preparation for selection is done by a process that is called "preafference" [39], Bressler Chapter. This process has been identified with the formation of a global pattern of synchronous oscillation [20, 30, 31] that is established through widespread exchanges of action potentials among sensory cortices and the limbic system. The global pattern elicits and modulates the attractor landscapes of all sensory cortices simultaneously, thereby preparing them for the range of expected outcomes following each act of observation on the environment.

This global state of preparedness in anticipation constitutes a metastable state [Kelso Chapter]. The incoming sensory stimuli select one basin of attraction from the landscape in each sensory cortex. This exclusive choice can be referred to as spontaneous symmetry breaking, as distinct from the explicit symmetry breaking of the evoked potential [33]. The transition from high-dimensional chaos to more stable dynamics with the selection of a lower-dimensional attractor may correspond to dynamic logic [Perlovsky Chapter], which describes transitions from vague and uncertain potential states not yet realized to a crisp and certain perception and cognition following detection of an anticipated stimulus, as an essential mechanism preceding higher cognitive functions.

The endogenous state transition in the formation of a wave packet having an AM pattern is the essence of an act of perception. Transmission of the AM pattern must be followed by another state transition that returns the cortex to the receiving state. The key evidence for the first transition is provided by measurements of the pattern of phase modulation of the beta or gamma carrier waves of successive bursts. Each AM pattern is accompanied by a spatiotemporal pattern of phase modulation (PM). The phase is defined at the dominant or mean carrier frequency in the immediate past. The onset of the transition is characterized by an apparent discontinuity in the phase resembling "phase slip", followed by emergence of a new carrier frequency, on average differing by $\pm$10–20 Hz. The spatial pattern of the phase has the form of a cone (Fig. 3, left) for which the apex is either maximal lead or maxmimal lag. The cone is displayed by concentric circles representing isophase contours. The cone appears because a state transition in a distributed medium does not occur everywhere instantaneously. Like the formation of a snowflake or raindrop it begins at a site of nucleation and spreads radially. The phase gradient in rad/m divided by the carrier frequency in rad/s gives the phase velocity in m/s. The observed phase velocities conform to the conduction velocities of propagated action potentials on axons running parallel to the cortical surfaces. The apical location and sign vary randomly between successive frames (Fig. 3, right). The spread depends on the existence of a small proportion of long axons among the predominantly short axons [42]. The sign

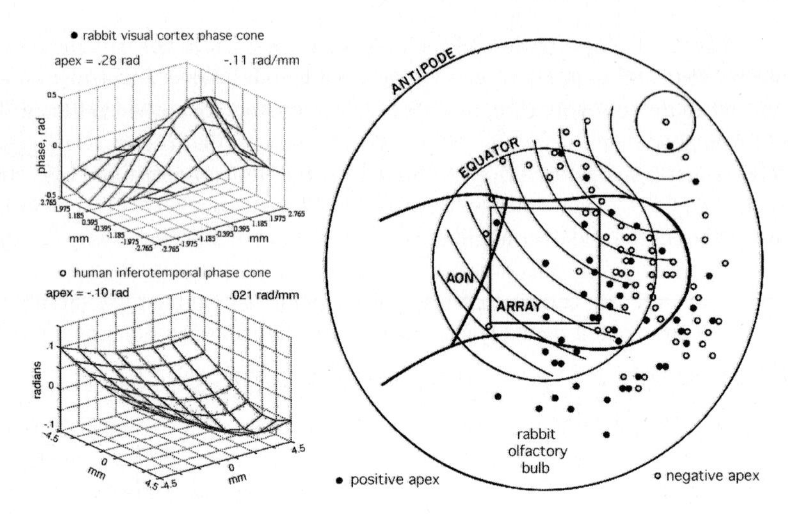

Fig. 3 Left: examples of phase cones from neocortical ECoG of human and rabbit. From [30]. **Right**: outline of rabbit bulb with square array superimposed on the opened spherical surface. The arcs show a representative phase cone; the dark symbols • show the locations of positive apices ("*explosion*"); the light symbols ^{o} show negative apices ("*implosion*"). From [24]

at the apex depends on whether the short axons (phase lead at the apex, explosion) or the long axons (phase lag at the apex, implosion) dominate the transmission.

These properties show that the state transition is an endogenous property of distributed cortical networks. It cannot be attributed to a localized "pacemaker" in the cortex, thalamus, or striatum. Then there is the second state transition that returns the cortex to the receiving state is intrinsic to the cortical background activity. There is no apparent phase discontinuity. Instead, the analytic amplitude may diminish to such a low level that the phase becomes undefined. If in this condition there is fresh input to the cortex, the possibility emerges of a new state transition. Recurrence of the episodic phase re-setting gives the cinematographic sequences of wave packets expressed in AM and PM frames. Each frame begins after the phase discontinuity manifests phase re-setting. The time lapse before the cone disappears demonstrates the duration of the frame.

Next the question is addressed, to what extent might the dynamics of the olfactory system hold for other sensory areas? The main components of the olfactory system are allocortex. This is an ancient three-layered neuropil that is found in all vertebrates in a variety of forms. The six-layered neocortex, found only in mammals, is far more complex and has widely varying forms and degrees of specialization. The relative simplicity of allocortical structure and dynamics makes it a good platform from which to discern what properties might be fundamental in the dynamics of neocortex. The visual, auditory and somatic sensory cortices, all or which are neocortical, reveal the same basic modes of operation in perception as in olfaction: the genesis of broad-spectrum, aperiodic ECoG activity with statistically correlated spike activity; formation of successive wave packets with spatially coherent carrier waves; AM patterns that are classifiable with respect to learned conditioned stimuli and responses (CS and CR); and accompanying PM patterns in the form of a cone

(Fig. 3, left) with randomly varying location and sign of the apices. Neocortical AM and PM patterns are larger and more complex than allocortical patterns, and they lack the easy frame marker provided by respiration and sniffing (the bursts of gamma seen in Fig. 1, A). Otherwise the properties described in this and the following sections hold for both types of cortex. Paramount among these properties is the ubiquitous background activity, which holds the key to understanding cortical function and therefore higher cognitive function.

3 Self-organization by Positive Feedback: Mutual Excitation and Excitatory Bias

The pervasive *spontaneous* activity of cerebral cortex is easily observed both in spikes from axons and in field potentials from dendritic current seen in ECoG (Fig. 1). The source of this activity has been identified as mutual excitation among excitatory cells [14, 22]. In neocortex 80% of the neurons are pyramidal cells, which are excitatory, and 90% of their synapses are from intracortical neurons. The 10% of synapses from sources outside cortex are excitatory. Therefore, the overwhelming interaction among neocortical neurons is by mutual excitation among pyramidal cells. In the olfactory bulb the organization differs, because the equivalents of pyramidal cells (the mitral and tufted cells) are outnumbered >100:1 by the equivalent inhibitory interneurons (the internal granule cells). A dynamic balance is maintained by a large population of excitatory interneurons (the periglomerular cells in the outer layer bulb) [14]. These neurons transmit and receive by GABA, so they are commonly misidentified as inhibitory interneurons; however, their high content of chloride ions [56] makes their GABA-A synapses excitatory. Their sustained mutually excitatory activity is not the same as that of the "reverberatory circuits" of Hebb and Lashley for putative temporary storage of memories [2] in small nerve cell assemblies [49]. Instead their background activity is steady state and large scale.

The mechanism in periglomerular and other excitatory populations has been described using a network of ordinary differential equations (ODE) [14]. The topology of connections in neural populations is represented (Fig. 4) by K-sets of neurons of like kind. The simplest are populations of noninteractive neurons of two kinds: KOe excitatory or KOi inhibitory, which when synchronously activated behave as would an average neuron, but with state variables of wave and pulse densities in place of membrane currents and spike trains. A KIe population of interacting excitatory neurons such as the periglomerular cells can, by consideration of symmetry conditions, be reduced to a positive feedback loop, in which the dynamics of the KOe forward and feedback limbs is identical, because each neuron in the receiving state continually renews the population in the transmitting state. Each limb is described by weighted linear summation of wave density in dendrites and static nonlinear conversion of wave density to pulse density. The dendritic operation can be described by a 2nd order linear ODE with gain constants for the synapses. The solution of the ODE for impulse input conforms to the dendritic postsynaptic potential (PSP)

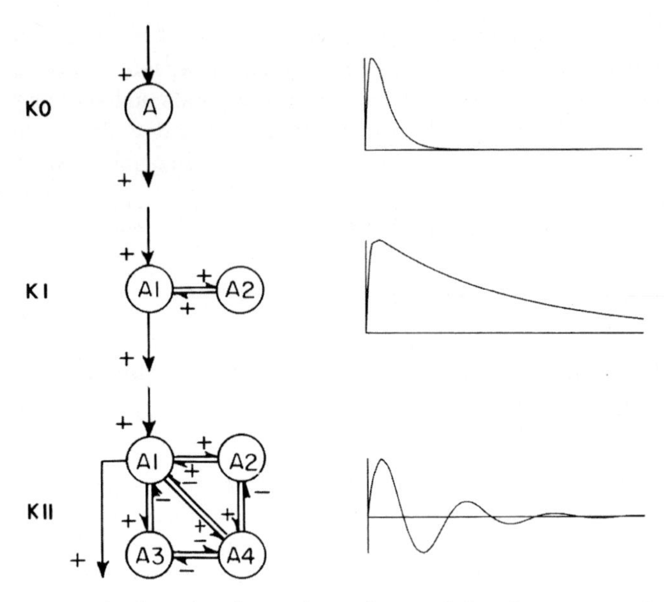

Fig. 4 KO represents the dynamics of a non-interactive population; the response to impulse input (single shock electrical stimulation) corresponds to the postsynaptic dendritic potential of single neurons. KI represents an interactive population of like kind: excitatory KIe or inhibitory KIi with a prolonged monotonic closed loop impulse response (averaged evoked potential AEP and PSTH). KII represents interaction between excitatory KIe and inhibitory KIi neuron populations with an oscillatory closed loop impulse response (AEP or PSTH). From [14], Fig. 1.6, p. 40

of single neurons to single-shock impulse input: a rapid exponential rise governed by synaptic delay and the cable properties of the dendrites, followed by exponential decay that is governed by passive membrane resistance and capacitance (Fig. 4, KO). In a population of neurons the output of the axons is observed in a post-stimulus time histogram (PSTH) of representative single neurons that has the same form and rate constants as the summed PSP of the dendrites.

Conversion from dendritic wave density to axonal pulse density at trigger zones is governed by an asymmetric sigmoid function [14, 15] (Fig. 5). The nonlinear thresholds of the axons determine lower asymptote. The neurons cannot fire when they are inhibited below threshold by input from inhibitory interneurons. The upper asymptote is determined by the axon refractory periods. The neurons cannot fire with excitatory input when they have already recently fired.

Piecewise linearization [14] of cortical dynamics is achieved by replacing the nonlinear function (Fig. 5) with a constant gain value given by the slope of the tangent to the sigmoid curve at a specified operating point for wave density input and pulse density output. The operating point in the steady state is at unity gain, when input amplitude is equal to output amplitude. The displacement from a steady state set point by impulse or sustained input can raise or lower the gain from unity. There are two stable set points at unity gain: one is found in the KIe set at normalized wave density $v > 0$ (Fig. 5, A). The other set point is found at $v = 0$ in the KII set, which is formed by the interaction between a KIe set and a KIi set. The mutually

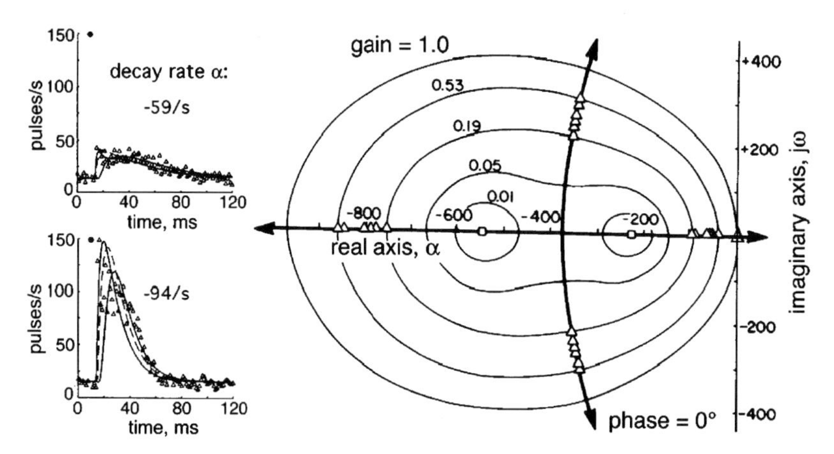

Fig. 5 Wave-to-pulse conversion at trigger zones in populations is governed by a static nonlinearity, which is the asymmetric sigmoid curve that is shown with a high slope in arousal (Fig. 1, A) and a low slope in rest (Fig. 1, B). The pairs of curves asymptotic to zero show the derivative of the sigmoid curve, which gives the nonlinear gain. Linear approximation is by replacing the function with the tangent at the desired operating point, for which the slope approximates the fixed feedback gain. A : KIe set, B: KIIob set. Equations are given in [1, 14, 49]

inhibitory KIi set has a stable set point only at zero gain giving a flat ECoG, so a KIi set is nonfunctional without an excitatory bias from one or more KIe sets. The maximal asymptote, Q_m, is determined by refractory periods; however, as indicated by Fig. 5, Q_m varies with the level of arousal under limbic control. Two levels are shown in Fig. 5. The ECoG with arousal state is seen in Fig. 1, A. The ECoG in the rest state is seen in Fig. 1, B.

The impulse response of the closed loop KIe system (Fig. 1, KI) is given by the averaged, single-shock evoked potential (AEP) in the wave density mode (ECoG recording) or by the post-stimulus time histogram (PSTH) in the pulse density mode (time-locked averages of spike train recording from single neurons, Fig. 6, left, or multiunit recording – not shown). The utility of the impulse response lies in the information it gives about the state of the system and its dynamic properties, because it contains all frequencies, and so it reveals the characteristic frequencies of the neural population in its current near-linear operating state. These properties are displayed in a state space diagram (Fig. 6, right) that is derived by linearization of the two coupled 2nd-order nonlinear ODE that model KIe dynamics in the vicinity of the prestimulus operating point (Fig. 5, A). Averaging of the impulse responses (Fig. 6, left) from repeated stimulation at regular time intervals is necessary to remove the microscopic background activity treated as *noise* and extract the mesoscopic *signal*, which is the trajectory by which the system returns to its prestimulus state after each impulse.

Linear analysis is the most powerful tool available to neurodynamicists [6, 7, 12, 34, 44, 63]. In the variant using root locus analysis the characteristic frequencies of the linearized system are represented as closed loop poles (small triangles Δ in Fig. 6, right), and the characteristic forbidden "anti-resonant" frequencies are shown as zeroes ((unknown char) in Fig. 6, right) in the roots of 2 coupled

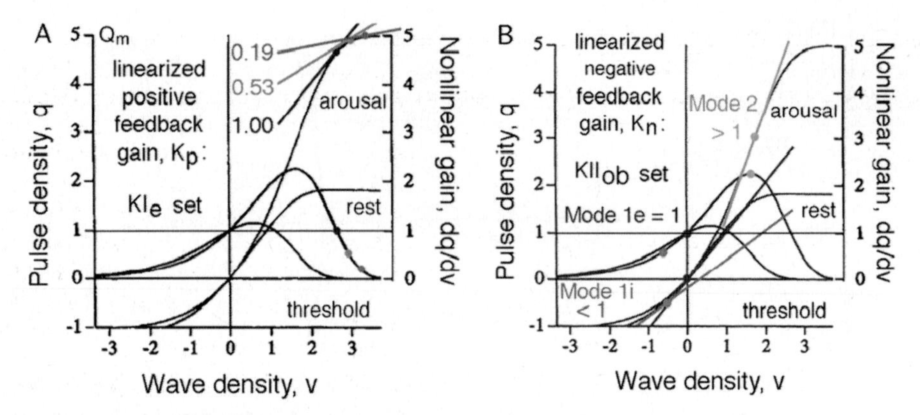

Fig. 6 **Left**: Impulse responses of a periglomerular cell on excitation of the primary olfactory nerve at 6 intensities. Data (Δ) in averaged post-stimulus time histograms were fitted with solutions to a matrix of four 1st order ODE. **Right**: Root loci (*dark lines*) of the changes in poles as a function of gain (elliptical contours), k_p. At zero gain the poles (Δ) lie at the open loop values (⊓), rise rate = -530/s, decay rate = -230.sl . The left-right arrows indicate the direction of change in the real rate constants as gain contours (*ellipses*) increase to infinity. The up-down arrows show the direction of approach of the complex conjugate roots to zeroes in the right half of the complex plane (not shown). From [14], Fig. 5.10 p. 289, Fig. 5.13 p. 292

2nd-order ODE representing the closed loop dynamics. The open loop poles in the feedback ODE become the zeroes in the closed loop, small-signal, linearized 4th- order transfer function (the symbols (unknown char) on the left half of the negative real axis in Fig. 6, right).

A very useful experimental control parameter is the intensity of the impulse (the product of amplitude and duration). Changing the stimulus intensity causes the characteristic frequencies in the response to change. The changes in frequency are translated graphically into the most useful physiological parameters, which is the strengths of interaction (gains) between populations of neurons. The frequencies (roots) change along the heavy lines (root loci), and the gains are calculated as the values for linearized closed loop gain (the intersections of the root loci with the contours in Fig. 6, right). When the intensity is raised or lowered, the characteristic frequencies of the open loop components of the system (the synapses and their strengths) are not changed, but the functional properties of the system are altered by the amplitude-dependent nonlinearity at the trigger zones that is described by the asymmetric sigmoid curve (Fig. 5, A). Changing the response amplitudes by changing the input intensity changes the characteristic frequencies of the closed loop poles and zeroes by changing the set points at which the tangent slope is calculated. With systematic step-wise change in stimulus intensity, the changed values of the closed loop poles inscribe the root loci (orthogonal to the gain contours).

These root loci (at zero radians phase in positive feedback and at p radians phase in negative feedback) quantify the degree of stability of the system and specify how a change in the input can bring about either greater stability with resistance to state transition (poles moving to the left away from the imaginary axis), or to lesser stability and an increase in likelihood of a transition to a new state (poles moving to the right toward the imaginary axis). The postulate to be considered is that the formation of an AM pattern during an act of perception might take place when poles shift to the right of the imaginary axis, giving exponential terms with positive rates of increase.

The arrowheads on the root loci in Fig. 6 indicate the direction in which gain increases with decreasing response amplitude. On the one hand, strong impulse that drives an increase in impulse intensity increases response amplitude and decreases feedback gain, which increases the decay rate to the prestimulus background level. The decrease in KIe feedback gain with increased input intensity is due to the refractory periods; neurons that are challenged to fire cannot do so if they have recently already fired. On the other hand, extrapolation of the decay rate of the PSTH to zero input (response threshold) gives zero decay rate. Such a response would be a step function [14]); it cannot be observed, because it has zero amplitude. The extrapolation of the decay rates to zero and the feedback gain to unity at threshold using this root locus demonstrates that the population is self-stabilized at its steady state excitatory output by refractory periods without need for inhibition. The pole at the origin of the complex plane (Fig. 6, right, large Δ) conforms to a zero eigenvalue (zero rate of change, an infinite time constant).

That pole represents the functional steady state of the KIe population that would hold in the absence of input: steady state excitatory pulse density output, the background activity. The state is stable, because excitatory input raises the output but decreases the gain, so the population activity tends toward the steady state level. Decreased excitatory input or increased inhibitory input does the reverse. Therefore the pole represents a point attractor, which differs from the state of zero activity under deep anesthesia or in death, in that the mutual excitation provides sustained, stable, steady state excitatory bias within itself and to other neural populations in the olfactory system. The non-zero point attractor governs the KIe population dynamics and modulates all other populations that receive the excitatory bias.

Direct experimental observation of the pole at the origin is obviously not possible, because the impulse response has zero amplitude and zero decay rate. Moreover, every KIe population has on-going input from other populations, including the periglomerular population getting both sensory input and centrifugal input from other parts of the forebrain, so the observed steady state gives a pole that is displaced to the negative side of zero frequency on the real axis of the complex plane (gain = 0.53).

The root locus method requires piecewise linearization of the dynamics. The technique depends on the fact that the impulse input does not change the system structure. Instead the input imposes an extra term, a perturbation superimposed on the background activity that reveals the dynamics by the trajectory of relaxation back to the prestimulus state: the *explicit breakdown of symmetry*, in contrast to the *spontaneous breakdown of symmetry* [33] that, as will be shown, characterizes perception. The same pair of 2nd-order ODE is applicable across the whole range of piecewise linearization but with a different value of the gain parameter for each level of input. More generally, populations of neurons that share the properties of thresholds, refractory periods, and mutual excitation can be described as governed by a non-zero point attractor, which describes the process by which excitatory populations maintain non-zero steady state background activity without need for inhibition to prevent runaway excitatory discharge. Thereby mutual excitation provides the excitatory bias in cortex that is essential for maintaining "spontaneous" background activity. This activity in many circumstances has been shown to have approximately a Gaussian noise amplitude histogram and a power-law power spectral density with slope near −2 (1/f^2, "brown noise" [55]).

4 Measuring and Modeling Oscillations Using Root Loci in Piecewise Linearization

The introduction of inhibitory neurons (Fig. 4, KII) is required not for stabilization of neural networks but for oscillations, most notably in the olfactory system for oscillations in the gamma range, 30–80 Hz (Fig. 1, A and Fig. 4 KII, Fig. 7, left). The specification of a central frequency in this range, 40 Hz, is immediately apparent from measurement of the open loop time constants (Fig. 4 KO) of the PSP or PSTH, the impulse responses of the non-interactive excitatory and inhibitory neurons comprising the olfactory populations. The exponential rise times average near 1.25 ms; the exponential decay times are near 5 ms. One cycle of oscillation requires two passages around the loop, giving a total delay of 25 ms (the wave

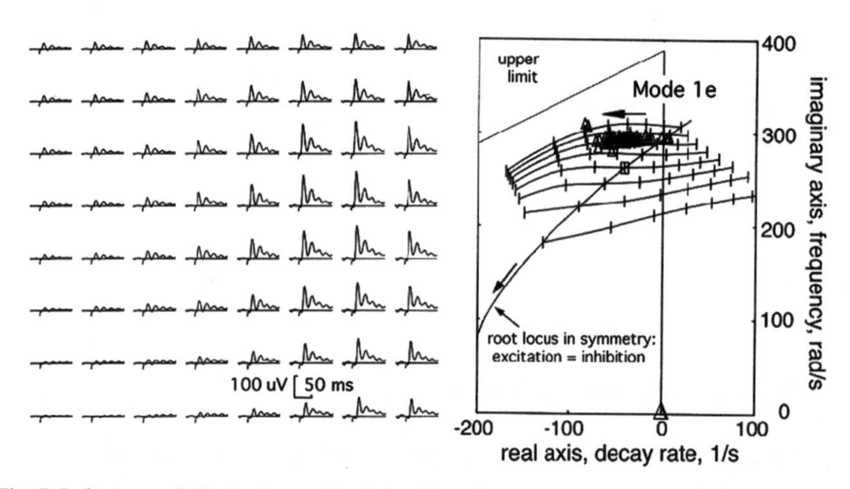

Fig. 7 **Left**: A set of 64 simultaneously derived impulse responses from an 8×8 4×4 mm array of electrodes for recording ECoG (the square frames in Fig. 3 left and the open square inset in Fig. 2, right) gave different amplitudes at different locations on single-shock stimulation of the olfactory nerve. The upward shift in baseline was provided by the KIe periglomerular neurons with positive gain k_p (shown by the PSTH in Fig. 6, left). Oscillations all had the same frequency, with decay rates proportional to amplitude. From [14], Fig. 4.27 p. 221. **Right**: the poles given by the frequencies and decay rates of the dominant damped cosine wave that was fitted to each of the 64 impulse responses gave root locus Mode 1e for the KII set with maintenance of unity gain (Fig. 5, B). The leftward arrows indicate the increased decay rate of the evoked gamma oscillation at sites with increased response amplitude due to gain reduction by refractory periods. From [14], Fig. 6.86 p. 361

duration of 40 Hz). However, the excitatory neurons excite each other, and the inhibitory neurons inhibit each other, as well as interacting in negative feedback (Fig. 1, KII). These two types of positive feedback — mutual excitation in the KIe population and mutual inhibition in the KIi population — modulate the frequency of negative feedback over the gamma range by variation in the 4 types of forward gain: excitatory or inhibitory synapses on excitatory or inhibitory neurons. If the four gains are set equal and changed concurrently, the symmetry of mutual excitation and mutual inhibition results in pole-zero cancellation. The root locus (phase = p radians in negative feedback) is shown for the upper half of the complex plane as the arc ("symmetry" in Fig. 7, right) rising with increasing negative feedback gain from the open loop pole (shown on Fig. 6 at a = –230/s) on the negative real axis and crossing the imaginary axis near 300 rad/s (50 Hz). The centered small rectangle (unknown char) identifies the average state or rest point that is observed for impulse responses at near-threshold impulse input by single-shock stimulation of the primary olfactory nerve [14]. This route of input to the bulb is called "orthodromic" because the direction of transmission conforms to the normal input of action potentials to the bulb from the sensory receptors in the nose. In contrast, electric excitation of the output path of the bulb, the lateral olfactory tract formed by axons of mitral cells, is called "antidromic", because the evoked action potentials travel into the bulb abnormally, that is, in the opposite direction to normal propagation (Fig. 8).

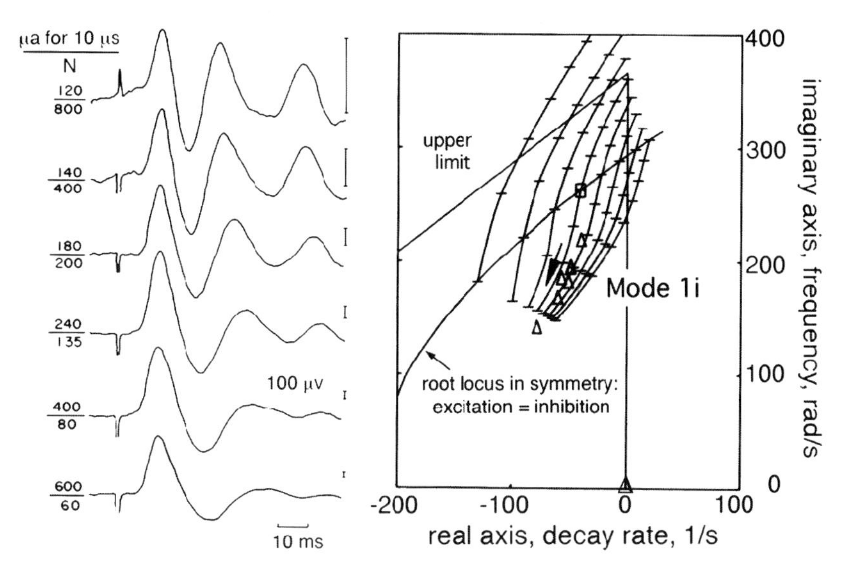

Fig. 8 **Left**: antidromic single-shocks to the lateral olfactory tract evoked impulse responses in the bulb, which bypassed the periglomerular neurons and the excitatory bias they contributed to the input. With increasing stimulus intensity the frequency decreased with little change in decay rate, giving a vertical root locus. **Right**: root loci from solving ODE for impulse responses predicting frequencies and decay rates. The same Mode 1i root loci are derived from impulse responses obtained with fixed impulse intensity while reducing the background activity by administration of barbiturate anesthesia. That is, the frequency of oscillation is determined by the ratio of evoked activity to background activity, when the evoked potential exceeds the background. From [14], Fig. 6.11, p. 374

Mode 1e shows the increased stability imposed by the bulbar mechanism in response to irrelevant or noisy input. The upper line shows the limit imposed by the value of the feedback gain in the KIe population that gives a pole at the origin of the complex plane for unity positive feedback gain, $k_p = 1$. That is, when the bombarding input to cortex abates, the cortex can settle toward its internal steady state amplitude of background activity governed by the non-zero attractor represented by the pole at the origin of the complex plane. Then the oscillatory pole tends toward a higher frequency and faster envelope of decay of the impulse response, but only to the limit of the diagonal line shown as the "upper limit" in the steady state. As with the pole at the origin, oscillator y impulse responses at the limit are not observable.

The impulse responses of the olfactory system (the averaged evoked potentials AEP) for which the peak amplitudes do not exceed the amplitude range of the ongoing background ECoG conform to additivity and proportionality on paired-shock testing, showing that there is a small-signal near-linear range in the dynamics of the major components of the central olfactory system. The arrow downward to the left in Fig. 7 shows the reduction in frequency and increase in decay rate when all feedback gains are reduced equally ("symmetry") by increased input intensity or by reduction of the background bias from KIe sets.

This symmetric root locus under the condition of equality of excitatory and inhibitory positive feedback is rarely seen in experimental root loci derived from evoked potentials, because the mutual excitation (KIe) and mutual inhibition (KIi) are rarely equal. When mutual excitation predominates, the root locus is horizontal (Mode 1e). Stimulation of the input tract to the bulb evokes potentials in which the decay rate is proportional to response amplitude over the spatial distribution of input but the frequency is constant (the horizontal root loci in Fig. 7, right). When mutual inhibition dominates, the root locus can approach vertical (Mode 1i. Fig. 8, right). Antidromic stimulation of the output tract of the bulb, the lateral olfactory tract, activates the inhibitory interneurons directly by dendrodendritic synapses, giving a strong inhibitory bias to the oscillations (Fig. 8, Mode 1i) that is not compensated by KIe input as it is in orthodromic input (Fig. 7, left), because the periglomerular cells are not activated by antidromic stimulation.

Mode 1e and 1i root loci are seen only when the amplitude of the single-shock evoked potentials exceeds the range of the background ECoG amplitudes. This is because the thresholds of the axons during recurrent inhibition limit the amplitude of the evoked activity that exceeds the spontaneous range. When neurons are inhibited below their thresholds, they cannot fire to signal the magnitude of inhibitory input. The thresholds block axonal output that signals excess dendritic inhibition; the block effectively reduces the feedback gain in the KII loop [14], Fig. 5.28, p. 335. In contrast, the refractory periods (not the thresholds) of the axons limit the upper range of background activity and the superimposed evoked potentials (Fig. 5, $k_n > 1$).

When the input intensity is fixed to give evoked potentials for which the peak-to-peak amplitudes do not exceed the range of background activity, the frequency and decay rate of successive averaged evoked potentials are high with low amplitude and high with high amplitude (downward-right arrow, Fig. 9). The variations provide evidence for spontaneous fluctuations in feedback gain and set point,

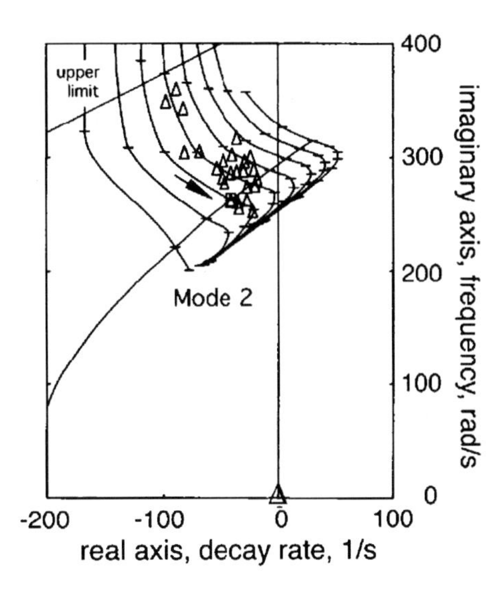

Fig. 9 Closed loop root loci in the upper half of the complex plane at p radians are revealed by the spontaneous variation of the oscillatory impulse responses in the small signal, near-linear range. The small triangles show the characteristic frequencies of the evoked potentials fitted with damped cosines. The arrow shows the direction of decreased decay rate and frequency with increased response amplitude (the reverse direction compared with Mode 1e in Fig. 7, right). From [14], Fig. 6.21, p. 374

owing to spontaneous fluctuations in the levels of the background activity. These relations differ markedly from those of Mode 1e and Mode 1i. Mode 2 root loci from solving the ODE to fit the data (Δ) cross the imaginary axis with increasing amplitude, which implies runaway excitation by positive feedback between amplitude and gain. However, the root loci turn back toward the imaginary axis and converge to a crossing point close to 250 radians/s (40 Hz). That convergence to a complex conjugate pair of poles on the imaginary axis predicts the existence of a limit cycle attractor for the KII set in the middle of the gamma range. The state is predicted to be stable, because further increase in amplitude is prevented by decreased feedback gain. In piecewise linear analysis of observations based on time averages that flatten the background activity (AEP and PSTH) this result predicts bistability: convergence of cortex either to a point attractor or a limit cycle attractor.

5 Definition of a Cortical Order Parameter Related to Information Content of Cortical Activity

The carrier frequency is everywhere the same across the olfactory bulb and likewise across each sensory neocortex, but it tends to vary in time. Owing to this frequency modulation (FM) the study of spatial patterns is optimally done using the instantaneous frequency derived from the Hilbert transform [20, 21, 39], which gives better temporal resolution of AM and PM patterns than does decomposition using the Fourier transform [20, 21]. The digitized values of 64 ECoG from an 8×8 array yield the conic PM pattern of phase modulation (PM) at the instantaneous frequency (Fig. 3, left), for which the isophase contours form equidistant circles around an apex (Fig. 3, right). This phenomenon is also found in all areas of sensory neocortex, which implies that communication within cortex is with a finite velocity across its surface that is imposed by the necessity for action potentials acting at synapses to provide for phase locking of oscillations at the time-varying carrier frequency. Any form of time averaging across multiple measurements of phase differences between two points must converge to zero phase lag [28] no matter how great the distance between the points.

The analytic amplitude, $A_j(t)$, at the j-th electrode of the 64 electrodes in the array co-varies widely with time. The variation is reflected in the spatial average, $\underline{A}(t)$ (red/grey curve in Fig. 10, A). When $\underline{A}(t)$ approaches zero, the phase becomes undefined. At each spatial location the rate of increase in phase (the instantaneous frequency) transits to a new value that is shared by waves in all other locations, independently of response amplitude. The spatially coherent oscillation re-synchronizes near a new average frequency, typically differing by ± 10–20 Hz from the average frequency in the prior frame. During the transition the spatial standard deviation of phase differences, $SD_X(t)$, fluctuates around high values (blue/grey curve in Fig. 10, A). There are two reasons. One reason is that at very low analytic amplitudes, the errors of measurement of phase are very large when the phase is indeterminate. The other reason is that the transition is not simultaneous but dispersed in accord with the phase cone but within the range ± p/4 (± 45°), the half-power value. Therefore the peaks $SD_X(t)$ provide a marker in the ECoG for locating widely synchronized jumps signifying distributed state transitions in the cortex.

This pattern of inverse relation between $SD_X(t)$ and $\underline{A}(t)$ occurs both in aroused subjects actively engaged in perception [5, 18, 20, 21, 29, 48] and in subjects at rest (Fig. 10, A). The pattern is simulated in a simple way [22]. The transfer function in the positive feedback loop modeling the dynamics of the KIe set can be approximated by a transcendental exponential term having the same form as a 1-D diffusion process. The autocorrelation, cross correlation and interval histograms conform to those of a Poisson process with refractory periods. The pulse trains of the component neurons can be modeled with random numbers. The sum of a large number

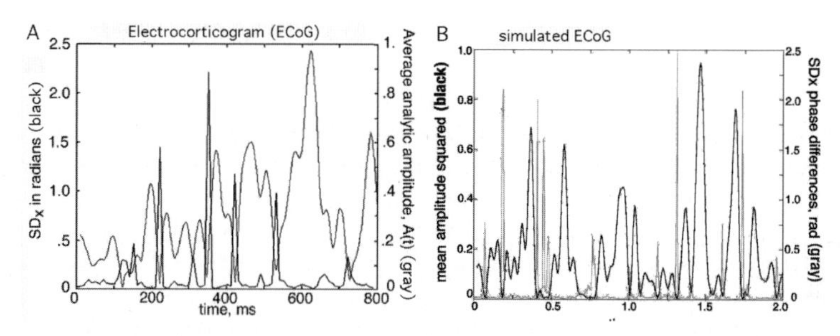

Fig. 10 (**A**) this is an example of ECoG from a human subject at rest, showing the background activity. When the spatial standard deviation of phase differences, $SD_X(t)$, falls to a low value, the mean analytic amplitude, $\underline{A}(t)$, increases to a maximum during the period of relatively constant frequency that follows a state transition [21]. (**B**) this pattern is simulated by passing brown noise $1/f^2$ through a band pass filter in the gamma range. Mean analytic amplitude and spatial variance of analytic phase co-vary inversely [22]

of identical, independently distributed random numbers converges to a power density spectrum with $1/f^2$ slope in log-log coordinates [55]. When this *brown noise* is passed through a band pass filter in the gamma range, the amplitude and phase vary in the manner shown in Fig. 10, B. The cortex provides the pass band at the characteristic frequency the negative feedback relation between the pyramidal (or mitral) cells and the inhibitory interneurons (or internal granule cells) (Fig. 7, right). The result is seen in the fluctuations of the gamma band activity in the ECoG; the amplitude histogram is close to Gaussian; the envelope of the activity conforms to the Rayleigh distribution [14], p. 148, Fig. 3.13c; and the power spectral density in log-log coordinates has a slope near –2. The spikes in $SD_X(t)$ recur erratically in the theta range, as revealed by correlation of ECoG with $SD_X(t)$ in both the real ECoG [20, 21, 26] and the simulation [22]. Therefore the spikes in $SD_X(t)$ in Fig. 10, A can be viewed as intrinsic to the cortical dynamics. The brown noise is generated by KIe dynamics (Figs. 4 and 6); the null spikes emerge when the noise is passed through the KII band pass filter (Fig. 4 and 7). The null spikes are not imposed by the sensory input; instead they reveal episodic intrinsic silencing of the background activity. That momentary silencing may enable perception for the following reason. The cortex passes undergoes a state transition only in the presence of suprathreshold sensory input. The threshold is materially reduced for stimuli for which a Hebbian assembly already exists with its attendant attractor and basin of attraction. The threshold may be still further reduced by the brief abatement in cortical background noise.

The analytic amplitude, $A_j(t)$, also varies with spatial recording location, giving the spatial AM pattern that is represented by a 64×1 vector, $\mathbf{A}(t)$. The vector is normalized by dividing each value of A_j (t) by the mean amplitude, $\underline{A}(t)$. While the AM pattern at every digitizing step is accompanied by a PM pattern, only the patterns that last 3 to 5 cycles of the sustained mean frequency and its phase cone are readily identified, measured, and classified with respect to stimulus categories. The AM patterns with classification of greatest statistical significance are those at the peak of $\underline{A}(t)$, when the rate of increase in the analytic phase (the instantaneous frequency) is nearly constant and the spatial variance in phase, $SD_X(t)$, is low. These are the AM patterns that are classifiable with respect to the CS given to subjects in whom the ECoG is being recorded (Fig. 2, right).

The normalized vector, $\mathbf{A}(t)$, representing AM patterns is adopted as the order parameter for cortex, because it measures the kind and degree of structure in the cortical activity. However, for descriptive purposes a scalar index is needed to show relations of the order parameter to other variables, in particular to the ECoG amplitude. The index is derived as follows. The normalized $\mathbf{A}(t)$ specifies a point in 64-space that represents the AM pattern. Similar patterns form a cluster of points that manifest an attractor in cortical state space. Multiple clusters reveal an attractor landscape. The change from one basin of attraction to another across a boundary (separatrix) is shown by a trajectory of successive digitized points for $\mathbf{A}(t)$, as cortical state shifts from one cluster to another cluster. The absolute rate of change in the order parameter, $D_e(t) = |\mathbf{A}(t) - \mathbf{A}(t\text{-}1)|$, is given by the Euclidean distance between successive points. $D_e(t)$ corresponds to the absolute value of the numerical derivative of the order parameter. It has high values during state transitions and decreases

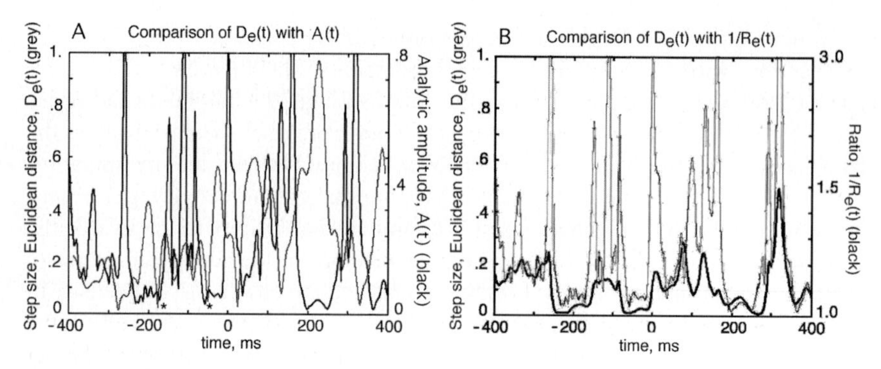

Fig. 11 (**A**): The rate of change in the order parameter, $D_e(t)$, falls to a low value well before a major increase in average analytic amplitude, $\underline{A}(t)$. (**B**): the index of synchrony, $R_e(t)$, (shown here as its reciprocal, $1/R_e(t)$) rises to near unity before the AM patterns stabilize (low $D_e(t)$) and well before $\underline{A}(t)$ increases, showing that increased synchrony is not the basis for increased analytic amplitude in frames of ECoG. From [20]

to low values when a new AM pattern emerges with convergence to an attractor, followed by a major increase in $\underline{\mathbf{A}}$(t) (Fig. 11, A). Therefore D_e(t) varies inversely with the degree of stability of the pattern specified by the vectorial order parameter [20] The degree of order also varies in proportion to the rate of free energy dissipated in cortical activity, which is manifested in the square of the mean ECoG analytic amplitude, $\underline{A}^2$(t). The best available index for locating in time the classifiable AM patterns in the ECoG [18] is the ratio, $H_e(t) = \underline{A}^2(t) / D_e(t)$. Atmanspacher and Scheingraber [3] define this quantity as the *pragmatic information* carried by the wave packet in the beta or gamma range. It is the ratio of the rate of dissipation of free energy to the rate of increase in the order parameter. H_e(t) also provides a unique scalar value to index each classifiable AM pattern as an inequivalent ground state [33].

Desynchronization is commonly associated with low-voltage fast activity, while synchronization is associated with high-voltage slow activity. An increase in mean ECoG analytic amplitude might occur with more synchrony, more total dendritic current, or both. The two factors are distinguished in beta-gamma ECoG records recorded with a high-density 8×8 array by calculating the temporal standard deviation, SD_T(t), of the average waveform in a moving window twice the duration of the wavelength of the center frequency of the pass band and dividing it by the average of the 64 $\underline{SD_T}$(t). When there is no synchrony, the ratio, $R_e(t) = SD_T(t) / \underline{SD_T}(x)$, approaches $1/n^{.5}$, where n is the number of time steps in the window. R_e(t) equals unity when there is complete synchrony (identical instantaneous frequency and phase), whether or not there are AM patterns. R_e(t) is used to show that the major increases in $\underline{A}$(t) during AM pattern formation are not attributable to increases in the level of synchrony (Fig. 11, B), which occur before amplitude increases.

Every classifiable AM pattern has a phase cone. In the olfactory bulb there appears to be only one phase cone at a time for the duration of the gamma burst. When there is more than one distinguishable component of the time-varying carrier frequency, each peak frequency has its phase cone with the same location and sign of the apex and the same phase velocity. The dual phase cones manifest the same broad-spectrum event. In neocortex there are multiple overlapping phase cones at all times, giving the appearance of a pan of boiling water. The frequencies, signs and locations of the phase cones differ as much as they vary between successive events [21]. The distributions of the durations are power-law, and likewise of the diameters within the limits of measurement. The distributions of carrier frequencies and more generally of the power spectral densities in both time and space [21, 22, 27, 30] of neocortex are power-law. These multiple power-law distributions of the derived state variables suggest that neocortex is stabilized in a form of self-organized criticality (SOC) as first described by Bak, Wiesenfeld and Tang [4] and in more detail by Jensen [38].

The classic object of study of SOC is a sand pile fed from above as at the bottom of an hourglass. The steady drip of sand creates a conic pile that increases slope to a critical angle of repose that is held thereafter by repeated avalanches. Likewise a pan of water brought to a boil holds a constant temperature by forming bubbles of steam. An open system of interacting elements, whether grains of sand, water molecules or neurons, evolves to a stable steady state that is far from equilibrium. That global state is maintained at pseudo-equilibrium by repeated adjustments: avalanches of sand, bubbles of steam, or state transitions among groups of neurons. The records of the changes appear chaotic and give temporal and spatial spectra with $1/f^{\alpha}$ forms implying self-similarity across wide scales of time and space. Likewise the ECoG gives records that appear chaotic, with temporal and spatial power spectral densities that conform to $1/f^{\alpha}$, where the exponent α has been calculated as ranging $1 < \alpha < 3$ [20, 21, 22, 27, 51]. In all three systems the appearance of "noise" is illusory. When viewed from the proper perspective, informative structures in the "noise" become clear. The critical parameter that is maintained by SOC in cortex is proposed to be the mean firing rate of the neurons comprising the interactive populations in cortex, which is regulated homeostatically everywhere in cortex by the refractory periods. The mechanism is represented by a point attractor expressed in the pole at the origin of the complex plane (Figs. 6,7,8 and 9, right). Like avalanches the times and locations of cones are predictable not locally but only in the average. They overlap so that any neuron may participate in multiple cones simultaneously. The sizes and durations of cones give histograms that are power-law. The smallest and briefest cones are the most numerous. Their means and SD change in proportion to the size of the measuring windows [22, 30]. The distributions show that the neural patterns may be self-similar across multiple scales [12, 36, 37, 45, 51]. These functional similarities indicate that neocortical dynamics is scale-free [10, 19, 62]: the largest events are in the tail of a continuous distribution and share the same mechanism of onset and the same brief transit time of onset despite their large size.

6 A Proposed Phase Diagram to Represent the Dynamics of Cerebral Cortex

The classic phase diagram of a substance such as water at thermodynamic equilibrium (Fig. 12, A from Blauch [8]) serves as a platform from which to construct a diagram for cortex as a system operating far from equilibrium, yet maintaining a conditionally stable steady state (Fig. 12, B). The classic system is static and

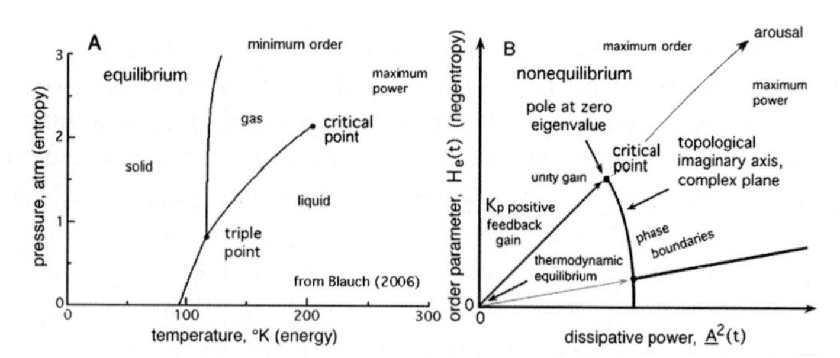

Fig. 12 (**A**) conventional phase diagram at equilibrium, from [8]. (**B**) by analogy to physical matter, a phase diagram is proposed for cortical dynamics. Far from equilibrium the levels of order and power are conceived to increase together with increased arousal (Fig. 1). The critical point and imaginary axis are taken from the complex plane (Fig. 9). Arousal from deep anesthesia (the "open loop" state with flat ECoG, KO in Fig. 4) leads to stabilization at a pseudo-equilibrium that is maintained by the refractory periods of the excitatory interactions among pyramidal (mitral) cells. Transformation of A to B is by translating the origin in B to the critical point shown in A and constructing two new orthogonal dimensions for change: the rate of pragmatic information increase (negentropy) as a scalar index of order on the ordinate and the rate of dissipation of free energy (power) on the abscissa

closed. The level of energy displayed on the abscissa is indexed by temperature. The degree of order displayed on the ordinate is measured by pressure, volume or entropy. Maximum order occurs at minimum energy. Three phases are separated by phase boundaries that meet at a triple point. However, the phase boundary between liquid and gas ends at a *critical point*, beyond which the phase of constituents is undefined.

Brains are open, dynamic systems that continually dissipate free energy [33, 60] in burning glucose; brains constitute 5% of body mass yet consume 20% of basal metabolic rate. Two new state variables are required: the rate of energy dissipation (power) on the abscissa, and the rate of increase in information (negentropy) on the abscissa, indexing the order emergent in dissipative structures [52]. In the dynamic display the equilibrium at criticality appears as a point at the origin of the new state variables. Equilibrium is approached in brain death and in the non-interactive state imposed by deep anesthesia (Fig. 4, KO). With recovery from anesthesia the degree of order and the rate of energy increase together (the diagonal line, Fig. 12, B).

Brain temperature is unsuitable as an index of power, because birds and mammals use homeostatic feedback to hold brain temperatures within a narrow range. A useful index of power is the mean square analytic amplitude, $\underline{A}^2(t)$, because ECoG amplitude depends on current density across the relatively fixed extracellular specific resistance of cortical tissue. The square of current or voltage is proportional to the power expended by neurons generating the dendritic potentials. This dissipation is the basis for imaging brain activity by measuring cerebral blood flow using fMRI, SPECT, PET, etc., because 95% of brain energy is consumed by oxidative metabolism in dendrites, only 5% in axons.

The degree of order on the ordinate is indexed by the scalar ratio specifying pragmatic information, $H_e(t)$. This state variable makes explicit the premise that order in dissipative structures increases with increasing power [52]. Cortex indexCortexbreak operates as an open system far from equilibrium (Fig. 12, B), yet it is self-stabilized while transiting through a collection of states by processes variously denoted as orbiting in metastability [42, 57], chaotic itinerancy [40]; transient coherence in unitarily inequivalent ground states [33]; and bifurcations [35] among hierarchical levels by neuropercolation [Kozma Chapter] through an extension of random graph theory [42]. In all formulations the magnitude of the background activity at which the cortex is stabilized can vary; it generally co-varies with the level of arousal (Fig. 1). In cortex the stability is maintained by the refractory periods, so that no matter how great the increase in arousal, the output of the KIe populations is self-stabilized at unity feedback gain. The pole at the origin of the complex plane (Δ at the origin in Figs. 6,7,8 and 9) corresponds to the critical point (Δ) on the diagonally upward line (Fig. 12, B), so that the metastability of cortex can be described as an instance of self-organized criticality (SOC). Thereby the steady state can be described as a phase in a system at pseudo-equilibrium, and the state transition can be labeled as a phase transition. The demarcation by the imaginary axis that is accessible to piecewise linear analysis between stable and unstable states (Figs. 6,7,8 and 9) in the upper half of the complex plane is seen to correspond to a phase boundary (Fig. 12, B) between bistable receiving and transmitting states (Fig. 13). A second phase boundary that is inaccessible to linear analysis separates a phase domain of self-sustained complex partial seizure [16].

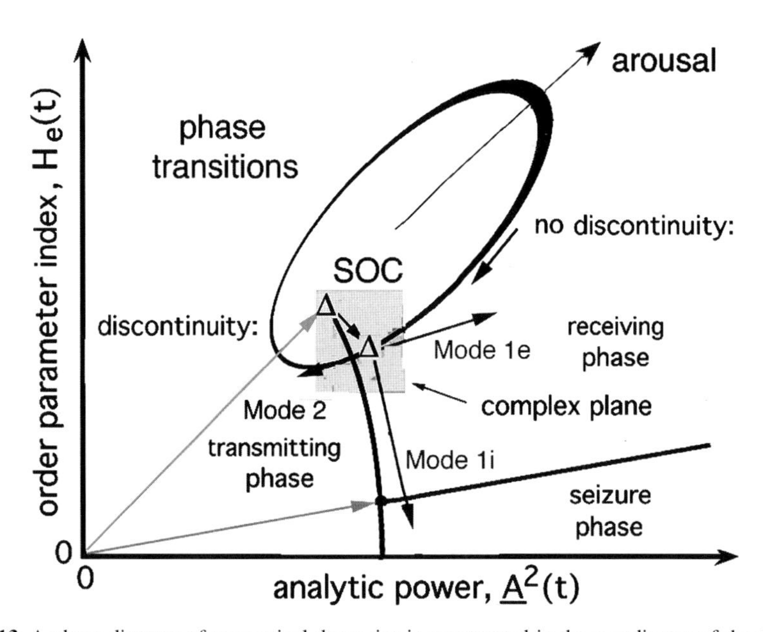

Fig. 13 A phase diagram of neocortical dynamics is constructed in the coordinates of the degree of order given by the rate of change in the order parameter, $H_e(t)$, as the dependent variable and the average analytic power, $\underline{A}^2(t)$, as the independent variable. The central grey area shows the linearized subspace of the complex plane in Figs. 6,7,8 and 9

These properties lead to a view of perception as a cyclical process (ellipse in Fig. 13). The cycle begins with cortex in a receiving state with an attractor landscape already established by learning from input and selected by the limbic system in preafference [39]. The pole at the origin representing the stable, non-zero point attractor of the KIe set (Δ at the origin in Figs. 6,7,8 and 9, right) specifies the location of the critical point (upper Δ) maintained by SOC, which is located on the 45° line that represents increasing order with increasing power of neural activity in accord with Prigogine's [52] "order from disorder" in formation of "dissipative structures" and Haken's [35] "slaving principle". Three phases are conceived in a simplified approach: receiving, transmitting, and seizure. The set point (Fig. 13, lower Δ) corresponding to the center small rectangle (unknown char) in Figs. 7,8 and 9 is postulated not to lie on the upwardly diagonal line but below the line. It is offset in sensory cortices by the background input from sensory receptors, and in other cortices by background input from other parts of the brain. The displacement (small downward arrow between triangles Δ — Δ in Fig. 13) reflects gain reduction below unity in for the smallest observable impulse responses (Fig. 6, A). The displacement represents decreased order and increased power.

The root locus in Mode 1e is represented by the nearly horizontal rightward arrow from the set point (lower Δ) indicating increased power and small increase in order with orthodromic evoked potentials above the range of background activity. The downward orange arrow indicates the root locus in Mode 1i, showing the increased power and decreased order accompanying impulse responses involving strong inhibition. The leftward magenta arrow indicates the root locus in Mode 2, which crosses the phase boundary (imaginary axis) in a phase transition. Mode 2 appears diametrically opposed to Mode 1e in showing a decrease in power with a small decrease in order, tending toward instability.

The most remarkable feature of the diagram in Fig. 13 is the description of the phase transition in an act of perception as *beginning with decreased power and decreased order*. This feature is counterintuitive, because perception is usually conceived as resulting from the impact of a sensory volley that increases power and imposes increased order by the injection of information from a stimulus. Certainly that takes place, but the phase diagram adds a different dimension to the story, as follows. The event that initiates a phase transition is an abrupt decrease in the analytic power of the background activity to near zero, as is shown to occur in Fig. 10, A, and as is simulated in B. This reduction induces a brief state of indeterminacy, in which the amplitude of the ECoG is near zero and phase of the ECoG is undefined. If a stimulus-induced volley arrives at or just before this state, then the cortex in accord with Mode 2 can be driven by the input across the phase boundary. The Mode 2 root loci in Fig. 9 show that exogenous input (as distinct from endogenous activity) increase amplitude and also instability by bringing the cortex closer and across the imaginary axis. The response amplitude depends not on the input amplitude but on the intrinsic state of the cortex, specifically the degree of reduction in the power and order of the background brown noise.

If the phase transition occurs, it re-sets the carrier frequency of oscillation through a discontinuity in analytic phase. Then the oscillation converges to a shared frequency (Fig. 9) as shown by the reduction in $SD_X(t)$ in Fig. 10, A. Thereafter in succession follow re-synchronization as shown by increased $R_e(t)$ in Fig. 11, B; increased order as a stable pattern emerges with decreased $D_e(t)$ in Fig. 11, A; and lastly increased mean power $\underline{A}^2(t)$ in Fig. 11, A. After power has peaked, the cortex returns to its receiving state by a second phase transition without a discontinuity in the oscillatory phase. According to this hypothesis the cycle for the opportunity to undergo a phase transition repeats aperiodically at an intrinsic rate in the theta range. The actualization of a phase transition requires the presence of a surge of input to the cortex. In the experimental conditioning paradigm used in this study, the input surge is a sensory volley brought to the cortex by a behavioral act of observation, so the gamma ellipse in Fig. 13 can properly be called a representation of the neural trajectory of an action-perception cycle [17, 19, 28]. Conceivably the input surge can also be provided in sensory cortices by corticocortical transmission from the limbic system. Such a phase transition may underlie the formation of beta wave packets in a "virtual" action-perception cycle.

7 Conclusions and Summary

The significance of the reduction in power prior to initiation of a phase transition can be seen in the fact that an expected conditioned stimulus typically is a very weak signal that is embedded in contextual noise. The afferent volley activates the sensory cortex and actualizes the attractor landscape that expresses the several possible categories of CS that are expected during an intentional act of observation. A Hebbian nerve cell assembly that was formed in prior learning governs each attractor. The relatively few action potentials sent by the relevant receptors can ignite one of the cell assemblies, which can direct the cortex into the corresponding basin of attraction. That moment of selection can be conceived as done optimally when the background activity is quenched, and the output of a relevant Hebbian assembly has maximal signal:noise ratio in selecting a basin to which to guide the entire sensory cortex. When the background power then increases, it is imprinted with the AM pattern provided by an attractor in the landscape. The power is not provided by the sensory input; it is provided by the intrinsic mutual excitation. The null spike may be likened to the eye of a hurricane—a transient interlude of silence that opens the possibility for a change in direction of movement. The next null spike quenches the cortical activity and releases the dynamics from the attractor as the AM pattern disappears.

The null spike in the band pass filtered brown noise activity is conceived as a *shutter* [43] that blanks the intrinsic background. At very low analytic amplitude when the analytic phase is undefined, it may be that the system trajectory approaches a singularity that enhances the likelihood for a small extrinsic sensory input to re-set the background activity in a new frame. The null spike does not qualify as a gate, because the sensory input is not blocked or withheld, and the AM pattern in the following frame, if any, is formed by reorganization of existing activity through attractor selection, not by the driving of cortical activity by input as in the "information processing" model. The null spike is not a scanner [61], because the selection of a basin of attraction is by competition among cell assemblies in state space and not by a search for a module in cortex. It is not really a clock, because the recurrence of null spikes is aperiodic in a limited range, and the formation of an AM pattern requires an exogenous factor, the input as from a sniff or a saccade [58]. Gating, scanning, and clock timing [46] are better conceived as mesoscopic thalamocortical operations Bressler Chapter], while the precise timing of the phase transition is performed within the neural populations that are to construct the wave packet carrying an AM pattern. Exploration will require development of a mathematical foundation for the null spike, which does not currently exist. Two promising bases for research are foreseen in random graph theory [9, 42] and quantum field theory [33, 60]. They are approaches widely used to describe processes by which microscopic events might be up-scaled into macroscopic patterns in complex systems.

Figure 13 offers the basis for more complicated graph in which to synthesize and display further data and concepts derived from studies of the nonlinear dynamics of cortex. This simple form can be elaborated to include additional data from drug studies. For example, induction of anesthesia by barbiturates is by enhancement of inhibition that is replicated by the root locus in Mode 1i. The same root loci are obtained by reducing the background activity with barbiturate or by increasing the impulse intensity (Fig. 8). At the extreme level of flattening the ECoG, the gain k_p = 0, and a point attractor with zero amplitude is revealed [32] giving the open loop impulse response (Fig. 4, KO). An instability is manifested in barbiturate spindles: brief bursts of oscillation in the theta range [14]. Stages of sleep might be represented by yet other phases.

Brief tetanization of the lateral olfactory tract at supra-maximal intensity can induce a complex partial seizure with *absence* and 3/s spike-and-wave [16]. This autonomous pattern of activity implies the existence of the lower phase boundary that separates the receiving and transmitting states from an epileptic domain with very high power and minimal order, which is accessed paradoxically by raising the activity of mutually inhibitory neurons to high intensity. Remarkably the seizure spike repetition rate is also in the same theta range as the null spike. The seizure spike-and-wave would form a much larger ellipse (not shown).

The normal ellipse might be regarded as a 2-D projection of a torus, which manifests a quasi-periodic attractor, in which the low frequency of burst repetition in the theta range is represented by the diameter of the torus, and the high frequency of the burst carrier wave in the beta or gamma range is represented by the thickness of the torus. The torus might in turn be regarded as a projection of a helix extending into the time domain in 3-D. Multiple ellipses might serve to represent high-dimensional chaotic attractor landscapes [32], dynamic memory systems in neuropercolation theory [56], and multiple ground states modeled by dissipative quantum field theory [16, 60]. More immediately, a solid mathematical foundation is needed beyond ODE, in order to derive and describe more precisely the forms of the neural action-perception cycles and the phase boundaries that they cross, for which ODE are not fully suitable.

In summary, the olfactory bulb, nucleus and cortex constitute a semi-autonomous system in each cerebral hemisphere that receives odorant information, categorizes an odor percept by generalization and abstraction, and transmits the percept broadly through the forebrain. The form of the percept is a frame constructed and carried by a wave packet with an aperiodic "chaotic" carrier oscillation in the gamma range. The content is expressed in a spatial pattern of amplitude modulation (AM) of the carrier wave that is determined by modified synapses among bulbar mitral cells that form Hebbian nerve cell assemblies in reinforcement learning. Each cell assembly for a discriminated odorant guides the bulb to an AM pattern constructed by a "chaotic" attractor. The collection of learned categories of odors is retained in the form of an attractor landscape. Each attractor is surrounded by a basin of attraction; the processes of generalization and abstraction occur when the system converges to the attractor, regardless of where in a basin a stimulus places the system. Three requirements met by the olfactory system in percept formation are a state transition from a receiving state to a transmitting state, another to return, and a third for repeated sampling of the olfactory environment: a shutter.

Dynamics in the visual, auditory and somatic perceptual systems share these properties. In this Chapter evidence from ECoG of allocortex and neocortex serves to describe the neural mechanisms that generate and stabilize cortical carrier waves; that enable the state transitions; and that provide a shutter to terminate old frames and initiate new ones in cinematographic sequences in every sensory system. It is postulated that a percept forms a wave packet only when the endogenous background activity briefly abates in a "null spike", which provides the high signal:noise ratio that a burst of cortical activity that is driven by input relayed from sensory receptors needs to select and initiate a new frame. The aperiodic sequence of null spikes provides the shutter that is a necessary though not sufficient condition for the cinematographic process of perception.

Acknowledgment I am grateful to Mark D. Holmes and Ceon Ramon, University of Washington, Seattle WA and Sampsa Vanhatalo, University of Helsinki, Finland for data and modeling using 64 channels of the 256-channel recording System 200 provided by Don Tucker, Electrical Geodesics Inc., Eugene OR. For critical insights I am grateful to physicists Giuseppe Vitiello, Salerno University and Michael Stringer, Northwestern University, and to mathematicians Robert Kozma, Memphis University and Béla Bollobás, Cambridge University. A preliminary version of this Chapter appears in Proc. IJCNN'07 [23].

References

1. *Neurodynamics. An Exploration of Mesoscopic Brain Dynamics.* London: Springer, 2001.
2. D. J. Amit. The hebbian paradigm reintegrated: Local reverberations as internal representations. *Behavioral and Brain Science*, 18:617–657, 1995.
3. H. Atmanspacher and H. Scheingraber. Pragmatic information and dynamical instabilities in a multimode continuous-wave dye laser. *Canadian Journal of Physics*, 68:728–737, 1990.
4. P. Bak, C. Tang C, and K. Wiesenfeld. Self-organized criticality: an explanation of 1/f noise. *Physical Review Letters*, 59:364–374, 1987.
5. J. M. Barrie, W. J. Freeman, and M. Lenhart. Modulation by discriminative training of spatial patterns of gamma eeg amplitude and phase in neocortex of rabbits. *Journal of Neurophysiology*, 76:520–539, 1996.

6. E. Basar. Eeg - brain dynamics. *Amsterdam: Elsevier*, 1980.
7. E. Basar. Brain function and oscillations. *Berlin: Springer-Verlag*, 1998.
8. D. N. Blauch. Chemistry experiments & exercises: Phase changes. 2006.
9. B. Bollobás. Random graphs, cambridge studies in advanced mathematics 2nd ed. *Cambridge UK: Cambridge University Press*, 1985/2001.
10. B. Bollobás and O. Riordan. *Results on scale-free random graphs*. Weinhiem: Wiley-VCH, 2003.
11. S. L. Bressler and J. A. S. Kelso. Cortical coordination dynamics and cognition. *Trends in Cognitive Science*, 5:26–36, 2001.
12. C. L. Chapman, P. D. Bourke, and J. J. Wright. Spatial eigenmodes and synchronous oscillation: coincidence detection in simulated cerebral cortex. *Journal of Mathematical Biology*, 45:57–78,, 2005.
13. J. D. Emery and W. J. Freeman. Pattern analysis of cortical evoked potential parameters during attention changes. *Physiololgy and Behavior*, 4:67–77, 1969.
14. W. J. Freeman. Mass action in the nervous system. *New York: Academic Press*, 1975.
15. W. J. Freeman. Nonlinear gain mediating cortical stimulus-response relations. *Biological Cybernetics*, 33:237–247, 1979.
16. W. J. Freeman. Petit mal seizure spikes in olfactory bulb and cortex caused by runaway inhibition after exhaustion of excitation. *Brain Research Reviews*, 11:259–284, 1986.
17. W. J. Freeman. *Societies of Brains. A Study in the Neuroscience of Love and Hate*. Mahwah NJ: Lawrence Erlbaum Assoc., 1995.
18. W. J. Freeman. Origin, structure, and role of background eeg activity. part 3. neural frame classification. *Clinical. Neurophysiology*, 116(5):1118–1129, 2005.
19. W. J. Freeman. Definitions of state variables and state space for brain-computer interface. part 1. multiple hierarchical levels of brain function. *Cognitive Neurodynamics*, 1(1): 3–14, 2006.
20. W. J. Freeman. Origin, structure, and role of background eeg activity. part 1. phase. *Clinical. Neurophysiology*, 115:2077–2088, 2006.
21. W. J. Freeman. Origin, structure, and role of background eeg activity. part 2. amplitude. *Clinical. Neurophysiology*, 115:2089–2107, 2006.
22. W. J. Freeman. Origin, structure, and role of background eeg activity. part 4. neural frame simulation. *Clinical. Neurophysiology*, 117(3):572–589, 2006.
23. W. J. Freeman. Cortical aperiodic 'clock' enabling phase transitions at theta rates. *Proceedings, International Joint Conference on Neural Networks (IJCNN)*, 2007.
24. W. J. Freeman and B. Baird. Relation of olfactory eeg to behavior: Spatial analysis. *Behavioral Neuroscience*, 101:393–408, 1987.
25. W. J. Freeman and B. C. Burke. A neurobiological theory of meaning in perception. part 4. multicortical patterns of amplitude modulation in gamma egg. *International Journal of Bifurcation and Chaos*, 13:2857–2866, 2003.
26. W. J. Freeman, B. C. Burke, and M. D. Holmes. Aperiodic phase re-setting in scalp eeg of beta-gamma oscillations by state transitions at alpha-theta rates. *Human Brain Mapping*, 19(4):248–272, 2003.
27. W. J. Freeman, B. C. Burke, M. D. Holmes, and S. Vanhatalo. Spatial spectra of scalp eeg and emg from awake humans. *Clinical. Neurophysiology*, 114:1055–1060, 2003.
28. W. J. Freeman, G. Gaál, and R. Jornten. A neurobiological theory of meaning in perception. part 3. multiple cortical areas synchronize without loss of local autonomy. *International Journal of Bifurcation and Chaos*, 13:2845–2856, 2003.
29. W. J. Freeman and K. Grajski. Relation of olfactory eeg to behavior: Factor analysis. *Behavioral Neuroscience*, 100:753–763, 1987.
30. W. J. Freeman, M. D. Holmes, G. A. West GA, and S. Vanhatalo. Fine spatiotemporal structure of phase in human intracranial egg. *Clinical Neurophysiology*, 117: 1228–1243, 2006.
31. W. J. Freeman and L. J. Rogers. A neurobiological theory of meaning in perception. part 5. multicortical patterns of phase modulation in gamma eeg. *International Journal of Bifurcation and Chaos*, 13:2867–2887, 2003.

32. W. J. Freeman and W. Schneider. Changes in spatial patterns of rabbit olfactory eeg with conditioning to odors. *Psychophysiology*, 19:44–56, 1982.
33. W. J. Freeman and G. Vitiello. Nonlinear brain dynamics as macroscopic manifestation of underlying many-body field dynamics. *Physics of Life Reviews*, 3:93–118, 2006.
34. E. Gordon. Integrative neuroscience. *Sydney: Harwood Academic*, 2000.
35. H. Haken. *What can synergetics contribute to the understanding of brain functioning?* C. Uhl (Ed.) Berlin: Springer-Verlag, 1999.
36. R. C. Hwa and T. Ferree. Scaling properties of fluctuations in the human electroencephalogram. *Physics Review*, 66, 2002.
37. L. Ingber. Statistical mechanics of multiple scales of neocortical interactions. *Nunez PL (ed.) Neocortical Dynamics and Human EEG Rhythms, New York: Oxford University Press*, pp. 628–681, 1995.
38. H. J. Jensen. Self-organized criticality: Emergent complex behavior in physical and biological systems. *New York: Cambridge University Press*, 1998.
39. L. M. Kay and W. J. Freeman. Bidirectional processing in the olfactory-limbic axis during olfactory behavior. *Behavioral Neuroscience*, 112:541–553, 1998.
40. J. A. S. Kelso. Dynamic patterns: The self-organization of brain and behavior. *Cambridge: MIT Press*, 1995.
41. R. Kozma and W. J. Freeman. Chaotic resonance: Methods and applications for robust classification of noisy and variable patterns. *International Journal of Bifurcation and Chaos*, 10:2307–2322, 2001.
42. R. Kozma, M. Puljic, P. Balister, B. Bollobás, and W. J. Freeman. Phase transitions in the neuropercolation model of neural populations with mixed local and non-local interactions. *Biological Cybernetics*, 92:367–379, 2005.
43. K. S. Lashley. *Brain Mechanisms and Intelligence.* Chicago IL: University of Chicago Press, 1929.
44. D. T. J. Liley, M. P. Dafilis, and P. J. Cadusch. A spatially continuous mean field of electrocortical activity. *Network: Computational Neural Systems*, 13:67–113, 2002.
45. K. Linkenkaer-Hansen, V. M. Nikouline, J. M. Palva, and R. J. Iimoniemi. Long-range temporal correlations and scaling behavior in human brain oscillations. *Journal of Neuroscience*, 15:1370–1377, 2001.
46. M. S. Matell and W. H. Meck. Neuropsychological mechanisms of interval timing behavior. *BioEssays*, 22(1):94–103, 2000.
47. M. Merleau-Ponty. Phenomenology of perception. *New York: Humanities Press*, 1945/1962.
48. F. W. Ohl, H. Scheich, and W. J. Freeman. Change in pattern of ongoing cortical activity with auditory category learning. *Nature*, 412:733–736, 2001.
49. J. Orbach. The neuropsychological theories of lashley and hebb. *Psycoloquy*, 10(29), 1999.
50. J. Panksepp. Affective neuroscience: The foundations of human and animal emotions. *Oxford UK: Oxford University Press*, 1998.
51. E. Pereda, A. Gamundi, R. Rial, and J. Gonzalez. Non-linear behavior of human eeg – fractal exponent versus correlation dimension in awake and sleep stages. *Neuroscence Letters*, 250:91–94, 1998.
52. I. Prigogine. From being to becoming: Time and complexity in the physical sciences. *San Francisco. W. H. Freeman*, 1980.
53. G. Viana Di Prisco and W. J. Freeman. Odor-related bulbar eeg spatial pattern analysis during appetitive conditioning in rabbits. *Behavioral Neuroscience*, (99):962–978, 1985.
54. O. Sacks. In the river of consciousness. *New York Book Revue*, 51(1), 2004.
55. M. Schroeder. Fractals, chaos, power laws. *San Francisco: W. H. Freeman*, 1991.
56. L. Siklós, M. Rickmann, F. Joó, W. J. Freeman, and J. R. Wolff JR. Chloride is preferentially accumulated in a subpopulation of dendrites and periglomerular cells of the main olfactory bulb in adult rats. *Neuroscience*, 64:165–172, 1995.
57. C. A. Skarda and W. J. Freeman. How brains make chaos in order to make sense of the world. *Behavioral and Brain Science*, 10:161–195, 1987.
58. L. W. Stark, C. M. Privitera, H. Yang, M. Azzariti, Y. F. Ho, T. Blackmon, and D. Chernyak. Representation of human vision in the brain: How does human perception recognize images? *Journal of Electronic Imaging*, 10(1):123–151, 2001.

59. I. Tsuda. Towards an interpretation of dynamic neural activity in terms of chaotic dynamical systems. *Behavioral and Brain Sciences*, 24:793–810, 2001.
60. G. Vitiello. My double unveiled. *Amsterdam: John Benjamins*, 2001.
61. W. G. Walter. *The Living Brain*. New York: W. W. Norton, 1953.
62. X. F. Wang and G. R. Chen. Complex networks: small-world, scale-free and beyond. *EEE Circuits and Systems*, 31:6–20, 2003.
63. J. J. Wright, C. J. Rennie, G. J. Lees, P. A. Robinson, P. D. Bourke, C. L. Chapman, E. Gordon, and D. L. Rowe. Simulated electrocortical activity at microscopic, mesoscopic and global scales. *Journal of Neuropsychopharmacology*, 28:S80–S93, 2003.

Toward a Complementary Neuroscience: Metastable Coordination Dynamics of the Brain

J. A. Scott Kelso and Emmanuelle Tognoli

Abstract Metastability has been proposed as a new principle of behavioral and brain function and may point the way to a truly complementary neuroscience. From elementary coordination dynamics we show explicitly that metastability is a result of a symmetry breaking caused by the subtle interplay of two forces: the tendency of the components to couple together and the tendency of the components to express their intrinsic independent behavior. The metastable regime reconciles the well-known tendencies of specialized brain regions to express their autonomy (segregation) and the tendencies for those regions to work together as a synergy (integration). Integration $\sim$ segregation is just one of the complementary pairs (denoted by the tilde ($\sim$) symbol) to emerge from the science of coordination dynamics. We discuss metastability in the brain by describing the favorable conditions existing for its emergence and by deriving some predictions for its empirical characterization in neurophysiological recordings.

Key words: Brain · Metastability · The complementary nature · Coordination indexCoordinationbreak dynamics · Consciousness

1 Prolegomenon

This essay starts with some *considerata* for science in general and cognitive computational neuroscience, in particular. It then focuses on a specific, empirically grounded model of behavioral and brain function that emanates from the theoretical framework of coordination dynamics. This model contains a number of attractive properties, one of which, metastability, has been acclaimed as a new principle of brain function. The term metastability is on the rise; it is well-known in physics and has been embraced by a number of well-known neuroscientists. As we explain, it is not the word itself that matters, but rather what the word means for understanding brain and cognitive function. In coordination dynamics, metastability is not a concept or an idea, but a fact that arises as a result of the observed self-organizing nature of both brain and behavior. Specifically, metastability is a result of broken symmetry in the relative phase equation that expresses the coordination

L. I. Perlovsky and R. Kozma, *Neurodynamics of Cognition and Consciousness.*

between nonlinearly coupled (nonlinear) oscillators. The latter design is motivated by empirical evidence showing that the structural units of the brain which support sensory, motor and cognitive processes typically express themselves as oscillations with well-defined spectral properties. According to coordination dynamics, nonlinear coupling among heterogeneous components is necessary to generate the broad range of brain behaviors observed, including pattern formation, multistability, switching (sans "switches"), hysteresis and metastability. Metastable coordination dynamics reconciles the well-known tendencies of specialized brain regions to express their autonomy, with the tendencies for those regions to work together as a synergy. We discuss metastability in the brain by describing the favorable conditions existing for its emergence and by deriving some predictions for its empirical characterization in neurophysiological recordings. A brief dialogue follows that clarifies and reconciles the present approach with that of W. Freeman. Finally, we explore briefly some of the implications of metastable coordination dynamics for perception.and thinking.

2 Toward a Complementary Science

Up until the time of Bohr, Heisenberg and Pauli, scientists debated over whether light, sound and atomic scale processes were more basically particle-like or wave-like in character. Philosophy spoke of thesis and antithesis, of dialectic tension, of self and not self, of the qualitative and the quantitative, the objective and the subjective, as if they were either/or divisions. This tendency to dichotomize, to divide the world into opposing categories appears to be a 'built in' property of human beings, arising very early in development and independent of cultural background [1]. It is, of course, central to the hypothetico-deductive method of modern science which has made tremendous progress by testing alternative hypotheses, moving forward when it rejects alternatives. Or so it seems.

For Bohr, Pauli and Heisenberg, three giants of 20th century science and chief architects of the most successful theory of all time, it became abundantly clear that sharp dichotomies and contrarieties must be replaced with far more subtle and sophisticated complementarities. For all of nature, human nature (and presumably human brains) included. Probably Pauli [2] expressed it best:

> "To us the only acceptable point of view appears to be one that recognizes both sides of reality—the quantitative and the qualitative, the physical and the psychical—as compatible with each other. It would be most satisfactory of all if physics and psyche could be seen as complementary aspects of the same reality" (p. 260).

The remarkable developments of quantum mechanics demonstrating the essential complementarity of both light and matter should have ushered in not just a novel epistemology but a generalized complementary science. However, they did not. Thinking in terms of contraries and the either/or comes naturally to the human mind. Much harder to grasp is the notion that contraries are complementary, *contraria sunt complementa* as Bohr's famous coat of arms says. One step in this direction might be if complementary aspects and their dynamics were found not just at the level of the subatomic processes dealt with by quantum mechanics, but at the level of human brains and human behavior dealt with by coordination dynamics.

3 Toward a Complementary Brain Science

How might a complementary stance impact on understanding the brain? The history of brain research over the last few centuries is no stranger to dichotomy: it contains two conflicting theories of how the human brain works (see [3] for an excellent treatment). One theory stresses that the brain consists of a vast collection of distinct regions each localizable in the cerebral cortex and each capable of performing a unique function. The other school of thought looks upon the brain not as a collection of specialized centers, but as a highly integrated organ. In this view, no single function can be the sole domain of any unique part of the cortex. Obeying the old dictum, the holistic brain is greater than and different from the sum of its parts. Like debates on nature versus nurture, learning versus innateness, reductionism versus holism, these two conflicting views of how the brain works have shed more heat than light. Yet surprisingly, the two either-or contrasts still survive. In modern parlance, researchers ask if the brain is "segregated" into its parts or "integrated" as a whole, if information is represented in a modular, category-specific way or in a distributed fashion in which many distinct areas of the brain are engaged, each one representing many different kinds of information.

In the last 20 years or so some new ideas about brain organization have emerged that may provide deeper insight into the human mind, both individual and collective. One step in this direction is by Sungchui Ji [4]. In promoting his "complementarist" epistemology and ontology, Ji draws on the biology of the human brain, namely the complementary nature of its hemispheric specializations. For Ji, the left and right hemispheres have relatively distinct psychological functions and "ultimate reality," as perceived and communicated by the human brain, is a complementary union of opposites [4]. This is a picture painted with a very broad brush. On a much finer grained scale, Stephen Grossberg [5] in a paper entitled "The Complementary Brain'' has drawn attention to the complementary nature of brain processes. For example, the visual system is divided by virtue of its sensitivity to different aspects of the world, form and motion information being carried by ventral and dorsal cortical pathways. For Grossberg, working memory order is complementary to working memory rate, color processing is complementary to luminance processing and so forth. Grossberg believes that the brain is organized this way in order to process complementary types of information in the environment. For him, a goal of future research is to study more directly how complementary aspects of the physical world are translated into complementary brain designs for coping with this world.

If the brain, like the physical world, is indeed organized around principles of complementarity, why then do we persist in partitioning it into contraries? What is it that fragments the world and life itself? Is it the way nature is? Or is it us, the way we are? (see how pernicious the either/or is!). Of course, this age-old question goes back thousands of years and appears again and again in the history of human thought, right up to the present [6, 7]. Outside quantum mechanics, however, no satisfactory answer from science has emerged. Motivated by new empirical and theoretical developments in coordination dynamics, the science of coordination, Kelso and Engstrøm [7] have offered an answer, namely that the reason the mind fragments the world into dichotomies (and more important how opposing tendencies are reconciled) is deeply connected to the way the human brain works, in particular its *metastable coordination dynamics* (e.g., [8, 9, 10, 11, 12, 13, 14]). Let's summarize some of the general aspects of coordination dynamics, before focusing in on its core mathematical form.

4 Coordination Dynamics of the Brain: Multistability, Phase Transitions and Metastability

From being on the periphery of the neurosciences for fifty years and more, brain dynamics is steadily inching toward center stage. There are at least four reasons for this. One is that techniques at many levels of description now afford an examination of both structure and function in real time: from gene expression to individual neurons to cellular assemblies and on to behavior, structures and their interrelation. Two is that slowly and surely the concepts, methods and tools of self-organizing dynamical systems are taking hold. It is twenty years since a review article in *Science* laid out the reasons why [15]. Three is that dynamics is a language for connecting events from the genetic to the mental [12]. Dynamics is and must be filled with content, each level possessing its own descriptions and quasi-autonomy (everything is linked, from a particle of dust to a star). Four is that empirical evidence indicates that dynamics appear to be profoundly linked to a broad range of disorders ranging from Parkinson's disease to autism and schizophrenia.

The theory of coordination dynamics is based on a good deal of empirical evidence about how brains are coordinated in space and time. One key set of results is that neurons in different parts of the brain oscillate at different frequencies (see [16, 17] for excellent reviews). These oscillations are coupled or "bound" together into a coherent network when people attend to a stimulus, perceive, think and act [18, 19, 20, 21, 22, 23, 24, 25, 26]. This is a dynamic, self-assembling process, parts of the brain engaging and disengaging in time, as in a good old country square dance. Such a coordinative mechanism may allow different perceptual features of an object, different aspects of a moving scene, separate remembered parts of a significant experience, even different ideas that arise in a conversation to be bound together into a coherent entity.

Extending notions in which the 'informational code' lies in the transient coupling of functional units, with physiological significance given to specific phase-lags realized between coordinating elements [27], we propose that phase relationships carry information, with multiple attractors (attracting tendencies) setting alternatives for complementary aspects to emerge in consciousness [28]. In the simplest case, oscillations in different brain regions can lock "in-phase", brain activities rising and falling together, or "anti-phase", one oscillatory brain activity reaching its peak as another hits its trough and vice-versa. In-phase and antiphase are just two out of many possible multistable phase states that can exist between multiple, different, specialized brain areas depending on their respective intrinsic properties, broken symmetry and complex mutual influence.

Not only does the brain possess many different phase relations within and among its many diverse and interconnected parts, but it can also switch flexibly from one phase relation to another (in principle within the same coalition of functional units), causing abrupt changes in perception, attention, memory and action. These switchings are literally "phase transitions" in the brain, abrupt shifts in brain states caused by external and internal influences such as the varying concentration of neuromodulators and neurotransmitter substances in cell bodies and synapses, places where one neuron connects to another.

Coordination dynamics affords the brain the capacity to lock into one of many available stable coordinative states or phase relations. The brain can also become unstable and switch to some completely different coordinative state. Instability, in this view, is a selection mechanism picking out the most suitable brain state for the circumstances at hand. Locking in and switching capabilities can be adaptive and useful, or maladaptive and harmful. They could apply as easily to the schizophrenic or obsessive-compulsive, as they could to the surgeon honing her skills.

A third kind of brain dynamic called metastability is becoming recognized as perhaps the most important of all for understanding ourselves. In this regime there are no longer any stable, phase and frequency synchronized brain states; the individual regions of the brain are no longer fully 'locked in' or interdependent. Nor, ironically enough, are they completely independent. According to a recent review [29]:

> Metastability is an entirely new conception of brain functioning where the individual parts of the brain exhibit tendencies to function autonomously **at the same time (emphasis ours)**as they exhibit tendencies for coordinated activity (Kelso, 1991; 1992; 1995; Bressler & Kelso, 2001; see also Bressler, 2003)

As the Fingelkurts's remark, metastability is an entirely new conception of brain organization, not merely a blend of the old. Individualist tendencies for the diverse regions of the brain to express themselves coexist with coordinative tendencies to couple and cooperate as a whole. In the metastable brain, local and global processes coexist as a complementary pair, not as conflicting theories. Metastability, by reducing the strong hierarchical coupling between the parts of a complex system while allowing them to retain their individuality leads to a looser, more secure, more flexible form of function that can promote the creation of new information. No dictator tells the parts what to do. Too much autonomy of the component parts means no chance of coordinating them together. On the other hand, too much interdependence and the system gets stuck, global flexibility is lost.

Metastability introduces four advantageous characteristics that neurocognitive models are invited to consider. First, metastability accommodates heterogeneous elements (e.g. brain areas having disparate intrinsic dynamics; brain areas whose activity is associated with the movement of body parts or events in the environment).

Second, metastability does not require a disengagement mechanism as when the system is in an attractor and has to switch to another state. This can be costly in terms of time, energy and information processing. In the metastable regime, neither stochastic noise nor parameter changes are necessary for the system to explore its patternings. Third, metastability allows the nervous system to flexibly browse through a set of possibilities (tendencies of the system) rather than adopting a single 'point of view'. Fourth, the metastable brain favors no extremes. Nor is it a "balance" of opposing alternatives. For example, it makes no sense to say the brain is 60% segregated and 40% integrated. Rather, metastability is an expression of the full complexity of the brain.

A number of neuroscientists have embraced metastability as playing a role in various neurocognitive functions, including consciousness (e.g. [14, 25, 30, 31, 32, 33, 34, 35, 36, 37]). As we explain below, it is not the word itself that matters, but what the word means for *understanding*. In coordination dynamics, metastability is not a concept or an idea, but a consequence of the observed self-organizing and pattern forming nature of brain, cognition and behavior [12, 15, 38, 39]. Specifically, metastability is a result of the broken symmetry of a system of (nonlinearly) coupled (nonlinear) oscillators called the *extended HKB model* [40]: *HKB* stands for Haken, Kelso and Bunz [41] and represents a core (idealized) dynamical description of coordinated brain and behavioral activity (see e.g. [42]). Importantly, it is the symmetry breaking property of the extended HKB model [40] that has led to metastability and the new insights it affords.

5 The Extended HKB Model

Etymologically, 'metastability', comes from the latin *'meta'* (beyond) and *'stabilis'* (able to stand). In coordination dynamics, metastability corresponds to a regime near a saddle-node or tangent bifurcation in which stable coordination states no longer exist (e.g., in-phase synchronization where the relative phase between oscillating components lingers at zero), but attraction remains to where those fixed points used to be ('remnants of attractor∼repellors'). This gives rise to a dynamical flow consisting of phase trapping and phase scattering. Metastability is thus the simultaneous realization of two competing tendencies: the tendency of the components to couple together and the tendency for the components to express their intrinsic independent behavior. Metastability was first identified in a classical model of coordination dynamics called the extended HKB [40], and later seen as a potential way by which the brain could operate [8, 10, 11, 12, 29, 30, 35, 43, 44].

The equation governing the coordination dynamics of the extended HKB model describes changes of the relative phase over time ($\dot{\phi}$) as:

$$\dot{\phi} = \delta\omega - a \sin\phi - 2b \sin(2\phi) + \sqrt{Q}\xi t\,. \tag{1}$$

where ϕ is the relative phase between two interacting components, a and b are parameters setting the strength of attracting regions in the system's dynamical landscape, $\sqrt{Q}\xi t$ is a noise term, and $\delta\omega$ is a symmetry breaking term arising from each component having its own intrinsic behavior. The introduction of this symmetry breaking term $\delta\omega$ (1) changes the entire dynamics (layout of the fixed points, bifurcation structure) of the original HKB system. It is the subtle interplay between the coupling term (k=b/a) in equation 1 and the symmetry breaking term, $\delta\omega$sghat gives rise to metastability.

The flow of the coordination dynamics across a range of values of $\delta\omega$ is presented in Fig. 1 for a fixed value of the coupling parameter, k = b/a=1 where a=1 and b=1). Stable fixed points (attractors) are presented as filled circles and unstable fixed points (repellors) as open circles. Note these fixed points refer to the coordination variable or order parameter: the relative phase (see Sect. 7 for further discussion of the order parameter concept). A fixed point of the coordination variable ϕ represents a steady phase- and frequency relationship between the oscillatory components or *phase-locking*. The surface shown in Fig. 1 defines three regions under the influence of the symmetry breaking term $\delta\omega$. In the first region present in the lower part of the surface, the system is multistable. Following the representative line labeled 1 in Fig. 1 from left to right, two stable fixed points (filled circles) are met which are the

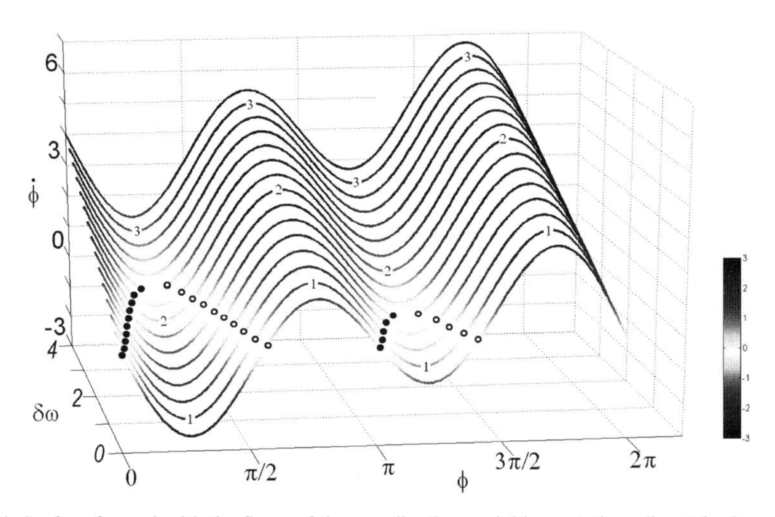

Fig. 1 Surface formed with the flows of the coordination variableg ϕ (in radians) for increasing values of $\delta\omega$ between 0 and 4. For this example, the coupling is fixed: a=1 and b=1. When $\dot{\phi}$ reaches zero (flow line becoming white), the system ceases to change and fixed point behavior is observed. Stable and unstable fixed points at the intersection of the flow lines with the isoplane $\dot{\phi}$=0 are represented as filled and open circles respectively. To illustrate the different regimes of the system, three representative lines labeled 1 to 3 fix $\delta\omega$ at increasing values. Following the flow line 1 from left to right, two stable fixed points (filled circles) and two unstable fixed points (open circles) exist. This flow belongs to the multistable (here bistable) regime. Following line 2 from left to right, one pair of stable and unstable fixed points is met on the left, but notice the complete disappearance of fixed point behavior on the right side of the figure. That is, a qualitative change (bifurcation; phase transition) has occurred. The flow now belongs to the monostable regime. Following line 3 from left to right, no stable or unstable fixed points exist yet coordination has not disappeared. This flow corresponds to the metastable regime, which is a subtle blend of coupling and intrinsic differences between the components

alternatives for the system to settle in. Which one, depends on the initial conditions and the size of the basin of attraction. In an intermediate region, following the line labeled 2 from left to right, one observes that the weakest attractor near anti-phase (right side) disappears after it collides with its associated repellor somewhere near $\delta\omega$=1.3, but the strongest attractor (left side) is still present as well as its repellor partner. Finally in the third region in the upper part of the surface, the regime becomes metastable. Following the line labeled 3 from left to right, no fixed points exist anymore (this part of the surface no longer intersects the isoplane $\dot{\phi}$=0 where the fixed points are located).

What does coordination behavior look like in the metastable regime? Although all the fixed points have vanished, a key aspect is that there are still some traces of coordination, 'ghosts' or 'remnants' of where the fixed points once were. These create a unique dynamics alternating between two types of periods which may be called dwell time and escape time. Escape times are observed when the trajectory of the coordination variable, relative phase, drifts or diverges from the horizontal. Dwell times are observed when the trajectory converges and holds (to varying degrees) around the horizontal. In Fig. 2c we show two locations for the dwell times: one that lingers a long time before escaping (e.g. Fig. 2c, annotation 1) slightly above the more stable in-phase pattern near 0 rad (modulo 2π), and the other that lingers only briefly (e.g. Fig. 2c, annotation 2) slightly above π (modulo 2π). The dwell time is reminiscent of the transient inflexions observed near the disappeared attractor-repellor pairs in the monostable regime (Fig. 2b, annotation 3). These inflexions recur over and over again as long as the system is maintained in the metastable regime, i.e. as long as it does not undergo a phase transition.

Despite the complete absence of phase-locked attractors, the behavior of the elements in the metastable regime is not totally independent. Rather, the dependence between the elements takes the form of dwellings (phase gathering) nearby the remnants of the fixed points and is expressed by concentrations in the histogram of the relative phase (see [12], Chap. 4). Can the brain make use of such a principle? In contrast to, or as a complement of theories of large-scale organization through linear phase-coupling [18, 19, 25, 45], our thesis is that the ability of the system to coordinate or compute without attractors opens a large set of possibilities. The classical view of phase-locked coordination prescribes that each recruited element looses its intrinsic behavior and obeys the dictates of the assembly. When such situations arise, from the functional point of view, individual areas cease to exert an influence for the duration of the synchronized state, and the pertinent spatial level of description of the unitary activity becomes the synchronous assembly itself. However, phylogenesis promoted specialized activity of local populations of neurons [30, 46, 47, 48, 49]. In theories proposing large-scale integration through phase synchronization, the expression of local activity can only exist when the area is not enslaved into an assembly, whereas in the metastable regime, the tendency for individual activity is more continually preserved (see also [35]).

As exemplified explicitly in the extended HKB model, a delicate balance between integration (coordination between individual areas) and segregation (expression of individual behavior) is achieved in the metastable regime [11, 12]. Excessive segregation does not allow the proper manifestation of cognition as seen for

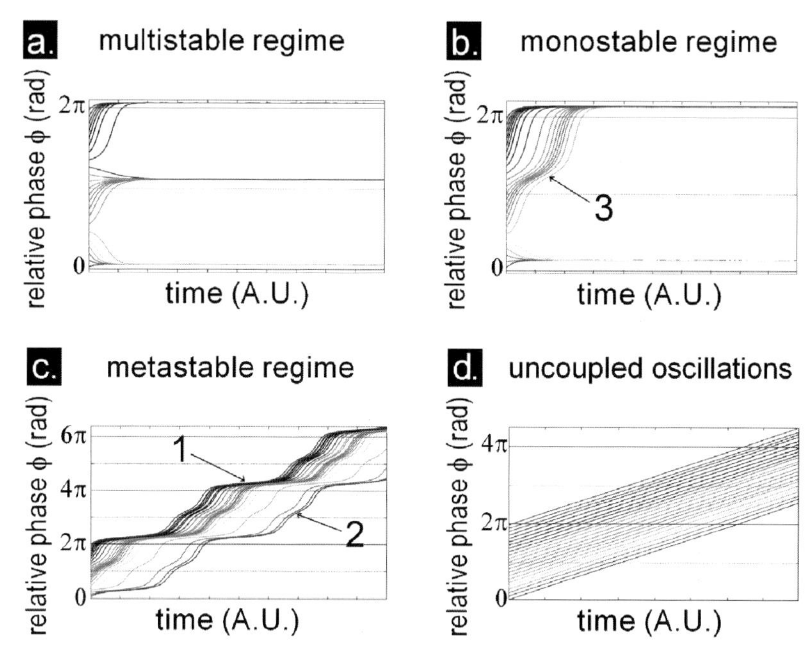

Fig. 2 Examples of trajectories of the coordination variable, relative phase ϕ arising from a range of initial conditions sampled between 0 and 2π radians, in the multistable (a), monostable (b) and metastable regimes (c) of the extended-HKB model. Trajectories in the multistable regime converge either to an attractor located slightly above 0 rad. modulo 2π or to another attractor located slightly above π rad. modulo 2π. In the monostable regime (a), trajectories converge to an attractor located slightly above 0 rad. modulo 2π. In the trajectories of relative phase for the metastable regime (c. unwrapped to convey continuity), there is no longer any persisting convergence to the fixed points, but rather a succession of periods of rapid drift (escape time) interspersed with periods inflecting toward, but not remaining on the horizontal (dwell time). Note dwells nearby 0 rad. modulo 2π in the metastable regime (e.g. dwell time at about 4π rad. annotated 1 in Fig 2c) and nearby π rad. modulo 2π (dwell time at about 3π rad. annotated 2 in c.) are reminiscent of the transient obtained for certain initial conditions in the monostable regime (b. annotation 3). For reference, the relative phase of uncoupled oscillators is displayed in (d.)

instance in autism and schizophrenia [50, 51, 52, 53, 54, 55]. On the other hand, excessive integration does not appear to be adaptive either. Resting states measured during cognitive idling are characterized by widespread oscillations across large cortical territories [56, 57, 58, 59, 60] that appear to block or inhibit the proper expression of a local areas' activity. Furthermore, propagation of synchronous activity leads to epileptic seizures [61, 62, 63, 64, 65] and is ultimately characterized by a total loss of cognition and consciousness once a certain mass of neurons is recruited. In a critical range between complete integration and complete segregation the most favorable situation for cognition is deemed to occur [44, 66, 67]. Studies of interareal connectivity both at the anatomical and functional level ([35, 68, 69], see also [37]) support this view by showing that measures of complexity reach a maximum when the balance between segregative and integrative forces is achieved. Note, however, that such measures are based upon stationarity assumptions whereas metastability in coordination dynamics is a 'stationary transient'. That is, the holding and releasing of the relative phase over time appears to be of a transient nature, but is actually quite stationary.

Another interesting feature related to the absence of attractors is the ability of the system to exhibit more than one coordination tendency in the time course of its life. This property is reminiscent of the multistable regime with attractors, with the difference that no transition is required to switch from one state to the other. Evidence of 'multistability' and spontaneous switching in perception and action abounds both at behavioral and brain levels (e. g., [39, 70, 71, 72, 73, 74, 75, 76, 77]). Aside from the multistable regime with attractors undergoing phase transition, the metastable regime is also suitable to explain those experimental results. The tendencies of the metastable regime toward the remnants of the fixed points readily implements spontaneous reversals of percepts and behaviors described in these studies [78]. From the perspective of coordination dynamics, the time the system dwells in each remnant depends on a subtle blend of the asymmetry of the components (longer dwelling for smaller asymmetry) and the strength of the coupling (longer dwelling for larger values of a or b). Such a mechanism provides a powerful means to instantiate alternating thoughts/percepts and their probability in both biological systems and their artificial models (e.g. alternating percepts of vase or faces in ambiguous Rubin figures, or alternative choices in the solving of a chess game).

Both a multistable regime with attractors and a metastable regime with attracting tendencies allow so-called perceptual and behavioral 'multistability'. Which attractor is reached in the multistable regime primarily depends on initial conditions. Once the system is settled into an attractor, a certain amount of noise or a perturbation is required to achieve a switching to another attractor. Or, if control parameters such as attention are modified, a bifurcation or phase transition may occur, meaning an attractor looses stability as the regime changes from multistable to monostable or vice-versa (see Ditzinger & Haken, [79]; [80] for excellent example of such modeling). In the metastable regime, successive visits to the remnants of the fixed points are intrinsic to the time course of the system, and do not require any external source of input. This is an important difference between between multistability and metastability and likely translates into an advantage in speed which is known to be an important constraint in neurocognitive systems [81] and a crucial aspect of their performance [82].

6 Metastability in the Brain

What is the anatomical basis in the brain for metastable coordination dynamics? As noted earlier, the fundamental requirements for metastability are the existence of coupled components each exhibiting spontaneous oscillatory behavior and the presence of broken symmetry. There are several spatial scales at which the collective behavior of the brain spontaneously expresses periodic oscillations [83, 84, 85, 86, 87] and represents the combined contribution of groups of neurons, the joint action of which is adequate to establish transfer of information [88, 89, 90]. The oscillatory activity of the brain may be captured directly via invasive neurophysiological methods such as LFP and iEEG, or indirectly from EEG scalp recordings (commonly at the price of estimating local oscillations by bandpass filtering of the signals). The coupling between local groups of neurons is supported by long-range functional connectivity [25, 91, 92]. Broken symmetry has several grounds to spring from, including the incommensurate characteristic frequencies of local populations of neurons [93] and their heterogeneous connectivity [94].

If the conditions required for metastable coordination in the brain are easily met, it remains to establish that the brain actually shows instances of operating in this regime. This empirical characterization encounters some difficulties. Before any attempt to find signatures of metastability, a first question is to identify from the continuous stream of brain activity some segments corresponding to individual regimes. In other words, it consists in finding the transitions between regimes, a task undertaken by only a few [95, 96]. Provided adequate measurement/estimation of local oscillations in the presence of noise and spatial smearing is possible, insights can be gained by identifying episodes of phase-locking (these forming states) and ascribing their interim periods as transitions (e.g. [25]). In the absence of ad hoc segmentation of the EEG, it remains as a possibility to use behavioral indices as cues to when brain transitions occur (e.g. [39, 73]). The case of metastability is evidently more difficult since the regime is stationary but not stable. Initial attempts have targeted the more recognizable dwell time as a quasi phase-locked event [29]. To gain understanding on the mechanism, it seems necessary to elaborate strategies that comprise the coordination pattern of the metastable regime in its entirety (dwell and escape time as an inseparable whole) and to establish criteria for the differentiation of state transitions and dwell $\sim$ escape regimes.

To identify and understand potential mechanisms, it is of prime importance to be able to distinguish between the different regimes..For instance, a transition between the metastable and the monostable regime could be a way the brain instantiates a process of decision among a set of possibilities. This amounts to the creation of information [97]. Figure 3 shows the isomorphism of simulated systems belonging to both regimes in their relative phases' trajectory. In this window of arbitrary size, a succession of states is shown in the multistable regime (right) separated by transitions. It differs from the metastable regime (left) by the presence of horizontal segments (stable relative phase) during the states and by sharp inflections of the relative phase at the onset and offset of transitions. The corresponding histograms of the relative phase cumulated over this period of time are similar as well. The ability to distinguish the multistable regime from the metastable regime in a non-segmented EEG depends critically on the precision of the estimation of the components' frequency and phase. Unfavorable circumstances are met since the EEG is a noisy, broadband signal [98], and because each component's frequency shifts when coupled in a metastable regime.

Other criteria might be sought for to distinguish between those regimes. State-transition regimes have been conceptually formulated and empirically verified by a line of studies initiated by Eckhorn et al. [18], Gray, Singer et al. [19]. The theory of 'transient cell assemblies' has gathered numerous empirical findings at the microscale [99, 100], mesoscale [91, 101, 102] and macroscale [71, 103, 104, 105].

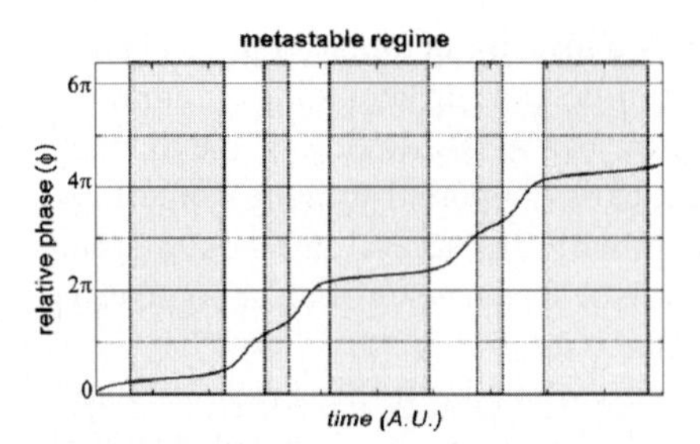

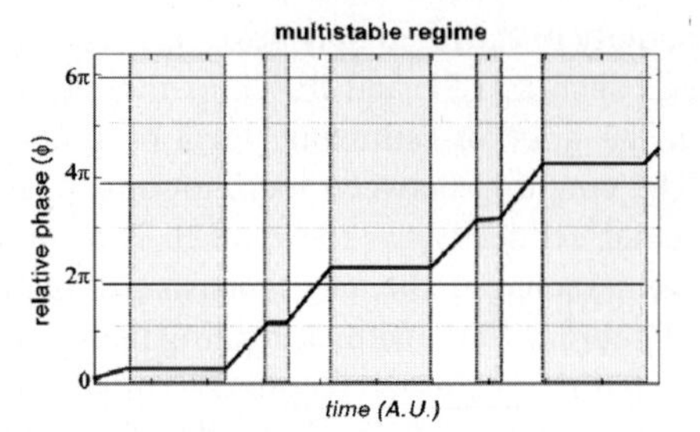

Fig. 3 Comparison of relative phase trajectories in the metastable and multistable regime for a temporal window of arbitrary size. Coordination (multistable regime) and tendency to coordinate (metastable regime) are shown in grey boxes. In the multistable regime (right), a succession of states (stable relative phase near 0 and pi radians) is interweaved with transitions. Horizontal segments are lost in the metastable regime (left) which only shows tendencies for synchronization toward inphase and antiphase. In a situation in which coordination is estimated from a broadband signal in the presence of noise, distinguishing between the two regimes may difficult. The transitions on the right however are induced by parametric change; the flow on the left is not

This set of studies relies on linear pairwise phase synchronization methods applied both to Single- and Multi-Unit Activity and Field Potentials. Whereas many studies have focused on the 'state' part of the state transition, an interesting feature is seen in the study by Rodriguez et al. [105] of coherent oscillations elicited by Mooney faces. Two periods of intense synchronization at 250 msec and 700 msec are separated by a period of intense desynchronization that the authors described as phase scattering. They suggest that phase scattering is a mechanism by which the brain realizes the transition from a coherent assembly to another assembly–both belonging to stable regimes. Such a mechanism is unnecessary in the succession of tendencies that are characteristic of metastable coordination dynamics.

In summary, the brain by virtue of its properties forms a suitable ground for metastability to take place. The characterization of metastable onsets however is a matter which will certainly require some methodological developments outside the linear approach of transient phase synchronization. In the meantime, indices of metastability are found in the distribution of dwell times near phase-locked states.

7 Clarifying Nonlinear Brain Dynamics: The Freeman-Kelso Dialogue[1]

Recently, the eminent neurophysiologist Walter Freeman published an article entitled "Metastability, instability and state transitions in neocortex" [34] that led to a discussion with the authors which we think may be useful for clarificatory purposes and to expand awareness of nonlinear brain dynamics. Here we highlight some of the

[1] With the blessing of Walter Freeman

main issues—FAQ about metastable neurodynamics, if you like–in part as a tribute to Freeman and his pioneering work and its relation to coordination dynamics.

First the concept itself. Freeman draws heavily from the solid state physics literature where he notes the concept of metastability has been in use for over 30 years. Although this is correct and many useful analogies have been made between brains and other kinds of cooperative phenomena in nature (e.g., [12, 38, 41, 106, 107]) notice here that metastability arises because of a specific symmetry breaking in the coordination dynamics. That is, intrinsic differences in oscillatory frequency between the components are sufficiently large that they do their own thing, while still retaining a tendency to coordinate together. Thus, the relative phase between the components drifts over time, but is occasionally trapped near remnants of the (idealized) coordinated states e.g., near 0 and π radians (cf. Fig. 2). As a consequence of broken symmetry in its coordination dynamics, both brain and behavior are able to exhibit a far more variable, plastic and fluid form of coordination in which tendencies for the components to function in an independent, segregated fashion coexist with tendencies for the system to behave in an integrated, coordinative fashion.

Second, Freeman inquires about the order parameter in coordination dynamics. Freeman himself pursues spatial patterns of amplitude which he understands as manifestations of new attractors that form through learning. It is these amplitude patterns of aperiodic carrier waves derived from high density EEG recordings that constitute his order parameter. Freeman regards these as evidence for cortical dynamics accessing nonconvergent attractors for brief time periods by state transitions that recur at rates in the theta range. Although we originally also used the physicist's terminology of order parameters (e.g. [39, 41] we now prefer the term "collective variable" or "coordination variable" as a quantity that characterizes the cooperative behavior of a system with very many microscopic degrees of freedom. Indeed, our approach is called coordination dynamics because it deals fundamentally with informationally meaningful quantities [28]. Coordination in living things is not (or not only) matter in motion. The spatiotemporal ordering observed is functional and task-specific. Because in our studies the variable that changes qualitatively under parametric change is the relative phase, relative phase is one of the key order parameters of biological coordination. Relative phase is the outcome of a nonlinear interaction among nonlinear oscillatory components, yet in turn reciprocally conditions or "orders" the behavior of those components. In a system in which potentially many things can be measured, the variable that changes qualitatively is the one that captures the spatiotemporal ordering among the components. This strategy of identifying order parameters or coordination variables in experimental data goes back to Kelso [74, 75] and the early theoretical modeling by Haken, Kelso & Bunz, [41]. Recent empirical and theoretical research contacts Freeman's work in that it shows that phase transitions can also arise through the amplitudes of oscillation [108]. Both routes are possible depending on the situation, e.g. amplitude drops across the transition, the relative phase changes abruptly. Again, this is all under parametric control. In coordination dynamics, you don't "know" you have a coordination variable or order parameter and control parameters unless the former change qualitatively at transitions, and the latter–when systematically varied–lead the system through transitions. Order parameters and control parameters in the framework of coordination dynamics are thus co-implicative and complementary An issue in the olfactory system concerns what the control parameters are, e.g. that might lead the system from a steady state to a limit cycle or to chaotic oscillations and itineracy. Both Freeman's and our approach appeal to Haken's [106] synergetics and so-called 'circular' or 'reciprocal causality': whether one adopts the term order parameters or coordination variables, both arise from the cooperation of millions of neurons and in turn are influenced by the very field of activity they create (see also [42, 109] for an elaboration and application of neural field theory).

Both Freeman's approach and coordination dynamics appeal to nonlinear coupling among neural oscillators as the basis for varying degrees of global integration. Both make significant use of basin attractor dynamics to interpret experimental data. In coordination dynamics, the latter constitutes a step of idealization that is necessary in order to understand what it means to break the symmetry of the coordination dynamics [40, 97, 110]. Both approaches nevertheless invoke symmetry breaking, coordination dynamics from the loss of attractors of the relative phase dynamics, and Freeman in the emergence of spatial patterns of amplitude and phase from EEG recordings by convergence to a selected *a posteriori* attractor in a landscape of *a priori* attractors. There are obvious parallels between the two bodies of work; both are testament to the richness of detail and power of nonlinear theory. Freeman envisages transient but definite access to a succession of basins of attraction. Metastable coordination dynamics, on the other hand, has a very precise meaning: it is not about states but a subtle blend of integrative and segregative *tendencies*. Notably, these integrative tendencies to phase gather and segregative tendencies to phase wrap can be shown at the level of coupled neural ensembles. Figure 4 illustrates a set of coupled neural ensembles each composed of one hundred Fitzhugh-Nagumo oscillators connected by a sigmoidal function which is the usual consequence of the summation of threshold properties at cell bodies (2). A small amount of random noise has been added only for illustrative effect. Looking from top to bottom, the neuronal firing activity of each ensemble (X,Y) is shown, followed by the individual oscillatory phases, their relative phase and respective phase plane trajectories indicating limit cycle properties, along with a simple mathematical representation.

$$\begin{aligned} \frac{dX}{dt} &= F(X) = \frac{\partial}{\partial t} S(X - Y) + noise \\ \frac{dY}{dt} &= F(Y) = \frac{\partial}{\partial t} S(Y - X) + noise \end{aligned} \tag{2}$$

The intent here is only to establish proof of concept. It is quite clear that the relative phase between the neural groups dwells near phi = 0, wanders and then returns, indicating nicely the transient metastable tendencies to integrate and segregate. As Fingelkurts and Fingelkurts [29] note:

> One may note that the metastabilty principle extends the Haken synergetics rules..... Metastabilty extends them to situations where there are neither stable nor unstable states, only coexisting tendencies (see (Kelso, 2002)

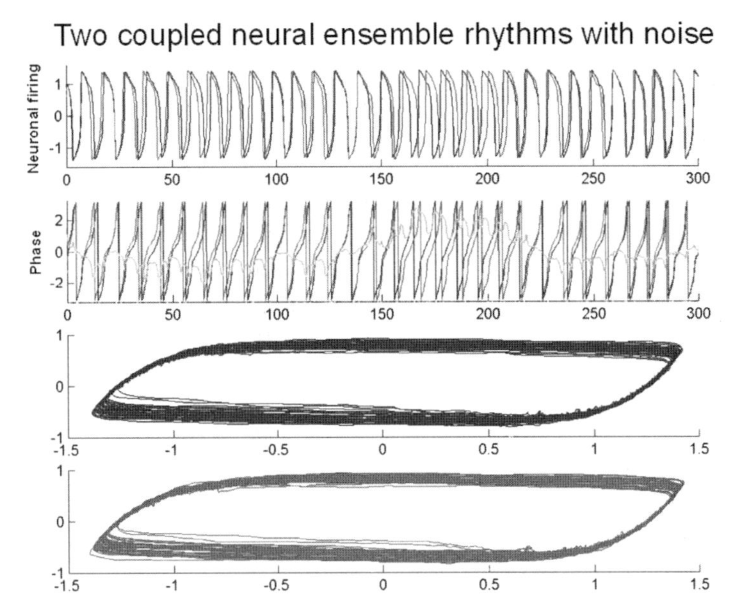

Fig. 4 A simulation of two coupled neural ensembles composed of an array of Fitzhugh-Nagumo oscillators (courtesy Viktor Jirsa). See text for description

8 A Short Afterthought

It has not escaped our attention that the metastable coordination dynamics of brain and behavior invites extension to the processes of thought and thinking. A case can be made that multistable perception and ambiguity offer test fields for understanding the self-organization of the brain. The perceptual system, when faced with a constant stimulus, may fluctuate between two or more interpretations. Through ambiguities, as Richard Gregory [111] remarks, we can investigate how the brain makes up its mind. One may speculate that when a naïve perceiver views a figure such as the hidden Dalmatian (Fig. 5), a series of mental phases and their associated brain states takes place. In a first stage, the naïve observer may attempt to group the blackened areas in various ways. There will be multistability and phase transitions. Eventually, he/she will correctly organize the picture and a monostable state will be reached in which the Dalmatian's picture is salient. Finally, the observer may think of the artist's work and consider simultaneously the fragmentation that allows the Dalmatian to disintegrate into the scene as well as the organization that hints its presence. A metastable dynamic will arise in which both the figure and its hiding texture will simultaneously be present in the mind. As Heisenberg noted fifty years ago [112]:

> "We realize that the situation of complementarity is not confined to the atomic world alone; we meet it when we reflect about a decision and the motives for our decision or when we have a choice between enjoying music and analyzing its structure" (p. 179)

Fig. 5 The hidden Dalmatian

Of course, Heisenberg, Bohr and Pauli's philosophy came from considerations in quantum mechanics. Here both the philosophy of the complementary nature and a complementary neuroscience are rooted in the metastable coordination dynamics of the brain.

Acknowledgment This work was supported by NIMH Grant MH42900, NINDS Grant NS48229 and ONR Grant N00014-05-1-0117.

References

1. Spelke, E.: The Baby Lab by Margaret Talbot. New Yorker, Sept. 4 (2006)
2. Pauli, W.: Writings on Physics and Philosophy. In: Enz, C.P., von Meyenn, K. (Eds.) Springer-Verlag, Berlin (1994)
3. Finger, S.: Origins of Neuroscience. Oxford, New York (1994)
4. Ji, S.: Complementarism: A Biology-Based Philosophical Framework to Integrate Western Science and Eastern Tao. Proc. of 16th International Congress of Psychotherapy (1995) 518-548
5. Grossberg, S.: The Complementary Brain: a Unifying View of Brain Specialization and Modularity. Trends in Cognitive Sciences 4 (2000) 233–246
6. Frayn, M.: The Human Touch: Our Part in the Creation of the Universe. Faber & Faber (2006)
7. Kelso, J. A. S., Engstrøm, D.: The Complementary Nature. MIT Press, Cambridge (2006)
8. Bressler, S. L., Kelso, J. A. S.: Cortical Coordination Dynamics and Cognition. Trends in Cognitive Sciences 5 (2001) 26–36
9. Jirsa, V. K., Kelso, J. A. S. (Eds.) Coordination Dynamics: Issues and Trends. Springer, Berlin (2004)
10. Kelso, J. A. S.: Behavioral and Neural Pattern Generation: the Concept of Neurobehavioral Dynamical System (NBDS) In: Koepchen, H. P., Huopaniemi, T. (Eds.) Cardiorespiratory and Motor Coordination, Springer-Verlag, Berlin (1991)
11. Kelso, J. A. S.: Coordination Dynamics of Human Brain and Behavior. Springer Proc. in Physics 69 (1992) 223–234
12. Kelso, J. A. S.: Dynamic Patterns: the Self-Organization of Brain and Behavior. MIT Press, Cambridge, MA (1995)

13. Tschacher, W., Dauwalder, J. P.: The Dynamical Systems Approach to Cognition: Concepts and Empirical Paradigms Based on Self-Organization, Embodiment and Coordination Dynamics. World Scientific, Singapore (2003)
14. Velazquez, J. L.: Brain, Behaviour and Mathematics: Are we Using the Right Approaches? Physica D: Nonlinear Phenomena 212(3–4) (2005) 161–182
15. Schöner, G., Kelso, J. A. S.: Dynamic Pattern Generation in Behavioral and Neural Systems. Science 239 (1988) 1513–1520
16. Basar, E.: Memory and Brain Dynamics: Oscillations Integrating Attention, Perception, Learning, and Memory. Conceptual Advances in Brain Research 7. CRC Press, Boca Raton (2004)
17. Buzsáki, G.: Rhythms of the Brain. Oxford University Press, Oxford (2006)
18. Eckhorn, R., Bauer, R., Jordan, W., Borsch, M., Kruse, W., Munk M., Reitboeck, HJ.: Coherent Oscillations : a Mechanism of Feature Linking in the Visual Cortex. Multiple Electrode Correlation Analyses in the Cat. Biological Cybernetics 60(2) (1988) 121–130
19. Gray, C. M., König, P., Engel, A. K., Singer, W.: Oscillatory Responses in Cat Visual Cortex Exhibit Inter-Columnar Synchronization which Reflects Global Stimulus Properties. Nature 338 (6213) (1989) 334–337
20. Munk, M. H. J., Roelfsema, P. R., König, P., Engel, A. K., Singer, W.: Role of Reticular Activation in the Modulation of Intracortical Synchronization. Science 272 (1996) 271–274
21. Bressler, S. L.: Interareal Synchronization in the Visual Cortex. Behavioral and Brain Research 76 (1996) 37–49
22. Steinmetz, P. N., Roy, A., Fitzgerald, P., Hsiao, S. S., Niebur, E., Johnson, K. O.: Attention Modulates Synchronized Neuronal Firing in Primate Somatosensory Cortex. Nature 404 (2000) 187–190
23. Mima, T., Matsuoka, T., Hallett, M.: Functional Coupling of Human Right and Left Cortical Motor Areas Demonstrated with Partial Coherence Analysis. Neurosci. Lett. 287 (2000) 93–96
24. Fries, P., Reynolds, J. H., Rorie, A. E., Desimone, R.: Modulation of Oscillatory Neuronal Synchronization by Selective Visual Attention. Science 291 (2001) 1560–1563
25. Varela, F. J., Lachaux, J. -P., Rodriguez, E. Martinerie, J.: The Brainweb: Phase Synchronization and Large-Scale Integration. Nature Reviews Neuroscience 2 (2001) 229–239
26. Brown, P., Marsden, J. F.: Cortical Network Resonance and Motor Activity in Humans. Neuroscientist 7 (2001) 518–527
27. König, P., Engel, A. K., Roelfsema, P. R., Singer, W.: Coincidence Detection or Temporal Integration. The Role of the Cortical Neuron Revisited. Trends Neurosci 19 (1996) 130–137
28. Kelso, J. A. S.: The Informational Character of SelfOrganized Coordination Dynamics. Human Movement Science 13 (1994) 393413
29. Fingelkurts, An.A., Fingelkurts, Al.A.: Making Complexity Simpler: Multivariability and Metastability in the Brain. International Journal of Neuroscience 114(7) (2004) 843–862
30. Bressler, S. L. Tognoli, E.: Operational Principles of Neurocognitive Networks. International Journal of Psychophysiology 60 (2006) 139–148
31. Edelman, G. M.: Naturalizing Consciousness: a Theoretical Framework. Proceedings of the National Academy of Science USA 100(9) (2004) 520–524
32. Edelman, G.: Second Nature: Brain Science and Human Knowledge. Yale University Press (2006)
33. Edelman, G., Tononi, G.: A Universe of Consciousness, New York, Basic Books (2000)
34. Freeman, W. J., Holmes, M. D.: Metastability, Instability, and State Transition in Neocortex. Neural Networks 18(5–6) (2005) 497–504
35. Friston, K. J.: Transients, Metastability, and Neuronal Dynamics. Neuroimage 5 (1997) 164–171
36. Koch, C.: Thinking about the Conscious Mind. A Review of Mind - An Introduction by John Searle. Science 306 (2004) 979–980
37. Sporns, O.: Complex Neural Dynamics. In: Jirsa, V. K., Kelso, J. A. S., (Eds.) Coordination Dynamics: Issues and trends. Springer-Verlag, Berlin (2004) 197–215
38. Haken, H.: Principles of Brain Functioning. Springer, Berlin (1996)

39. Kelso, J. A. S., Bressler, S. L., Buchanan, S., DeGuzman, G. C., Ding, M., Fuchs, A. Holroyd, T.: A Phase Transition in Human Brain and Behavior. Physics Letters A 169 (1992) 134–144
40. Kelso, J. A. S., DelColle, J., Schöner, G.: Action-Perception as a Pattern Formation Process. In: Jeannerod, M. (Ed.) Attention and Performance XIII. Erlbaum, Hillsdale, NJ (1990) 139–169
41. Haken, H., Kelso, J. A. S., Bunz, H.: A Theoretical Model of Phase Transitions in Human Hand Movements. Biological Cybernetics 51 (1985) 347–356
42. Jirsa, V. K., Fuchs, A., Kelso, J. A. S.: Connecting Cortical and Behavioral Dynamics: Bimanual Coordination. Neural Computation 10 (1998) 2019–2045
43. Perez Velazquez, J. L., Wennberg, R.: Metastability of Brain States and the many Routes to Seizures: numerous Causes, same Result. In: Recent Research Developments in Biophysics, 3: Transworld Research Network (2004) 25–59
44. Werner, G.: Metastability, Criticality and Phase Transitions in Brain and its Models. Biosystems (in press)
45. Bressler, S. L.: Interareal Synchronization in the Visual Cortex. Behavioural Brain Research 76 (1996) 37–49
46. Ebbesson, S. O. E.: Evolution and Ontogeny of Neural Circuits. Behavioral and Brain Science 7 (1984) 321–366
47. Deacon, T. W.: Rethinking Mammalian Brain Evolution. American Zoologist 30 (1990) 629–705
48. Jacobs, R. A., Jordan, M. I.: Computational Consequences of a Bias towards Short connections. Journal of Cognitive Neuroscience 4 (1992) 323–336
49. Chklovskii, D. B., Schikorski, T., Stevens, C. F.: Wiring optimization in cortical Circuits. Neuron 34(3) (2002) 341–347
50. Andreasen, N. C., Nopoulos, P., O'Leary, D.S., Miller, D. D., Wassink, T., Flaum, L.: Defining the Phenotype of Schizophrenia: Cognitive Dysmetria and its Neural Mechanisms. Biological Psychiatry 46 (1999) 908–920
51. Tononi, G., Edelman, G. M.: Schizophrenia and the Mechanisms of Conscious Integration. Brain Research Reviews 31 (2000) 391–400
52. Brock, J., Brown, C. C., Boucher, J., Rippon, G.: The Temporal Binding Deficit Hypothesis of Autism. Development and Psychopathology 14 (2002) 209–224
53. Niznikiewicz, M. A., Kubicki, M., Shenton, M. E.: Recent Structural and Functional Imaging Findings in Schizophrenia. Current Opinion in Psychiatry 16 (2003) 123–147
54. Welsh, J. P., Ahn, E. S., Placantonakis, D. G.: Is Autism Due to Brain Desynchronization? Int. J. Dev. Neurosci. 23 (2005) 253–263
55. Liang, M., Zhou, Y., Jiang, T., Liu, Z., Tian, L., Liu, H.; Hao, Y.: Widespread Functional Disconnectivity in Schizophrenia with Resting-State Functional Magnetic Resonance Imaging. Neuroreport 17(2) (2006) 209–213
56. Berger, H.: Ueber das Elektroenkephalogramm des Menschen. Archiv für Psychiatrie und Nervenkrankheiten 87 (1929) 527–570
57. Chatrian, G. E., Petersen, M. C., Lazarte, J. A.: The Blocking of the Central Wicket Rhythm and some Central Changes Related to Movement. Electroencephalogr Clinical Neurophysiology 11 (1959) 497–510
58. Chase, M. H., Harper, R. M.: Somatomotor and Visceromotor Correlates of Operantly Conditioned 12–14 Cycles Per Second Sensorimotor Cortical Activity. Electroencephalography and Clinical Neurophysiology 31 (1971) 85–92
59. Kuhlman, W. N.: Functional Topography of the Human Mu Rhythm. Electroencephalography and Clinical Neurophysiology 44 (1978) 83–93
60. Hughes, S. W., Crunelli, V.: Thalamic Mechanisms of EEG Alpha Rhythms and their Pathological Implications. The Neuroscientist 11 (2005) 357–372
61. Schiff, N. D., Plum, F.: The Role of Arousal and "Gating" Systems in the Neurology of Impaired Consciousness. Journal of Clinical Neurophysiology 17 (2000) 438–452
62. Glass, L.: Synchronization and Rhythmic Processes in Physiology. Nature 410 (2001) 277–284

63. Kostopoulos, G. K.: Involvement of the Thalamocortical System in Epileptic Loss of Consciousness. Epilepsia 42(30) (2001) 13–19
64. Blumenfeld, H., Taylor, J.: Why Do Seizures Cause Loss of Consciousness? The Neuroscientist 9(5) (2003) 301–310
65. Dominguez, L. G., Wennberg, R., Gaetz, W., Cheyne, D., Snead III, O. C., Perez Velazquez, J.L.: Enhanced Synchrony in Epileptiform Activity? Local Versus Distant Synchronization in Generalized Seizures. Journal of Neuroscience 25(35) (2005) 8077–8084
66. Atlan, H.: Entre le Cristal et la Fumée. Paris, Seuil (1979)
67. Chialvo, D. R.: Critical Brain Networks. Physica A 340(4) (2004) 756–765
68. Tononi, G., Sporns, O., Edelman, G. M.: A Measure for Brain Complexity: Relating Functional Segregation and Integration in the Nervous System. Proceedings of the National Academy of Science USA 91 (1994) 5033–5037
69. Tononi, G., Sporns, O., Edelman, G. M.: Complexity and Coherency: Integrating Information in the Brain. Trends in Cognitive Sciences 2 (1998) 474–484
70. Almonte, F., Jirsa, V.K., Large, E. W., Tuller, B.: Integration and Segregation in Auditory Streaming. Physica D 212 (2005) 137–159
71. Basar-Eroglu, C., Struber, D., Kruse, P., Basar, E., Stadler M.: Frontal Gamma-Band Enhancement during Multistable Visual Perception. International Journal of Psychophysiology 24 (1996) 113–125
72. Hock, H. S., Kelso, J. A. S., Schöner, G.: Bistability and Hysteresis in the Organization of Apparent Motion Patterns. Journal of Experimental Psychology: Human Perception Performance 19 (1993) 63–80
73. Keil, A., Muller, M. M., Ray, W. J., Gruber, T., Elbert, T.: Human Gamma Band Activity and Perception of a Gestalt. Journal of Neuroscience 19 (1999) 7152–7161
74. Kelso, J. A. S.: On the Oscillatory Basis of Movement. Bulletin of the Psychonomic Society 18 (1981) 63
75. Kelso, J. A. S.: Phase Transitions and Critical Behavior in Human Bimanual Coordination. American Journal of Physiology 246(6 Pt 2) (1984) R1000–1004
76. Kelso, J. A. S., DeGuzman, G. C. Holroyd, T.: Synergetic Dynamics of Biological Coordination with Special Reference to Phase Attraction and Intermittency. In: Haken, H., Koepchen, H. P. (Eds.) Rhythms in Physiological Systems. Springer Series in Synergetics, Vol. 55. Springer, Berlin (1991), 195–213
77. Tuller, B., Case, P., Ding, M., Kelso, J. A. S.: The Nonlinear Dynamics of Speech Categorization. Journal of Experimental Psychology: Human Perception and Performance 20 (1994) 1–14
78. Kelso, J. A. S., Case, P., Holroyd, T., Horvath, E., Raczaszek, J., Tuller, B., Ding, M.: Multistability and Metastability in Perceptual and Brain Dynamics. In: Kruse, P., Stadler, M. (Eds.) Ambiguity in Mind and Nature. Springer-Verlag, Heidelberg (1995) 159–185
79. Thorpe, S., Fize, D., Marlot, C.: Speed of Processing in the Human Visual System. Nature 381 (1996) 520–522
80. Jensen, A.R.: Why Is Reaction Time Correlated with Psychometric g? Current Directions in Psychological Science 2 (1993) 53–56
81. Ditzinger, T., Haken, H.: Oscillations in the Perception of Ambiguous Patterns. Biological Cybernetics 61 (1989) 279–287
82. Ditzinger, T., Haken, H.: The Impact of Fluctuations on the Recognition of Ambiguous Patterns. Biological Cybernetics 63 (1990) 453–456
83. Chen, Y., Ding, M., Kelso, J. A. S.: Task-Related Power and Coherence Changes in Neuromagnetic Activity during Visuomotor Coordination. Experimental Brain Research 148 (2003) 105–116
84. Chen, Y., Ding, M., Kelso, J. A. S.: Long Range Dependence in Human Sensorimotor Coordination. In: Rangarajan, G., Ding, M. (Eds.) Processes with Long-Range Correlations. Springer, Berlin (2003) 309–323
85. Freeman, W. J.: Neurodynamics: an Exploration in Mesoscopic Brain Dynamics. Springer, Berlin (2000)

86. Wright, J. J., Robinson, P. A., Rennie, C. J., Gordon, E., Bourke, P. D., Chapman, C. L., Hawthorn, N., Lees, G. J., Alexander, D.: Toward an Integrated Continuum Model of Cerebral Dynamics: the Cerebral Rhythms, Synchronous Oscillation and Cortical Stability. Journal of Biological and Information Processing Systems 63(1–3) (2001) 71–88
87. Buzsáki, G., Draguhn, A.: Neuronal Oscillations in Cortical Networks. Science 304(5679) (2004) 1926–1929
88. Braitenberg, V., Schuz, A.: Anatomy of the Cortex. Springer-Verlag (1991)
89. Douglas, R. J., Martin, K. A.: A Functional Microcircuit for Cat Visual Cortex. Journal of Physiology 440 (1991) 735–769
90. Buxhoeveden, D. P., Casanova, M. F.: The Minicolumnar Hypothesis in Neurosciences. Brain 125(5) (2002) 935–951
91. Bressler, S. L.: Large-Scale Cortical Networks and Cognition. Brain Research Reviews 20 (1995) 288–304
92. Sporns, O., Kötter, R.: Motifs in Brain Networks. PLoS Biology 2 (2004) 1910–1918
93. Freeman, W. J.: Making Sense of Brain Waves: The Most Baffling Frontier in Neuroscience. In: Parelus, P. Principe, J. Rajasekaran, S. (Eds) Biocomputing. Kluwer, New York (2001) 33–55
94. Jirsa, V. K., Kelso, J. A. S.: Spatiotemporal Pattern Formation in Continuous Systems with Heterogeneous Connection Topologies. Physical Review E 62(6) (2000) 8462–8465
95. Lehmann, D., Strik, W. K., Henggeler, B., Koenig, T., Koukkou, M.: Brain Electric Microstates and Momentary Conscious Mind States as Building Blocks of Spontaneous Thinking: I. Visual Imagery and Abstract Thoughts. International Journal of Psychophysiology 29(1) (1998) 1–11
96. Kaplan, A. Ya., Shishkin, S. L.: Nonparametric Method for the Segmentation of the EEG. Computer Methods and Programs in Biomed. 60(2) (1999) 93–106
97. Kelso, J. A. S.: The Complementary Nature of Coordination Dynamics: Self-Organization and the Origins of Agency. Journal of Nonlinear Phenomena in Complex Systems 5 (2002) 364–371
98. Pritchard, W. S.: The Brain in Fractal Time: 1/f-like Power Spectrum Scaling of the Human Electroencephalogram. Int. J. Neurosci. 66(1–2) (1992) 119–129
99. Engel, AK., König, P., Singer, W.: Direct Physiological Evidence for Scene Segmentation by Temporal Coding. Proceedings of the National Academy of Science USA 88 (1991) 9136–9140
100. Castelo-Branco, M., Goebel, R., Neuenschwander, S.: Neural Synchrony Correlates with Surface Segregation Rules. Nature 405 (2000) 685–689
101. Eckhorn, R., Obermüller, A.: Single Neurons Are Differently Involved in Stimulus-Specific Oscillations in Cat Visual Cortex. Experimental Brain Research 95(1) (1993) 177–182
102. Bressler, S. L., Coppola, R., Nakamura, R.: Episodic Multiregional Cortical Coherence at Multiple Frequencies during Visual Task Performance. Nature 366 (1993) 153–156
103. Tallon-Baudry, C., Bertrand, O., Fischer, C.: Oscillatory Synchrony between Human Extrastriate Areas during Visual Short-Term Memory Maintenance. J Neurosci 21 (2001) RC177(1–5)
104. Müller, M. M., Bosch, J., Elbert, T., Kreiter, A., Sosa, M., Valdes-Sosa, P., Rockstroh, B.: Visually Induced Gamma-Band Responses in Human Electroencephalographic Activity: a Link to Animal Studies. Experimental Brain Research 112(1) (1996) 96–102
105. Rodriguez, E., George, N., Lachaux, J. P., Martinerie, J., Renault, B., Varela, F.: Perception's Shadow: Long-Distance Synchronization in the Human Brain Activity. Nature 397 (1999) 430–433
106. Haken, H.: Synergetics: an Introduction. Springer Series in Synergetics. Springer, Berlin (1983)
107. Kelso, J. A. S., Haken, H.: New Laws to Be Expected in the Organism: Synergetics of Brain and Behavior. In: Murphy, M. O'Neill, L. (Eds.) What Is Life? The Next 50 Years. University Press, Cambridge (1995)

108. Assisi, C. G., Jirsa, V. K., Kelso, J. A. S.: Synchrony and Clustering in Heterogeneous Networks with Global Coupling and Parameter Dispersion, Physical Review Letters 94 (2005) 018106
109. Jirsa, V. K., Haken, H.: A Derivation of a Macroscopic Field Theory of the Brain from the Quasi-Microscopic Neural Dynamics. Physica D 99 (1997) 503–526
110. Kelso, J. A. S. Jeka, J. J.: Symmetry Breaking Dynamics of Human Multilimb Coordination. Journal of Experimental Psychology: Human Perception and Performance 18 (1992) 645–668
111. Gregory, R. L.: Editorial. Perception 29 (2000) 1139–1142
112. Heisenberg, W. Physics and Philosophy: The Revolution in Modern Physics. New York, Harper & Row (1958)

The Formation of Global Neurocognitive State

Steven L. Bressler

Abstract I propose in this chapter that the formation of global neurocognitive state in the cerebral cortex is central to the mammalian capacity for assessment of organismic state. I consider a putative mechanism for the formation of global neurocognitive state from interactions among interconnected cortical areas. In this model, each area makes a local assessment of its own current state, representing a partial assessment of organismic state, through the generation of packets of high-frequency oscillatory wave activity. The spatial amplitude modulation (AM) pattern of the wave packet is proposed to represent the expression of an area's current state in relation to the other areas with which it is interacting. Through their interactions, sets of cortical areas mutually constrain the AM patterns of their wave packets. It is proposed that this process leads to the manifestation of wave packets having cognitively consistent patterns, and the formation of globally unified consensual neurocognitive states.

1 Introduction

An essential function performed by the cerebral cortex is to dynamically assess the state of the mammalian organism on a moment-by-moment basis. This dynamic assessment plays a critical role in adaptive behavior by allowing the organism to perceive and act in a manner consistent with the context of the changing situation in which it exists [15]. It requires the monitoring of sensory inputs in multiple modalities that inform the cortex about the states of the external and internal environments, including the state of the musculature. The states of numerous brain systems, e.g. the preparatory, postural, and kinematic states of the motor system, must also be monitored. The assessments of these diverse states must further be integrated to create a global neurocognitive assessment state in the cortex from which judgements may be formed and actions planned and executed. Since the external and internal environments are in continual flux, the cortex must be highly flexible in its ability to monitor and integrate the plethora of data concerning organismic state to which it has access in order for the organism to behave adaptively.

It has long been recognized that no single area of the cerebral cortex serves as a central supervisor that performs all of these monitoring and integrative functions.

L. I. Perlovsky, R. Kozma, *Neurodynamics of Cognition and Consciousness.*

Although the functions of many areas of association cortex are considered to be primarily executive in function [37], there is no single "supreme" executive area. The cortex consists of a large number of areas profusely interconnected by long-range pathways in a complex topological structure [3, 24, 40, 41, 56, 73, 83, 87]. An important aspect of cortical connectivity is that each cortical area has a specialized topological position within the cortex, i.e. a unique pattern of interconnectivity with other cortical areas [10, 60]. To a large degree, the function of every cortical area, including executive areas, is determined by its unique patterning of long-range connectivity. Furthermore, in spite of the well-known cytoarchitectonic distinctiveness of certain areas [20, 21], the short-range interconnectivity of local circuits within cortical areas is generally similar throughout the cortex, implying that no area has a specialized monitoring function by virtue of its internal organization. In sum, these considerations suggest that cortical monitoring and integrative functions are a result of cooperative interaction among many distributed areas, and not the sole property of any one area or small group of areas.

The goal of this chapter is to attempt a reasonable explanation of how the cortex dynamically generates global neurocognitive states, representing the assessment of organismic state, from interactions among its areas. Evidence will be considered for the generation of local states by neuronal populations within cortical areas, and for the interaction of cortical areas in the generation of global neurocognitive states. The effect of short- and long-range patterning of interconnectivity within the cortex on the generation of local and global states will be considered [16, 72].

2 Coherent Local Cortical Domains and Their Interactions

This section considers the local organization of population activity in the cortex. There is ample support for the existence of locally coherent domains of high-frequency (13–100 Hz) oscillatory activity in the cortex. The evidence comes largely from recordings from the pial surface of sensory cortices of rabbits, cats, and monkeys by high-density electrode arrays [4, 7, 8, 25, 26, 31, 35]. The observed activity is spatially coherent, indicative of a common waveform, across spatial extents on the order of tens of square millimeters [27]. The spatial coherence arises largely as a result of mutually excitatory synaptic interactions among the excitatory neurons of the cortical area [22]. Within the spatial extent of a coherent domain, which has been termed a "wave packet" [27], the wave amplitude and phase are both spatially modulated.

The spatial pattern of amplitude modulation (AM) of the wave packet in sensory cortices has been found to correlate with the categorical perception of conditioned stimuli in the corresponding sensory modality [4, 31, 35, 58]. The spatial AM pattern changes reliably with behavioral context, conditioning, and the animal's cumulative experience of the sensory environment, but does not reflect the specific features of sensory stimuli. Wave packets have been observed to recur intermittently with rates in the delta-theta frequency range (2–7 Hz), each recurrence resulting from a state transition in the population activity of the local area [34]. The spatial pattern of phase modulation of the wave packet has not been found to correlate with perception or behavior, but is a useful indicator of the spatial extent of the wave packet. Since the spatial extent of coherent domains can exceed the size of the primary sensory areas [28, 30, 33], it remains an open question as to what degree the coherent domains are confined by the boundaries of cortical areas as defined anatomically.

Extrapolating from these considerations, it is reasonable to infer that each area of the cerebral cortex, whether sensory or non-sensory, generates a sequence of wave packets during normal cognitive behavior, and that interconnected areas concurrently generate wave packets that may result in their undergoing conjoint local state transitions. Generation of the spatially patterned wave packet arises from the local circuitry in each cortical area. The spatial AM pattern of the wave packet generated in an area depends on the interaction of pulse inputs from other areas with the local synaptic matrix acting as an associative memory formed by Hebbian learning [1, 53]. The spatial extent of the wave packet is limited by such factors as the spatial distributions of mutually excitatory connections and of the terminal arborizations of incoming axonal projection pathways.

From this perspective, the generation and transmission of spatially patterned activity between areas is central to cortical communication. Given that the anatomical interconnectivity of the cortex is largely bidirectional at the population level [24], areas typically receive feedback from the same areas to which they transmit, and in this sense cortical processing is considered to be "reentrant" [79]. By providing feedback inputs to the circuit elements in sets of interacting cortical areas, reentrant interactions mutually constrain the spatial AM patterns of wave packets generated by those areas.

In this process, each cortical area generates a spatial distribution of pulse activity whose density conforms to the spatial AM pattern of its wave packet. These pulse density patterns are transmitted between cortical areas over inter-areal fiber pathways, some of which have a degree of topographic specificity, but which in general have a high degree of divergence and convergence in their projection to other cortical areas. Thus wave packets undergo spatial integral transformations in transmission from one cortical area to another [4]. The pattern of postsynaptic potentials produced in an area by pulses transmitted over inter-areal fiber pathways depends further on the pattern of synaptic transmission strengths at the axon terminals that have been modified by transcortical Hebbian learning [36, 63]. The overall effect of multiple reentrant interactions among cortical areas is mutual constraint of the spatial AM patterns of the wave packets that they generate [12].

3 The Emergence of Global Neurocognitive State

In the awake state, cortical areas have been found to generate wave packets lasting on the order of 100–300 msec, interrupted by nonlinear state transitions [29]. Each time the wave packet is regenerated, its spatial AM pattern may change. The spatial AM pattern of the wave packet may be considered as an expression of a given area's own current state in relation to the other areas with which it is interacting. It is thus reasonable to suppose that, as reciprocally interconnected areas undergo reentrant interactions, the spatial AM patterns of the wave packets they generate are conjointly constrained. In that case, the wave packets in interconnected cortical areas may thereby undergo pattern constraint in conjunction with one another.

The conjunctive generation of wave packets by interconnected cortical areas represents a possible mechanism for the assessment of organismic state by the cortex. I hypothesize that the spatial AM pattern of the wave packet generated in each area represents an expression of the contribution by that area to the overall assessment of organismic state by the cortex. Each area's contribution arises from its own unique memory store, embodied in its local synaptic matrix, under the influence of inputs from other areas with which it interacts. Each area's spatial AM pattern may be expressed in conjunction with others in sets of interacting cortical areas as they conjointly undergo state transitions and co-generate wave packets. The mutual pattern constraint exerted by those areas is expected to promote the expression of consistent contributions to the state assessment, creating a "consensus" interpretation of available evidence concerning the state of the organism within the context of its current situation [15].

Although it is currently unknown what cellular mechanisms are responsible for the expression of consistent spatial AM patterns in interconnected cortical areas, there is evidence that they involve the (partial) phase synchronization of high-frequency oscillatory activity. The wave packet itself is a locally phase synchronized oscillatory event, and evidence suggests that high-frequency oscillatory activity in different cortical areas becomes phase synchronized in relation to cognitive state [13, 17, 18, 65, 75, 84]. The rapid emergence of large-scale patterns of phase-synchronized cortical sites at specific stages of cognitive task performance suggests that creation of the cognitive microstate [55] depends on the transient coordination of specific sets of areas by long-range oscillatory phase synchronization [9, 13, 14, 16]. It is thus likely that phase synchronization serves not only to coordinate local neuronal populations in wave packets within cortical areas, but also distant populations in different areas.

A neural process that produces long-range phase synchronization of neuronal populations in interconnected cortical areas has several attractive features for the creation of a global state of the cortex representing the ongoing assessment of organismic state. First, the possibility exists for both the transient inclusion and exclusion of cortical areas from large-scale phase-synchronized networks. Areas whose wave packets achieve the co-expression of consistent spatial AM patterns may be linked by long-range phase synchronization, whereas others may be excluded due to a lack of phase synchronization. Oscillatory activity in different cortical populations is called "coherent" when there exists a high level of phase synchronization between them, and "incoherent" when the degree of phase synchronization is low (Fig. 1). By a terminological coincidence, the term "coherence" has also been used to refer to the satisfaction of consistent cognitive constraints [76]. In this second sense, areas generating wave packets with consistent spatial AM patterns may be considered to express cognitive coherence with one another, and those with inconsistent patterns may be considered to express cognitive incoherence.

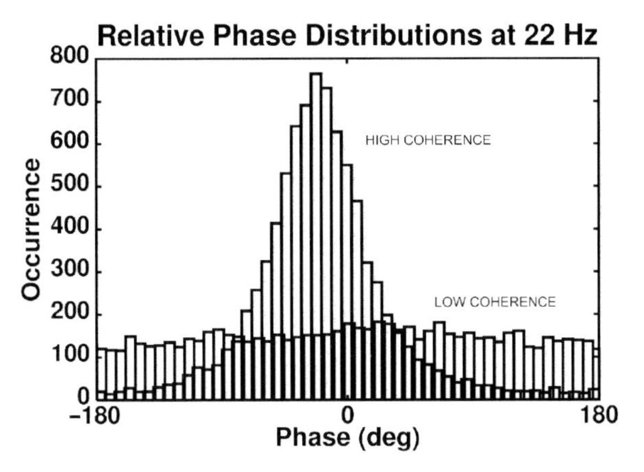

Fig. 1 Relative phase distributions between local field potentials recorded from two posterior parietal electrode sites in the macaque monkey. The 22 Hz frequency, in the mid-beta frequency range, represents the center frequency of oscillatory activity manifested by neuronal populations at both sites. The two distributions were computed from different time intervals as the monkey performed a visuomotor pattern discrimination task. The first distribution, labeled "HIGH COHERENCE", was from a prestimulus interval when oscillatory activity at 22 Hz was highly synchronized. The second distribution, labeled "LOW COHERENCE", was from a poststimulus interval when the phase synchronization at 22 Hz was very low. In general, high coherence indicates a highly concentrated relative phase distribution, and low coherence a weakly concentrated distribution

A second benefit of such a process is the possibility of maintaining the global coordination dynamics of the cortex in a state of metastability [10, 14]. A metastable system of interacting components preserves a balance between states of global coordination of the components and of component independence. In a system of

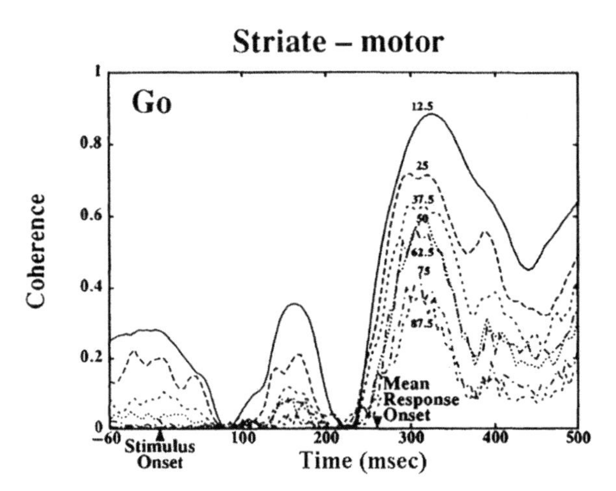

Fig. 2 Episodic increases in phase synchronization between local field potentials from electrodes sites in primary visual and primary motor cortical areas of the macaque monkey. Phase synchronization is measured by coherence, which varies from 0, representing no synchronization, to 1, representing total synchronization. Adapted from [13]

interacting oscillatory neuronal populations in interconnected cortical areas, metastability implies the coexistence of the tendency for these populations to be globally coordinated balanced by the counteracting tendency for them to be independent. A metastable system displays intermittency in coordination: the components may alternate between periods of coordination and uncoordination [14]. The concept of intermittent coordination dynamics is consistent with the experimentally observed intermittency of elevated phase synchronization among widely distributed cortical neuronal populations (Fig. 2), as well as the intermittent state transitions that separate wave packets [29, 32]. For both phenomena, the intermittency occurs roughly on the order of 100–300 msec.

4 Implications for the Study of Cognition and Consciousness

In essence, the proposal presented here is that an assessment of organismic state occurs in the cerebral cortex as the result of interactions among interconnected cortical areas, while they are receiving inputs from thalamic and other subcortical structures that convey the raw data about the internal and external milieus. By each cortical area making a local assessment of its own current state, it provides a partial assessment of the entire state of the organism. Depending on its position in the large-scale connectivity structure of the cortex [38, 56], each area expresses a different aspect of the total state.

A potential mechanism for the creation of a global neurocognitive state, representing the assessment of organismic state, involves the generation of packets of high-frequency oscillatory wave activity in cortical areas. The spatial AM pattern of the wave packet is proposed to represent the expression of an area's current state in relation to the other areas with which it is interacting. Through their interactions, cortical areas exert modulatory influences on one another that mutually constrain that expression. Cortical areas generate wave packets at rates of several times per second [29], each generation being initiated by a nonlinear state transition.

Mutual pattern constraint may lead sets of interacting cortical areas to achieve unified consensual states that are consistent according to cognitive coherence and incoherence relations [76]. It is proposed that such sets become coordinated in large-scale cortical networks by the phase synchronization of high-frequency oscillatory activity. Support for this idea comes from the observation of large-scale phase-synchronized cortical networks in relation to different cognitive functions [9, 65, 66, 75, 82, 84]. A example from the sensorimotor region of macaque monkeys monitoring the state of their limb position as part of a visuomotor task is shown in Fig. 3.

Depending on the location and extent of the areas involved, the emergence of unified consensual states may represent the assessment of organismic state in different cognitive forms [56]. Within perceptual systems, these states may lead to the recognition and identification of perceptual entities. Between perceptual and motor systems they may provide behavioral adaptation within the perception-action cycle [38]. Through integration across executive systems, they may establish

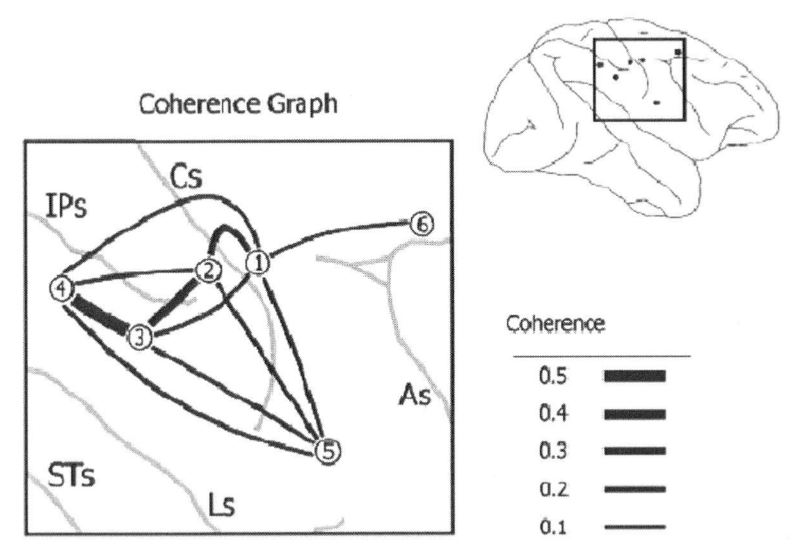

Fig. 3 Phase synchronized network in sensorimotor cortex of the macaque monkey. Lines connecting electrode sites represent significant coherence in the beta frequency range. Adapted from [17]

neurocognitive contexts [15] that allow the anticipation of impending events and the recognition of those events when they occur.

Of course, the assessment of organismic state is rarely perfect. A fundamental indeterminacy exists in the process of evaluating the states of sensory environments [12] and brain systems. Furthermore, sensory data may be incomplete, ambiguous, or contradictory. These considerations suggest that multiple clusters of consensual cortical areas may concurrently arise in the cortex, representing competitive assessments of organismic state, in much the same way as has been proposed for the competition among perceptual entities within [23] and between modalities [54]. Through increased consistency of expressed spatial AM patterns, as well as the recruitment of additional consistent areas, some clusters may be come better established over successive state transitions, whereas others may die out. Clusters may also coalesce if they express consistent interpretations of the organismic state. Through an ongoing process of recurring assessment, it is expected that the cortex may achieve neurocognitive states representing the globally consistent assessment of organismic state.

The framework proposed in this chapter is consistent with dynamical approaches in cognitive science that focus on the evolution of cognitive state on a sub-second time scale [5, 71]. The spatial AM pattern of a wave packet in a cortical area may be represented as a vector in high dimensional space [29]. As the wave packet of the area undergoes sequential generation, the vector follows a trajectory in state space, and its interactions with other areas modulates the state vector trajectory. Furthermore, the conjoint spatial AM patterns of interacting areas may also be represented as state vectors in high dimensional space. At the level of the global neurocognitive state, a cortical system trajectory is envisioned that is equivalent to cognitive state space trajectories described in the literature [45, 62, 70, 85].

The proposed assessment of organismic state is also similar to the appraisal process postulated to mediate the interplay of cognitive and emotional states [52]. The creation of global neurocognitive states in the cerebral cortex is likely to have important effects on subcortical brain structures with a wide variety of consequences [19]. Subcortical structures may feed influences back to the cortex by direct pathways using neurotransmitters or neuropeptides [59], as well as by more indirect paths involving circulating hormones. Several studies have demonstrated an interrelation between neurocognitive state and circulating hormones [48, 49, 64]. Thus, the cortical system trajectory described above is likely to be influenced by a number of sources elsewhere in the brain and body.

Disruption of the proposed organismic assessment process may be responsible for many of the symptoms observed in neuropsychiatric disorders. There is growing agreement that schizophrenia is a disease characterized by a disturbance of coordinated cognitive processes [2, 42, 57, 61, 78], and that this disruption occurs as a result of abnormal phase synchronization of high-frequency oscillatory neuronal population activity in the cortex [6, 11, 50, 51, 69, 74, 80, 81]. Evidence exists also for similar disruptions of phase synchronized cortical oscillations in Alzheimer's disease [47] and autism [86]. The impairment of organismic state assessment, in different forms, may be common to all these pathologies.

5 Conclusions

The global neurocognitive state discussed here as a means by which the organism is able to assess its own state may also be related to the state of conscious awareness. Although the process by which cortical areas become coordinated may largely be unconscious, it may possibly become conscious when a threshold is crossed in terms of the number of coordinated areas, the overall strength of coordination, or the recruitment of particular areas [44, 46, 67]. If, then, global neurocognitive assessment states are a prerequisite for consciousness, the latter should be considered as an emergent property of the interaction of distributed cortical areas [68, 77]. To view consciousness as a dynamic property of brain function that emerges from the interactions of cortical areas in relation to the assessment of organismic state suggests that it may endow mammals with enhanced survival value [39, 43]. By augmenting the assessment of organismic state, consciousness may provide mammals with greater flexibility in their ability to distinguish environmental entities and a greater behavioral repertoire. It further suggests that all mammalian species possess consciousness with a level indexed by the number of differentiated cortical areas and the complexity of their interconnectivity.

Acknowledgment Supported by grant MH64204 from the National Institute of Mental Health.

References

1. D. J. Amit. The hebbian paradigm reintegrated: Local reverberations as internal representations. *Behav Brain Sci*, 18:617–657, 1995.
2. N. C. Andreasen, P. Nopoulos, D. S. O'Leary, D. D. Miller, T. Wassink, and M. Flaum. Defining the phenotype of schizophrenia: cognitive dysmetria and its neural mechanisms. *Biol Psychiatry*, 46(7):908–20, 1999.
3. H. Barbas. Connections underlying the synthesis of cognition, memory, and emotion in primate prefrontal cortices. *Brain Res Bull*, 52(5):319–30, 2000.
4. J. M. Barrie, W. J. Freeman, and M. D. Lenhart. Spatiotemporal analysis of prepyriform, visual, auditory, and somesthetic surface eegs in trained rabbits. *J Neurophysiol*, 76(1):520–39, 1996.
5. R. D. Beer. Dynamical approaches to cognitive science. *Trends Cogn Sci*, 4:91–99, 2000.
6. C. A. Brenner, O. Sporns, P. H. Lysaker, and B. F. O'Donnell. Eeg synchronization to modulated auditory tones in schizophrenia, schizoaffective disorder, and schizotypal personality disorder. *Am J Psychiatry.*, 160(12):2238–40., 2003.
7. S. L. Bressler. *Spatio-temporal analysis of olfactory signal processing with behavioral conditioning*. PhD thesis, University of California, Berkeley, 1982.
8. S. L. Bressler. Spatial organization of eegs from olfactory bulb and cortex. *Electroencephalogr Clin Neurophysiol*, 57(3):270–276, Mar 1984.
9. S. L. Bressler. Large-scale cortical networks and cognition. *Brain Res Brain Res Rev*, 20(3):288–304, 1995.
10. S. L. Bressler. Understanding cognition through large-scale cortical networks. *Curr Dir Psychol Sci*, 11:58–61, 2002.
11. S. L. Bressler. Cortical coordination dynamics and the disorganization syndrome in schizophrenia. *Neuropsychopharmacology*, 28(Suppl 1):S35–9, 2003.
12. S. L. Bressler. Inferential constraint sets in the organization of visual expectation. *Neuroinformatics*, 2(2):227–38, 2004.
13. S. L. Bressler, R. Coppola, and R. Nakamura. Episodic multiregional cortical coherence at multiple frequencies during visual task performance. *Nature*, 366(6451):153–6, 1993.
14. S. L. Bressler and J. A. Kelso. Cortical coordination dynamics and cognition. *Trends Cogn Sci*, 5(1):26–36, 2001.
15. S. L. Bressler and A. R. McIntosh. The role of neural context in large-scale neurocognitive network operations. In V. Jirsa and A. R. McIntosh, editors, *Handbook of Brain Connectivity*. Springer-Verlag, 2007.
16. S. L. Bressler and E. Tognoli. Operational principles of neurocognitive networks. *Int J Psychophysiol*, 60(2):139–48. Epub 2006 Feb 21, 2006.
17. A. Brovelli, M. Ding, A. Ledberg, Y. Chen, R. Nakamura, and S. L. Bressler. Beta oscillations in a large-scale sensorimotor cortical network: directional influences revealed by granger causality. *Proc Natl Acad Sci U S A*, 101(26):9849–54. Epub 2004 Jun 21, 2004.
18. G. Buzsaki and A. Draguhn. Neuronal oscillations in cortical networks. *Science*, 304(5679):1926–9, 2004.
19. A. R. Damasio, T. J. Grabowski, A. Bechara, H. Damasio, L. L. Ponto, J. Parvizi, and R. D. Hichwa. Subcortical and cortical brain activity during the feeling of self-generated emotions. *Nat Neurosci.*, 3(10):1049–56., 2000.
20. J. DeFelipe, L. Alonso-Nanclares, and J. I. Arellano. Microstructure of the neocortex: comparative aspects. *J Neurocytol.*, 31(3–5):299–316., 2002.
21. J. DeFelipe, G. N. Elston, I. Fujita, J. Fuster, K. H. Harrison, P. R. Hof, Y. Kawaguchi, K. A. Martin, K. S. Rockland, A. M. Thomson, S. S. Wang, E. L. White, and R. Yuste. Neocortical circuits: evolutionary aspects and specificity versus non-specificity of synaptic connections. remarks, main conclusions and general comments and discussion. *J Neurocytol.*, 31(3–5): 387–416., 2002.
22. R. J. Douglas, C. Koch, M. Mahowald, K. A. Martin, and H. H. Suarez. Recurrent excitation in neocortical circuits. *Science*, 269(5226):981–5., 1995.

23. J. Duncan, G. Humphreys, and R. Ward. Competitive brain activity in visual attention. *Curr Opin Neurobiol.*, 7(2):255–61., 1997.
24. D. J. Felleman and D. C. Van Essen. Distributed hierarchical processing in the primate cerebral cortex. *Cereb Cortex.*, 1(1):1–47., 1991.
25. W. J. Freeman. *Mass action in the nervous system*. Academic Press. 2004: http://sulcus.berkeley.edu/MANSWWW/MANSWWW.html, 1975.
26. W. J. Freeman. A neurobiological theory of meaning in perception. part ii: Spatial patterns of phase in gamma eegs from primary sensory cortices reveal the dynamics of mesoscopic wave packets. *Int J Bifurcat Chaos*, 13:2513–2535, 2003.
27. W. J. Freeman. The wave packet: an action potential for the 21st century. *J Integr Neurosci*, 2(1):3–30, 2003.
28. W. J. Freeman. Origin, structure, and role of background eeg activity. part 2. analytic phase. *Clin Neurophysiol*, 115(9):2089–107, 2004.
29. W. J. Freeman. Origin, structure, and role of background eeg activity. part 4: Neural frame simulation. *Clin Neurophysiol*, 117(3):572–89. Epub 2006 Jan 25, 2006.
30. W. J. Freeman and J. M. Barrie. Analysis of spatial patterns of phase in neocortical gamma eegs in rabbit. *J Neurophysiol*, 84(3):1266–1278, 2000.
31. W. J. Freeman and B. C. Burke. A neurobiological theory of meaning in perception. part iv: Multicortical patterns of amplitude modulation in gamma eeg. *Int J Bifurcat Chaos*, 13: 2857–2866, 2003.
32. W. J. Freeman, B. C. Burke, and M. D. Holmes. Aperiodic phase re-setting in scalp eeg of beta-gamma oscillations by state transitions at alpha-theta rates. *Hum Brain Mapp*, 19(4): 248–272, 2003.
33. W. J. Freeman, M. D. Holmes, G. A. West, and S. Vanhatalo. Fine spatiotemporal structure of phase in human intracranial eeg. *Clin Neurophysiol*, 117(6):1228–1243, 2006.
34. W. J. Freeman and L. J. Rogers. A neurobiological theory of meaning in perception. part v. multicortical patterns of phase modulation in gamma eeg. *Int J Bifurc Chaos*, 13:2867–2887, 2003.
35. W. J. Freeman and B. W. van Dijk. Spatial patterns of visual cortical fast eeg during conditioned reflex in a rhesus monkey. *Brain Res*, 422(2):267–276, 1987.
36. J. M. Fuster. Network memory. *Trends Neurosci*, 20(10):451–9, 1997.
37. J. M. Fuster. Upper processing stages of the perception-action cycle. *Trends Cogn Sci.*, 8(4):143–5., 2004.
38. J. M. Fuster. The cognit: a network model of cortical representation. *Int J Psychophysiol*, 60:125–32, 2006.
39. J. A. Gray. The contents of consciousness: a neuropsychological conjecture. *Behav Brain Sci*, 18:659–722, 1995.
40. C. C. Hilgetag, M. A. O'Neill, and M. P. Young. Indeterminate organization of the visual system. *Science.*, 271(5250):776–7., 1996.
41. C. C. Hilgetag, M. A. O'Neill, and M. P. Young. Hierarchical organization of macaque and cat cortical sensory systems explored with a novel network processor. *Philos Trans R Soc Lond B Biol Sci.*, 355(1393):71–89., 2000.
42. R. E. Hoffman and T. H. McGlashan. Reduced corticocortical connectivity can induce speech perception pathology and hallucinated 'voices'. *Schizophr Res.*, 30(2):137–41., 1998.
43. W. James. Are we automata? *Mind*, 4:1–22, 1879.
44. E. R. John. From synchronous neuronal discharges to subjective awareness? *Prog Brain Res.*, 150:143–71., 2005.
45. J. A. S. Kelso. *Dynamic patterns*. MIT Press, 1995.
46. J. F. Kihlstrom. The cognitive unconscious. *Science.*, 237(4821):1445–52., 1987.
47. T. Koenig, L. Prichep, T. Dierks, D. Hubl, L. O. Wahlund, E. R. John, and V. Jelic. Decreased eeg synchronization in alzheimer's disease and mild cognitive impairment. *Neurobiol Aging*, 26:165–71, 2005.
48. S. Kuhlmann and O. T. Wolf. Arousal and cortisol interact in modulating memory consolidation in healthy young men. *Behav Neurosci.*, 120(1):217–23., 2006.
49. S. Kuhlmann and O. T. Wolf. A non-arousing test situation abolishes the impairing effects of

cortisol on delayed memory retrieval in healthy women. *Neurosci Lett.*, 399(3):268–72. Epub 2006 Feb 28., 2006.
50. J. S. Kwon, B. F. O'Donnell, G. V. Wallenstein, R. W. Greene, Y. Hirayasu, P. G. Nestor, M. E. Hasselmo, G. F. Potts, M. E. Shenton, and R. W. McCarley. Gamma frequency-range abnormalities to auditory stimulation in schizophrenia. *Arch Gen Psychiatry*, 56(11):1001–5, 1999.
51. L. Lee, L. M. Harrison, and A. Mechelli. The functional brain connectivity workshop: report and commentary. *Network.*, 14(2):R1–15., 2003.
52. M. D. Lewis. Bridging emotion theory and neurobiology through dynamic systems modeling. *Behav Brain Sci.*, 28(2):169–94; discussion 194–245., 2005.
53. M. Marinaro, S. Scarpetta, and M. Yoshioka. Learning of oscillatory correlated patterns in a cortical network by a stdp-based learning rule. *Math Biosci*, doi:10.1016/j.mbs.2006.10.001, 2007.
54. J. B. Mattingley, J. Driver, N. Beschin, and I. H. Robertson. Attentional competition between modalities: extinction between touch and vision after right hemisphere damage. *Neuropsychologia.*, 35(6):867–80., 1997.
55. J. L. McClelland, D. E. Rumelhart, and G. E. Hinton. The appeal of parallel distributed processing. In D. E. Rumelhart and J. L. McClelland, editors, *Parallel Distributed Processing: Explorations in the Microstructure of Cognition, Vol. 1*, pp. 3–44. MIT Press, 1986.
56. M. M. Mesulam. From sensation to cognition. *Brain.*, 121(Pt 6):1013–52., 1998.
57. A. Meyer-Lindenberg, J. B. Poline, P. D. Kohn, J. L. Holt, M. F. Egan, D. R. Weinberger, and K. F. Berman. Evidence for abnormal cortical functional connectivity during working memory in schizophrenia. *Am J Psychiatry*, 158(11):1809–17, 2001.
58. F. W. Ohl, H. Scheich, and W. J. Freeman. Change in pattern of ongoing cortical activity with auditory category learning. *Nature*, 412(6848):733–736, 2001.
59. J. Panksepp. At the interface of the affective, behavioral, and cognitive neurosciences: decoding the emotional feelings of the brain. *Brain Cogn.*, 52(1):4–14., 2003.
60. R. E. Passingham, K. E. Stephan, and R. Kotter. The anatomical basis of functional localization in the cortex. *Nat Rev Neurosci.*, 3(8):606–16., 2002.
61. W. A. Phillips and S. M. Silverstein. Convergence of biological and psychological perspectives on cognitive coordination in schizophrenia. *Behav Brain Sci.*, 26(1):65–82; discussion 82–137., 2003.
62. R. Port and T. van Gelder. *Mind as motion.* MIT Press, 1995.
63. F. Pulvermuller. Words in the brain's language. *Behav Brain Sci.*, 22(2):253–79; discussion 280–336., 1999.
64. M. Reuter. Impact of cortisol on emotions under stress and nonstress conditions: a pharmacopsychological approach. *Neuropsychobiology.*, 46(1):41–8., 2002.
65. P. R. Roelfsema, A. K. Engel, P. Konig, and W. Singer. Visuomotor integration is associated with zero time-lag synchronization among cortical areas. *Nature.*, 385(6612):157–61., 1997.
66. A. Schnitzler and J. Gross. Functional connectivity analysis in magnetoencephalography. *Int Rev Neurobiol.*, 68:173–95., 2005.
67. J. T. Serences and S. Yantis. Selective visual attention and perceptual coherence. *Trends Cogn Sci*, 10:38–45, 2006.
68. A. K. Seth, E. Izhikevich, G. N. Reeke, and G. M. Edelman. Theories and measures of consciousness: an extended framework. *Proc Natl Acad Sci U S A*, 103:10799–804, 2006.
69. K. M. Spencer, P. G. Nestor, M. A. Niznikiewicz, D. F. Salisbury, M. E. Shenton, and R. W. McCarley. Abnormal neural synchrony in schizophrenia. *J Neurosci.*, 23(19):7407–11., 2003.
70. M. Spivey. *The continuity of mind.* Oxford University Press, 2006.
71. M. J. Spivey and R. Dale. Continuous dynamics in real-time cognition. *Current Directions in Psychological Science*, 15:207–211, 2006.
72. O. Sporns, G. Tononi, and G. M. Edelman. Theoretical neuroanatomy: relating anatomical and functional connectivity in graphs and cortical connection matrices. *Cereb Cortex.*, 10(2): 127–41., 2000.
73. O. Sporns, G. Tononi, and G. M. Edelman. Theoretical neuroanatomy and the connectivity of the cerebral cortex. *Behav Brain Res.*, 135(1–2):69–74., 2002.

74. V. B. Strelets, V. Y. Novototsky-Vlasov, and J. V. Golikova. Cortical connectivity in high frequency beta-rhythm in schizophrenics with positive and negative symptoms. *Int J Psychophysiol.*, 44(2):101–15., 2002.
75. C. Tallon-Baudry, O. Bertrand, and C. Fischer. Oscillatory synchrony between human extrastriate areas during visual short-term memory maintenance. *J Neurosci.*, 21(20):RC177., 2001.
76. P. Thagard and K. Verbeurgt. Coherence as constraint satisfaction. *Cognit Sci*, 22:1–24, 1998.
77. E. Thompson and F. J. Varela. Radical embodiment: neural dynamics and consciousness. *Trends Cogn Sci.*, 5(10):418–425., 2001.
78. G. Tononi and G. M. Edelman. Schizophrenia and the mechanisms of conscious integration. *Brain Res Brain Res Rev.*, 31(2–3):391–400., 2000.
79. G. Tononi, O. Sporns, and G. M. Edelman. Reentry and the problem of integrating multiple cortical areas: simulation of dynamic integration in the visual system. *Cereb Cortex*, 2:310–335, 1992.
80. P. J. Uhlhaas, D. E. Linden, W. Singer, C. Haenschel, M. Lindner, K. Maurer, and E. Rodriguez. Dysfunctional long-range coordination of neural activity during gestalt perception in schizophrenia. *J Neurosci.*, 26(31):8168–75., 2006.
81. O. van der Stelt, A. Belger, and J. A. Lieberman. Macroscopic fast neuronal oscillations and synchrony in schizophrenia. *Proc Natl Acad Sci U S A.*, 101(51):17567–8. Epub 2004 Dec 15., 2004.
82. F. Varela, J. P. Lachaux, E. Rodriguez, and J. Martinerie. The brainweb: phase synchronization and large-scale integration. *Nat Rev Neurosci.*, 2(4):229–39., 2001.
83. J. Vezoli, A. Falchier, B. Jouve, K. Knoblauch, M. Young, and H. Kennedy. Quantitative analysis of connectivity in the visual cortex: extracting function from structure. *Neuroscientist.*, 10(5):476–82., 2004.
84. A. von Stein, P. Rappelsberger, J. Sarnthein, and H. Petsche. Synchronization between temporal and parietal cortex during multimodal object processing in man. *Cereb Cortex.*, 9(2):137–50., 1999.
85. L. M. Ward. *Dynamical cognitive science*. MIT Press, 2002.
86. J. P. Welsh, E. S. Ahn, and D. G. Placantonakis. Is autism due to brain desynchronization? *Int J Dev Neurosci.*, 23(2–3):253–63., 2005.
87. M. P. Young. The organization of neural systems in the primate cerebral cortex. *Proc Biol Sci.*, 252(1333):13–8., 1993.

Neural Dynamic Logic of Consciousness: the Knowledge Instinct

Leonid I. Perlovsky

Abstract The chapter discusses evolution of consciousness driven by the knowledge instinct, a fundamental mechanism of the mind which determines its higher cognitive functions and neural dynamics. Although evidence for this drive was discussed by biologists for some time, its fundamental nature was unclear without mathematical modeling. We discuss mathematical difficulties encountered in the past attempts at modeling the mind and relate them to logic. The main mechanisms of the mind include instincts, concepts, emotions, and behavior. Neural modeling fields and dynamic logic mathematically describe these mechanisms and relate their neural dynamics to the knowledge instinct. Dynamic logic overcomes past mathematical difficulties encountered in modeling intelligence. Mathematical mechanisms of concepts, emotions, instincts, consciousness and unconscious are described and related to perception and cognition. The two main aspects of the knowledge instinct are differentiation and synthesis. Differentiation is driven by dynamic logic and proceeds from vague and unconscious states to more crisp and conscious states, from less knowledge to more knowledge at each hierarchical level of the mind. Synthesis is driven by a hierarchical organization of the mind; it strives to achieve unity and meaning of knowledge: every concept finds its deeper and more general meaning at a higher level. These mechanisms are in complex relationship of symbiosis and opposition, and lead to complex dynamics of evolution of consciousness and cultures. Mathematical modeling of this dynamics in a population leads to predictions for the evolution of consciousness, and cultures. Cultural predictive models can be compared to experimental data and used for improvement of human conditions. We discuss existing evidence and future research directions.

1 The Knowledge Instinct

To satisfy any instinctual need—for food, survival, and procreation—first and foremost we need to understand what's going on around us. The knowledge instinct is an inborn mechanism in our minds, an instinctual drive for cognition which compels us to constantly improve our knowledge of the world.

Humans and higher animals engage in exploratory behavior, even when basic bodily needs, like eating, are satisfied. Biologists and psychologists have discussed

L. I. Perlovsky, R. Kozma, *Neurodynamics of Cognition and Consciousness.*

various aspects of this behavior. Harry Harlow discovered that monkeys as well as humans have the drive for positive stimulation, regardless of the satisfaction of drives such as hunger [1]; David Berlyne emphasized curiosity as a desire for acquiring new knowledge [2]; Leon Festinger discussed the notion of cognitive dissonance and human drive to reduce the dissonance [3]. Until recently, however, this drive for exploratory behavior was not mentioned among "basic instincts" on a par with instincts for food and procreation.

The fundamental nature of this mechanism became clear during mathematical modeling of workings of the mind. Our knowledge always has to be modified to fit the current situations. We don't usually see exactly the same objects as in the past: angles, illumination, and surrounding contexts are usually different. Therefore, our internal representations that store past experiences have to be modified; adaptation-learning is required. For example, visual perception (in a simplified way) works as follows [4, 5, 6]. Images of the surroundings are projected from the retina onto the visual cortex, while at the same time memories-representations of expected objects are projected on the same area of cortex. Perception occurs when actual and expected images coincide. This process of matching representations to sensory data requires modifications-improvement of representations.

In fact virtually all learning and adaptive algorithms (tens of thousands of publications) maximize correspondence between the algorithm internal structure (knowledge in a wide sense) and objects of recognition. Paul Werbos' chapter in this book discusses a fundamental role of reinforcement learning; the knowledge instinct is a reinforcement learning, when reinforcers include correspondence of internal mind representations to the surrounding world. Internal mind representations, or models, which our mind uses for understanding the world, are in constant need of adaptation. Knowledge is not just a static state; it is in a constant process of adaptation and learning. Without adaptation of internal models we would not be able to understand the world. We would not be able to orient ourselves or satisfy any of the bodily needs. Therefore, we have an inborn need, a drive, an instinct to improve our knowledge, and we call it *the knowledge instinct*. It is a foundation of our higher cognitive abilities, and it defines the evolution of consciousness and cultures.

2 Aristotle and Logic

Before we turn to mathematical description of the knowledge instinct, it is instructive to analyze previous attempts at mathematical modeling of the mind. Founders of artificial intelligence in the 1950s and 60s believed that mathematical logic was the fundamental mechanism of the mind, and that using rules of logic they would soon develop computers with intelligence far exceeding the human mind. Although this belief turned out to be wrong, still many people believe in logic. It plays a fundamental role in many algorithms and even neural networks, and we start from logic to analyze difficulties of mathematical modeling of the mind.

Logic was invented by Aristotle. Whereas multiple opinions may exist on any topic, Aristotle found general rules of reason that are universally valid, and he called this set of rules "logic". He was proud of this invention and emphasized, "Nothing in this area existed before us" (Aristotle, IV BCE, a). However, Aristotle did not think that the mind works logically; he invented logic as a supreme way of argument, not as a theory of the mind. This is clear from many Aristotelian writings, for example from "Rhetoric for Alexander" (Aristotle, IV BCE, b), which he wrote when his pupil, Alexander the Great, requested from him a manual on public speaking. In this book he lists dozens of topics on which Alexander had to speak publicly. For each topic, Aristotle identified two opposing positions (e.g. making peace or declaring war; using or not using torture for extracting the truth, etc.). Aristotle gives logical arguments to support each of the opposing positions. Clearly, Aristotle saw logic as a tool to argue for decisions that were already made; he did not consider logic as the fundamental mechanism of the mind. Logic is, so to speak, a tool for politicians. Scientists follow logic when writing papers and presenting talks, but not to discover new truths about nature.

To explain the mind, Aristotle developed a theory of Forms, which will be discussed later. During the centuries following Aristotle the subtleties of his thoughts were not always understood. With the advent of science, intelligence was often identified with logic. In the 19th century mathematicians striving for exact proofs of mathematical statements noted that Aristotelian ideas about logic were not adequate for this. The foundation of logic, since Aristotle (Aristotle, IV BCE), was the law of excluded middle (or excluded third): every statement is either true or false, any middle alternative is excluded. But Aristotle also emphasized that logical statements should not be formulated too precisely (say, a measure of wheat should not be defined with an accuracy of a single grain). He emphasized that language implies the adequate accuracy, and everyone has his mind to decide what is reasonable. George Boole thought that Aristotle was wrong, that the contradiction between exactness of the law of excluded third and vagueness of language should be corrected.

In this way formal logic, a new branch of mathematics was born. Prominent mathematicians contributed to the development of formal logic, including Gottlob Frege, Georg Cantor, Bertrand Russell, David Hilbert, and Kurt Gödel. Logicians discarded uncertainty of language and founded formal mathematical logic on the law of excluded middle. Many of them were sure that they were looking for exact mechanisms of the mind. Hilbert wrote, "The fundamental idea of my proof theory is none other than to describe the activity of our understanding, to make a protocol of the rules according to which our thinking actually proceeds." (See Hilbert, 1928). In the 1900 he formulated Entscheidungsproblem: to define a set of logical rules sufficient to prove all past and future mathematical theorems. This would formalize scientific creativity and define a logical mechanism for the entire human thinking.

Almost as soon as Hilbert formulated his formalization program, the first hole appeared. In 1902 Russell exposed an inconsistency of formal logic by introducing a set R as follows: *R is a set of all sets which are not members of themselves.* Is R a member of R? If it is not, then it should belong to R according to the definition, but if R is a member of R, this contradicts the definition. Thus either way leads to a contradiction. This became known as the Russell's paradox. Its jovial formulation is as follows: A barber shaves everybody who does not shave himself. Does the barber shave himself? Either answers to this question (yes or no) lead to a contradiction.

This barber, like Russell's set, can be logically defined but cannot exist. For the next 25 years mathematicians where trying to develop a self-consistent mathematical logic, free from paradoxes of this type. But in 1931, Gödel (see in Gödel, 1986) proved that it is not possible, and that formal logic is inexorably inconsistent and self-contradictory.

For a long time people believed that intelligence is equivalent to conceptual logical reasoning. Although it is obvious that the mind is not always logical, since the first successes of science many people came to identify the power of intelligence with logic. This belief in logic has deep psychological roots related to the functioning of the mind. Most of the mind processes are not consciously perceived. For example, we are not aware of individual neuronal firings. We become conscious about the final states resulting from perception and cognition processes; these are perceived by our minds as "concepts" approximately obeying formal logic. For this reason many people believe in logic. Even after Gödelian theory, founders of artificial intelligence still insisted that logic is sufficient to explain how the mind works.

Let us return to Aristotle. He addressed relationships between logic and the working of the mind as follows. We understand the world due to Forms (representations, models) in our mind. Cognition is a learning process in which a Form-as-potentiality (initial model) meets matter (sensory signals) and becomes a Form-as-actuality (a concept). Whereas Forms-actualities are logical, Forms-potentialities do not obey logic. Here Aristotle captured an important aspect of the working of the mind which has eluded many contemporary scientists. Logic is not a fundamental mechanism of the mind, but rather the result of mind's illogical operations. Later we describe the mathematics of dynamic logic, which gives a mathematical explanation for this process: how logic appears from illogical states and processes. It turns out that dynamic logic is equivalent to the knowledge instinct.

3 Mechanisms of the Mind

The basic mind mechanisms making up operations of the knowledge instinct are described mathematically in the next section. Here we give a conceptual preview of this description. Among the mind's cognitive mechanisms, the most directly accessible to consciousness are concepts. Concepts are like internal models of the objects and situations in the world. This analogy is quite literal, e.g., as already mentioned, during visual perception of an object, a concept-model in our memory projects an image onto the visual cortex, which is matched there to an image, projected from retina. This simplified description will be refined later.

Concepts serve to satisfy the basic instincts, which have emerged as survival mechanisms long before concepts. Current debates regarding instincts, reflexes, motivational forces, and drives, often lump together various mechanisms. This is inappropriate for the development of mathematical description of the mind mechanisms. I follow proposals (see Grossberg & Levine, 1987; Perlovsky 2006, for further references and discussions) to separate instincts as internal sensor mechanisms indicating the basic needs, from "instinctual behavior," which should be described by appropriate mechanisms. Accordingly, I use the word "instincts" to describe mechanisms of internal sensors: for example, when a sugar level in blood goes below a certain level an instinct "tells us" to eat. Such separation of instinct as "internal sensor" from "instinctual behavior" is only a step toward identifying all the details of relevant biological mechanisms.

How do we know about instinctual needs? Instincts are connected to cognition and behavior by emotions. Whereas in colloquial usage, emotions are often understood as facial expressions, higher voice pitch, exaggerated gesticulation, these are outward signs of emotions, serving for communication. A more fundamental role of emotions within the mind system is that emotional signals evaluate concepts for the purpose of instinct satisfaction. This evaluation does not take place according to rules or concepts (like in rule-systems of artificial intelligence), but according to a different instinctual-emotional mechanism, described first by Grossberg and Levine (1987); the role of emotions in the working of the mind is considered in this book in chapter by Daniel Levine. Below we describe emotional mechanisms for higher cognitive functions.

Emotions evaluating satisfaction or dissatisfaction of the knowledge instinct are not directly related to bodily needs. Therefore, they are "spiritual" emotions. We perceive them as harmony-disharmony between our knowledge and the world (between our understanding of how things ought to be and how they actually are in the surrounding world). According to Immanuel Kant [7] these are aesthetic emotions (emotions that are not related directly to satisfaction or dissatisfaction of bodily needs).

Aesthetic emotions related to learning are directly noticeable in children. The instinct for knowledge makes little kids, cubs, and piglets jump around and play fight. Their inborn models of behavior must adapt to their body weights, objects, and animals around them long before the instincts of hunger and fear will use the models for the direct aims of survival. In adult life, when our perception and understanding of the surrounding world is adequate, aesthetic emotions are barely perceptible: the mind just does its job. Similarly, we do not usually notice adequate performance of our breathing muscles and satisfaction of the breathing instinct. However, if breathing is difficult, negative emotions immediately reach consciousness. The same is true about the knowledge instinct and aesthetic emotions: if we do not understand the surroundings, if objects around do not correspond to our expectations, negative emotions immediately reach consciousness. We perceive these emotions as disharmony between our knowledge and the world. Thriller movies exploit the instinct for knowledge: their personages are shown in situations in which knowledge of the world is inadequate for survival.

Let me emphasize again, aesthetic emotions are not peculiar to art and artists, they are inseparable from every act of perception and cognition. In everyday life we usually do not notice them. Aesthetic emotions become noticeable at higher cognitive levels in the mind hierarchy, when cognition is not automatic, but requires conscious effort. Antonio Damasio's view [8] of emotions defined by visceral mechanisms, as far as discussing higher cognitive functions, seems erroneous in taking secondary effects for the primary mechanisms.

In the next section we describe a mathematical theory of conceptual-emotional recognition and understanding, which is the essence of neural cognitive dynamics. As we discuss, in addition to concepts and emotions, this theory involves the mechanisms of intuition, imagination, conscious, and unconscious. This process is intimately connected to an ability of the mind to think, to operate with symbols and signs. The mind involves a heterarchy of multiple levels. It is not a strict hierarchy, but a heterarchy, because it involves feedback connections throughout many levels; to simplify discussion we often refer to the mind stricture as a hierarchy. Hierarchy of multiple levels of cognitive mechanisms: knowledge instinct, concept-models, and emotions, operate at each level from simple perceptual elements (like edges, or moving dots), to concept-models of objects, to relationships among objects, to complex scenes, and up the hierarchy... toward the concept-models of the meaning of life and purpose of our existence. Hence the tremendous complexity of the mind, yet relatively few basic principles of the mind organization explain neural evolution of this system.

4 Neural Modeling Fields

Neural Modeling Fields (NMF) is a neural architecture that mathematically implements the mechanisms of the mind discussed above. It is a multi-level, hetero-hierarchical system [9]. The mind is not a strict hierarchy; there are multiple feedback connections among adjacent levels, hence the term hetero-hierarchy. At each level in NMF there are concept-models encapsulating the mind's knowledge; they generate so-called top-down neural signals, interacting with input, bottom-up signals. These interactions are governed by the knowledge instinct, which drives concept-model learning, adaptation, and formation of new concept-models for better correspondence to the input signals.

This section describes a basic mechanism of interaction between two adjacent hierarchical levels of bottom-up and top-down signals (fields of neural activation). Sometimes it will be more convenient to talk about these two signal-levels as an input to, and output from, a (single) processing-level. At each level, output signals are concepts recognized in (or formed from) input signals. Input signals are associated with (or recognized, or grouped into) concepts according to the models and the knowledge instinct at this level. This general structure of NMF corresponds to our knowledge of neural structures in the brain; still, in this chapter we do not map mathematical mechanisms in all their details to specific neurons or synaptic connections.

At a particular hierarchical level, we enumerate neurons by the index $n = 1, \ldots N$. These neurons receive bottom-up input signals, $\mathbf{X}(n)$, from lower levels in the processing hierarchy. $\mathbf{X}(n)$ is a field of bottom-up neuronal synapse activations, coming from neurons at a lower level. Each neuron has a number of synapses. For generality, we describe each neuron activation as a set of numbers, $\mathbf{X}(n) = \{X_d(n), d = 1, \ldots D\}$. Top-down, or priming signals, to these neurons are sent by concept-models, $\mathbf{M}_h(\mathbf{S}_h, n)$, and we enumerate models by the index $h = 1, \ldots H$. Each model is characterized by its parameters, $\mathbf{S}_h$; in the neuron structure of the brain they are encoded by strength of synaptic connections. Mathematically, we describe them as a set of numbers, $\mathbf{S}_h = \{S^a_h, a = 1, \ldots A\}$. Models *represent* signals in the following way. Say, signal $\mathbf{X}(n)$ is coming from sensory neurons activated by object h, characterized by a model $\mathbf{M}_h(\mathbf{S}_h, n)$ and parameter values $\mathbf{S}_h$. These parameters may include position, orientation, or lighting of an object h. Model $\mathbf{M}_h(\mathbf{S}_h, n)$ predicts a value $\mathbf{X}(n)$ of a signal at neuron *n*. For example, during visual perception, a neuron n in the visual cortex receives a signal $\mathbf{X}(n)$ from retina and a priming signal $\mathbf{M}_h(\mathbf{S}_h, n)$ from an object-concept-model h. A neuron *n* is activated if both bottom-up signal from lower-level-input and top-down priming signal are strong. Various models compete for evidence in the bottom-up signals, while adapting their parameters for better match as described below. This is a simplified description of perception. The most benign everyday visual perception uses many levels from retina to object perception. The NMF premise is that the same laws describe the basic interaction dynamics at each level. Perception of minute features, or everyday objects, or cognition of complex abstract concepts is due to the same mechanism described in this section. Perception and cognition involve models and learning. In perception, models correspond to objects; in cognition, models correspond to relationships and situations.

The knowledge instinct drives learning, which is an essential part of perception and cognition. Learning increases a similarity measure between the sets of models and signals, $L(\{\mathbf{X}\}, \{\mathbf{M}\})$. The similarity measure is a function of model parameters and associations between the input bottom-up signals and top-down, concept-model signals. For concreteness I refer here to an object perception using a simplified terminology, as if perception of objects in retinal signals occurs in a single level.

In constructing a mathematical description of the similarity measure, it is important to acknowledge two principles (which are almost obvious) [10]. First, the visual field content is unknown before perception occurred and second, it may contain any of a number of objects. Important information could be contained in any bottom-up signal; therefore, the similarity measure is constructed so that it accounts for all bottom-up signals, $\mathbf{X}(n)$,

$$L(\{\mathbf{X}\}, \{\mathbf{M}\}) = \prod_{n \in N} l(\mathbf{X}(n)). \tag{1}$$

This expression contains a product of partial similarities, $l(\mathbf{X}(n))$, over all bottom-up signals; therefore it forces the mind to account for every signal (even if one term in the product is zero, the product is zero, the similarity is low and the knowledge instinct is not satisfied); this is a reflection of the first principle. Second, before perception occurs, the mind does not know which object gave rise to a signal from a particular retinal neuron. Therefore a partial similarity measure is constructed so that it treats each model as an alternative (a sum over models) for each input neuron signal. Its constituent elements are conditional partial similarities between signal $\mathbf{X}(n)$ and model $\mathbf{M}_h$, $l(\mathbf{X}(n)|h)$. This measure is "conditional" on object h being present, therefore, when combining these quantities into the overall similarity measure, L, they are multiplied by r(h), which represent a probabilistic measure of object h actually being present. Combining these elements with the two principles noted above, a similarity measure is constructed as follows:

$$L(\{X\}, \{M\}) = \prod_{n \in N} \sum_{h \in H} r(h) l(X(n)|h). \tag{2}$$

The structure of (2) follows standard principles of the probability theory: a summation is taken over alternatives, h, and various pieces of evidence, n, are multiplied. This expression is not necessarily a probability, but it has a probabilistic structure. If learning is successful, it approximates probabilistic description and leads to near-optimal Bayesian decisions. The name "conditional partial similarity" for $l(\mathbf{X}(n)|h)$ (or simply $l(n|h)$) follows the probabilistic terminology. If learning is successful, $l(n|h)$ becomes a conditional probability density function, a probabilistic measure that signal in neuron n originated from object h. Then L is a total likelihood of observing signals $\{\mathbf{X}(n)\}$ coming from objects described by models $\{\mathbf{M}_h\}$. Coefficients r(h), called priors in probability theory, contain preliminary biases or expectations, expected objects h have relatively high r(h) values; their true values are usually unknown and should be learned, like other parameters $\mathbf{S}_h$.

We note that in probability theory, a product of probabilities usually assumes that evidence is independent. Expression (2) contains a product over n, but it does not assume independence among various signals $\mathbf{X}(n)$. Partial similarities $l(n|h)$ are structured in a such a way (described later) that they depend on differences between signals and models; these differences are due to measurement errors and can be considered independent. There is a dependence among signals due to models: each model $\mathbf{M}_h(\mathbf{S}_h,n)$ predicts expected signal values in many neurons n.

During the learning process, concept-models are constantly modified. Here we consider a case when functional forms of models, $\mathbf{M}_h(\mathbf{S}_h,n)$, are all fixed and learning-adaptation involves only model parameters, $\mathbf{S}_h$. More complicated structural learning of models is considered in [11, 12]. From time to time a system forms a new concept, while retaining an old one as well; alternatively, old concepts are sometimes merged or eliminated. This requires a modification of the similarity measure (2); the reason is that more models always result in a better fit between the models and data. This is a well known problem, it is addressed by reducing similarity (2) using a "skeptic penalty function," p(N,M) that grows with the number of models M, and this growth is steeper for a smaller amount of data N. For example, an asymptotically unbiased maximum likelihood estimation leads to multiplicative $p(N,M) = \exp(-N_{par}/2)$, where N_{par} is a total number of adaptive parameters in all models (this penalty function is known as Akaike Information Criterion, see [9] for further discussion and references).

The knowledge instinct demands maximization of the similarity (2) by estimating model parameters $\mathbf{S}$ and associating signals with concepts. Note that all possible combinations of signals and models are accounted for in expression (2). This can be seen by expanding a sum in (2), and multiplying all the terms; it would result in H^N items, a very large number. This is the number of combinations between all signals (N) and all models (H).

This very large number of combinations was a source of difficulties (that we call combinatorial complexity, CC) for developing intelligent algorithms and systems since the 1950s. The problem was first identified in pattern recognition and classification research in the 1960s and was named "the curse of dimensionality" [13]. It seemed that adaptive self-learning algorithms and neural networks could learn solutions to any problem "on their own" if provided with a sufficient number of training examples. It turned out that training examples should encompass not only all individual objects that should be recognized, but also objects in the context, that is combinations of objects. Self-learning approaches encountered CC of learning requirements.

Rule-based systems were proposed in the 1960s to solve the problem of learning complexity. An initial idea was that rules would capture the required knowledge and eliminate a need for learning [14]. However, in presence of variability the number of rules grew; rules depended on other rules, combinations of rules had to be considered and rule systems encountered CC of rules. Beginning in the 1980s, model-based systems were proposed. They used models that depended on adaptive parameters. The idea was to combine advantages of learning-adaptivity and rules by using adaptive models. The knowledge was encapsulated in models, whereas unknown aspects of particular situations was to be learned by fitting model parameters (see [15] and discussions in [9, 16]). Fitting models to data required selecting data subsets corresponding to various models. The number of subsets, however, is combinatorially large (N^M as discussed above). A general popular algorithm for fitting models to data, multiple hypotheses testing [17], is known to face CC of computations. Model-based approaches encountered computational CC (N and NP complete algorithms).

It turns out that CC is related to the most fundamental mathematical result of the 20th c., Gödel's theory of inconsistency of logic [18, 19]. Formal logic is based on the "law of excluded middle," according to which every statement is either true or false and nothing in between. Therefore, algorithms based on formal logic have to evaluate every variation in data or models as a separate logical statement (hypothesis). CC of algorithms based on logic is a manifestation of the inconsistency of logic in finite systems. Multivalued logic and fuzzy logic were proposed to overcome limitations related to the law of excluded third [20]. Yet the mathematics of multivalued logic is no different in principle from formal logic, "excluded third" is substituted by "excluded n+1." Fuzzy logic encountered a difficulty related to the degree of fuzziness. Complex systems require different degrees of fuzziness in various subsystems, varying with the system operations; searching for the appropriate degrees of fuzziness among combinations of elements again would lead to CC. Is logic still possible after Gödel? A recent review of the contemporary state of this field shows that logic after Gödel is much more complicated and much less logical than was assumed by the founders of artificial intelligence. The problem of CC remains unresolved within logic [21].

Various manifestations of CC are all related to formal logic and Gödel theory. Rule systems relied on formal logic in a most direct way. Self-learning algorithms and neural networks relied on logic in their training or learning procedures, every training example was treated as a separate logical statement. Fuzzy logic systems relied on logic for setting degrees of fuzziness. CC of mathematical approaches to theories of the mind is related to the fundamental inconsistency of logic.

5 Dynamic Logic

5.1 Mathematical Formulation

NMF solves the CC problem by using dynamic logic [10, 22, 23, 24]. An important aspect of dynamic logic is matching vagueness or fuzziness of similarity measures to the uncertainty of models. Initially, parameter values are not known, and uncertainty of models is high; so is the fuzziness of the similarity measures. In the process of learning, models become more accurate, and the similarity measure more crisp, the value of the similarity increases. This is the mechanism of dynamic logic.

Mathematically it is described as follows. First, assign any values to unknown parameters, $\{\mathbf{S}_h\}$. Then, compute association variables f(h|n),

$$f(h|n) = r(h) l(\mathbf{X}(n)|h) / \sum_{h' \in H} r(h') l(\mathbf{X}(n)|h'). \tag{3}$$

Equation (3) looks like the Bayes' formula for a posteriori probabilities; if l(n|h) in the result of learning become conditional likelihoods, f(h|n) become Bayesian probabilities for signal n originating from object h. The dynamic logic of NMF is defined as follows,

$$\begin{aligned} df(h|n)/dt = {} & f(h|n) \sum_{h' \in H} \{\delta_{hh'} - f(h'|n) \cdot \\ & [\partial \ln l(n|h')/\partial \mathbf{M}_{h'}] \, \partial \mathbf{M}_{h'}/\partial \mathbf{S}_{h'} \cdot d\mathbf{S}_{h'}/dt, \end{aligned} \tag{4}$$

$$d\mathbf{S}_h/dt = \sum_{n \in N} f(h|n) \left[\partial \ln l(n|h)/\partial \mathbf{M}_h\right] \partial \mathbf{M}_h/\partial \mathbf{S}_h, \tag{5}$$

here

$$\delta_{hh'} \text{ is } 1 \text{ if } h = h', \; 0 \text{ otherwise.} \tag{6}$$

Parameter t is the time of the internal dynamics of the MF system (like a number of internal iterations). These equations define the neural dynamics of NMF.

Gaussian-shape functions can often be used for conditional partial similarities,

$$l(n|h) = G(\mathbf{X}(n)|\mathbf{M}_h(\mathbf{S}_h, n), \mathbf{C}_h). \tag{7}$$

Here G is a Gaussian function with mean $\mathbf{M}_h$ and covariance matrix $\mathbf{C}_h$. Note, a "Gaussian assumption" is often used in statistics; it assumes that signal distribution is Gaussian. This is not the case in (7): here signal is not assumed to be Gaussian. Equation (7) is valid if *deviations* between the model **M** and signal **X** are Gaussian; these deviations usually are due to many random causes and are therefore Gaussian. If they are not Gaussian, appropriate functions could be used. If there is no information about the functional shapes of conditional partial similarities, still (7) is a good choice, it is not a limiting assumption: a weighted sum of Gaussians in (2) can approximate any positive function, like similarity.

Covariance matrices, $\mathbf{C}_h$, in (7) are estimated like other unknown parameters, as shown in (5). Initially they are set to large values, corresponding to uncertainty in the knowledge of models, $\mathbf{M}_h$. As parameter values and models improve, covariances are reduced to intrinsic differences between models and signals (due to sensor errors, or model inaccuracies). As covariances get smaller, similarities get crisper, closer to delta-functions; association variables (3) get closer to crisp $\{0, 1\}$ values, and dynamic logic solutions converge to crisp logic. This process of concurrent parameter improvement and convergence of similarity to a crisp logical function is an essential part of dynamic logic. This is the mechanism of dynamic logic defining the neural dynamics of NMF.

The dynamic evolution of fuzziness from large to small is the reason for the name "dynamic logic." Mathematically, this mechanism helps avoiding local maxima during convergence [9, 22, 25], and psychologically it explains many properties of the mind, as discussed later. Whichever functional shapes are used for conditional partial similarities, they ought to allow for this process of matched convergence in parameter values and similarity crispness. The brain might use various mechanisms for realizing the dynamic logic process at various stages. The chapter in this book by Emilio Del-Moral-Hernandez considers neural networks with recursive processing elements (RPE), which might implement dynamic logic through their chaotic dynamics. In RPE neural networks high vagueness of dynamic logic states and the knowledge instinct dissatisfaction corresponds to high value of the parameter p and to chaotic wide searches. Low vagueness, successful perception and cognition, and the knowledge instinct satisfaction corresponds to low p and ordered dynamics. Similar correspondence between dynamic logic, the knowledge instinct, and chaotic dynamic might be applicable to discussions in Walter Freeman chapter in this book. Dynamic logic seems to correspond to transitions from highly chaotic to lower chaotic states in cortical neural activity.

The following theorem was proved [9].

Theorem. Equations (3) through (6) define a convergent dynamic NMF system with stationary states defined by $\max\{\mathbf{S}_h\}L$.

It follows that the stationary states of a NMF system are the maximum similarity states satisfying the knowledge instinct. When partial similarities are specified as probability density functions (pdf), or likelihoods, the stationary values of parameters $\{\mathbf{S}_h\}$ are asymptotically unbiased and efficient estimates of these parameters [26]. A computational complexity of dynamic logic is linear in N.

In plain English, this means that dynamic logic is a convergent process. It converges to the maximum of similarity, and therefore satisfies the knowledge instinct.

Several aspects of NMF convergence are discussed in later sections. If likelihood is used as similarity, parameter values are estimated efficiently (that is, in most cases, parameters cannot be better learned using any other procedure). Moreover, as a part of the above theorem, it is proven that the similarity measure increases at each iteration. The psychological interpretation is that the knowledge instinct is satisfied at each step: a NMF system with dynamic logic *enjoys* learning.

Let us emphasize again, the fundamental property of dynamic logic is evolution from vague, uncertain, fuzzy, unconscious states to more crisp, certain, conscious states.

5.2 Example of Operation

Operations of NMF are illustrated in Fig. 1 using an example of detection and tracking of moving objects in clutter [27]. Tracking is a classical problem, which becomes combinatorially complex in clutter when target signals are below the clutter level. Solving this problem is usually approached by using multiple hypotheses tracking algorithm [28], which evaluates multiple hypotheses about which signals came from which of the moving objects, and which from clutter. This standard approach is well-known to face CC [9], because large numbers of combinations of signals and models have to be considered. Figure 1 illustrates NMF neurodynamics while solving this problem.

Figure 1(a) shows true track positions, while Fig. 1(b) shows the actual data available for detection and tracking. It contains 6 sensor scans on top of each other (time axis is not shown). The data set consists of 3000 data points, 18 of which

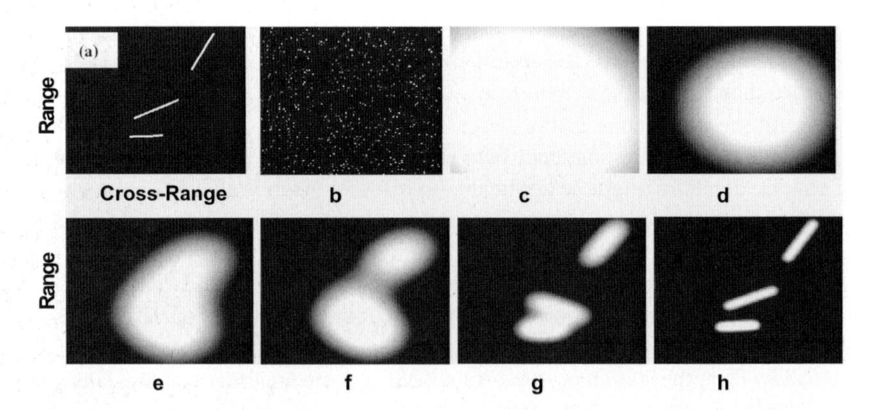

Fig. 1 Detection and tracking objects below clutter using NMF: (**a**) true track positions; (**b**) actual data available for detection and tracking. Evolution of the NMF neural network driven by the knowledge instinct is shown in (**c**)–(**h**), where (**c**) shows the initial, uncertain, model and (**h**) shows the model upon convergence after 20 iterations. Converged model (**h**) are in close agreement with the truth (**a**) Performance improvement of about 100 in terms of signal-to-clutter ratio is achieved due to dynamic logic evolution from vague and uncertain models to crisp and certain

belong to three moving objects. In this data, the target returns are buried in clutter, with signals being weaker than clutter (by factor of 2). Figures 1(c)–1(h) illustrate evolution of the NMF models as they adapt to the data during iterations. Figure 1(c) shows the initial vague-fuzzy model, while Fig. 1(h) shows the model upon convergence at 20 iterations. Between (c) and (h) the NMF neural network automatically decides how many model components are needed to fit the data, and simultaneously adapts the model parameters, including target track coefficients. There are two *types* of models: one uniform model describing clutter (it is not shown), and linear track models with large uncertainty. In (c) and (d), the NMF neural network fits the data with one model, and uncertainty is somewhat reduced. Between (d) and (e) NMF decides that it needs two models to "understand" the content of the data. Fitting with 2 tracks continues until (f); between (f) and (g) a third track is added. Iterations stop at (h), when similarity stops increasing. Detected tracks closely correspond to the truth (a). In this case NMF successfully detected and tracked all three objects and required only 10^6 operations, whereas a straightforward application of multiple hypotheses tracking would require $\mathrm{H}^N \sim 10^{1500}$ operations. This number, larger than the size of the Universe and too large for computation, prevents past algorithms from solving this problem. NMF overcoming this difficulty achieved about 100 times improvement in terms of signal-to-clutter ratio. This improvement is achieved by using dynamic evolution from vague and uncertain models to crisp and certain (instead of sorting through combinations).

6 Conscious, Unconscious, and Differentiation

NMF dynamics described above satisfy the knowledge instinct and improve knowledge by evolving vague, uncertain models toward crisp models, which maximize similarity between models and data. This process of knowledge accumulation, driven by the instinct for knowledge, proceeds in the minds of every member in a society and constitutes an essential aspect of cultural evolution. Vague and uncertain models are less accessible to consciousness, whereas crisp and concrete models are more conscious.

Most of the mind's operations are not accessible to consciousness. We definitely know that neural firings and connections cannot be perceived consciously. In the foundations of the mind there are material processes in the brain inaccessible to consciousness. Jung suggested that conscious concepts are developed by the mind based on genetically inherited structures, archetypes, which are inaccessible to consciousness [29, 30]. Grossberg [4] suggested that only signals and models attaining a resonant state (that is signals matching models) can reach consciousness. It was further detailed by Taylor [31]; he related consciousness to the mind being a control mechanism of the mind and body. A part of this mechanism is a prediction model. When this model predictions differ from sensory observations, this difference may reach a resonant state, which we are conscious about. To summarize the above analyses, the mind mechanisms, described in NMF by dynamic logic and fuzzy models, are not accessible to consciousness. Final results of dynamic logic processes, resonant states characterized by crisp models and corresponding signals are accessible to consciousness. Increase in knowledge and improved cognition results in better, more diverse, more differentiated consciousness.

How did the evolution of cognition and consciousness proceed? What was the initial state of consciousness: an undifferentiated unity or a "booming, buzzing confusion" [32]? Or, let us take a step back in evolutionary development and ask, what is the initial state of pre-conscious psyche? Or, let us move back even further toward the evolution of sensory systems and perception. Why would an evolution result in sensor organs? Obviously, such an expensive thing as a sensor is needed to achieve specific goals: to sense the environment with the purpose to accomplish specific tasks. Evolution of organisms with sensors went together with an ability to utilize sensory data.

In the process of evolution, sensory abilities emerged together with perception abilities. A natural evolution of sensory abilities could not result in a "booming, buzzing confusion," but must result in evolutionary advantageous abilities to avoid danger, attain food, etc. Primitive perception abilities (observed in primitive animals) are limited to a few types of concept-objects (light-dark, warm-cold, edible-nonedible, dangerous-attractive...) and are directly "wired" to proper actions (Walter Freeman and Robert Kozma chapters in this book refer to this primitive intelligence as low dimensional chaos; high dimensional chaos appears with higher intelligence). When perception functions evolve further, beyond immediate actions, it is through the development of complex internal model-concepts, which unify simpler object-models into a unified and flexible model of the world. Only at this point of possessing relatively complicated differentiated concept-models composed of a large number of sub-models, can an intelligent system experience a "booming, buzzing confusion" if it faces a new type of environment. A primitive system is simply incapable of perceiving confusion: It perceives only those 'things' for which it has concept-models. If its perceptions do not correspond to reality, it does not experience confusion, but simply ceases to survive. When a baby is born, it undergoes a tremendous change of environment, most likely without much conscious confusion. The original state of consciousness is undifferentiated unity. It possesses a single modality of primordial undifferentiated Self-World.

The initial unity of psyche limited the abilities of the mind, and further development proceeded through the differentiation of psychic functions or modalities (concepts, emotions, behavior); they were further differentiated into multiple concept-models, etc. This accelerated adaptation. Differentiation of consciousness began millions of years ago. It accelerated tremendously in our recent past, and still continues today [29, 33, 34].

In pre-scientific literature about mechanisms of the mind there was a popular idea of homunculus, that is, a little mind inside our mind which perceived our perceptions and made them available to our mind. This naive view is amazingly close to the actual scientific explanation. The fundamental difference is that the scientific explanation does not need an infinite chain of homunculi inside homunculi. Instead, there is a hierarchy of the mind models with their conscious and unconscious aspects. The conscious differentiated aspect of the models decreases at higher levels in the hierarchy, and they are more uncertain and fuzzy. At the top of the hierarchy there are most general and important models of the meaning of our existence (which we discuss later); these models are mostly unconscious.

Our internal perceptions of consciousness are due to Ego-model. This model essentially consists of, or has access to other model parts that are available to consciousness. It is the mind mechanism of what used to be called "homunculus." It "perceives" crisp conscious parts of other models, in the same way that models of perception "perceive" objects in the world. The properties of consciousness as we experience them, such as continuity and identity of consciousness, are due to properties of the Ego-model, [10]. These properties of unity, continuity, and identity, are the reasons to assume existence of this model. What is known about this "consciousness"-model? Since Freud, a certain complex of psychological functions was called Ego. Jung considered Ego to be based on a more general model or archetype of Self. Jungian archetypes are psychic structures (models) of a primordial origin, which are mostly inaccessible to consciousness, but determine the structure of our psyche. In this way, archetypes are similar to other models, e.g., receptive fields of the retina are not consciously perceived, but determine the structure of visual perception. The Self archetype determines our phenomenological subjective perception of ourselves, and in addition, structures our psyche in many different ways, which are far from being completely understood. An important phenomenological property of Self is the perception of uniqueness and in-divisibility (hence, the word individual).

According to Jung, conscious concepts of the mind are learned on the basis of inborn unconscious psychic structures, archetypes, [29]. Contemporary science often equates the mechanism of concepts with internal representations of objects, their relationships, situations, etc. The origin of internal representations-concepts is from two sources, inborn archetypes and culturally created models transmitted by language [12].

In preceding sections we described dynamic logic operating at a single hierarchical level of the mind. It evolves vague and unconscious models-concepts into more crisp and conscious. Psychologically this process was called by Carl Jung *differentiation* of psychic content [29].

7 Hierarchy and Synthesis

In previous sections we described a single processing level in a hierarchical NMF system. As we mentioned, the mind is organized in an approximate hierarchy. For example, in visual cortex, this approximate hierarchy is well studied [4, 5]. Not every two models are in hierarchical relationships (above-below or same level, more or less general, etc.). Also, feedback loops between higher and lower levels contradict to strict hierarchical ordering. Nevertheless, for simplicity, we will talk about the mind as a hierarchy (in terms of generality of models and the directions of bottom-up and top-down signal flows). At each level of the hierarchy there are input signals from lower levels, models, similarity measures (2) emotions (which are changes in similarity), and actions. Actions include adaptation, i.e., behavior satisfying the knowledge instinct. This adaptation corresponds to the maximization of similarity, as described mathematically by equations (3) through (6). An input to each level is a set of signals X(n), or in neural terminology, an input field of neuronal activations. The result of dynamic logic operations at a given level are activated models, or concepts h recognized in the input signals n; these models along with the corresponding instinctual signals and emotions may activate behavioral models and generate behavior at this level.

The activated models initiate other actions. They serve as input signals to the next processing level, where more general concept-models are recognized or created. Output signals from a given level, serving as input to the next level, could be model activation signals, a_h, defined as

$$a_h = \sum_{n \in N} f(h|n). \tag{8}$$

As defined previously in (3) f(h|n) can be interpreted as a probability that signal n came from object h; and a_h is interpreted as a total activation of the concept h from all signals. Output signals may also include model parameters. The hierarchical NMF system is illustrated in Fig. 2. Within the hierarchy of the mind, each concept-model finds its mental meaning and purpose at a higher level (in addition to other purposes). For example, consider a concept-model "chair." It has a "behavioral" purpose of initiating sitting behavior (if sitting is required by the body), this is the "bodily" purpose at the same hierarchical level. In addition, "chair" has a "purely mental" purpose at a higher level in the hierarchy, a purpose of helping to recognize a more general concept, say of a "concert hall," which model contains rows of chairs.

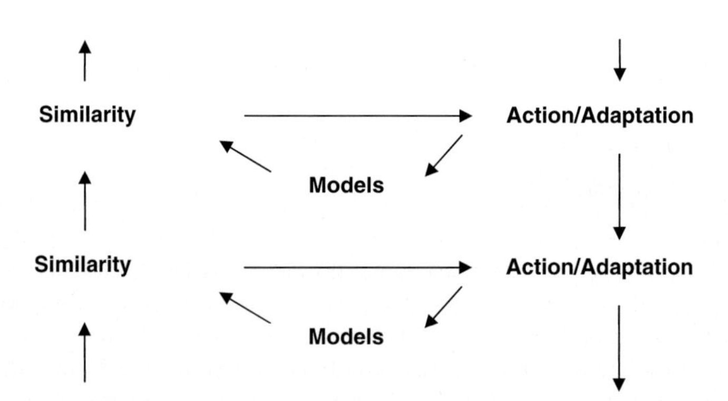

Fig. 2 Hierarchical NMF system. At each level of a hierarchy there are models, similarity measures, and actions (including adaptation, maximizing the knowledge instinct - similarity). High levels of partial similarity measures correspond to concepts recognized at a given level. Concept activations are output signals at this level and they become input signals to the next level, propagating knowledge up the hierarchy. Each concept-model finds its mental meaning and purpose at a higher level

Models at higher levels in the hierarchy are more general than models at lower levels. For example, if we consider the vision system, models at the very bottom of the hierarchy correspond (roughly speaking) to retinal ganglion cells and perform similar functions; they detect simple features in the visual field. At higher levels, models correspond to functions performed at V1 and higher up in the visual cortex, that is detection of more complex features such as contrast edges, their directions, elementary moves, etc. Visual hierarchical structures and models have been studied in detail [4, 5], and these models can be used in NMF. At still higher cognitive levels, models correspond to objects, to relationships among objects, to situations, and relationships among situations, etc. [10]. At still higher levels, even more general models reside, corresponding to complex cultural notions and relationships such as family, love, friendship, and abstract concepts such as law, rationality, etc. The contents of these models correspond to the wealth of cultural knowledge, including the writings of Shakespeare and Tolstoy. Mechanisms of the development of these models are reviewed in the next section. According to Kantian analysis [35], at the top of the hierarchy of the mind are models of the meaning and purpose of our existence, unifying our knowledge, and the corresponding behavioral models aimed at achieving this meaning. Chapter by Robert Kozma in this book considers a related neural architecture, K-sets, describing a hierarchy of the mind.

From time to time, as discussed, a system forms a new concept or eliminates an old one. Many pattern recognition algorithms and artificial neural networks lack this important ability of the mind. It can be modeled mathematically in several ways; adaptive resonance theory (ART) uses vigilance threshold [36], which is similar to a threshold for a similarity measure [9]. A somewhat different mechanism of NMF works as follows. At every level, the system always keeps a reserve of vague (fuzzy) inactive concept-models (with large covariance, C, 7). They are inactive in that their parameters are not adapted to the data; therefore their similarities to signals are low. Yet, because of a large fuzziness (covariance) the similarities are not exactly zero. When a new signal does not fit well into any of the active models, its similarities to inactive models automatically increase (because first, every piece of data is accounted for and second, inactive models are vague-fuzzy and potentially can "grab" every signal that does not fit into more specific, less fuzzy, active models). When the activation signal a_h of (8) for an inactive model, h, exceeds a certain threshold, the model is activated. Similarly, when an activation signal for a particular model falls below a threshold, the model is deactivated. Thresholds for activation and deactivation are set usually based on information existing at a higher hierarchical level (prior information, system resources, numbers of activated models of various types, etc.). Activation signals for active models at a particular level $\{a_h\}$ form a "neuronal field," which serve as input signals to the next level, where more abstract and more general concepts are formed, and so on along the hierarchy toward higher models of meaning and purpose.

Models at a higher level act as "eyes" perceiving the models at a lower level. Each higher level in the hierarchy is the "mind of a homunculus" perceiving the meaning of what was recognized at a lower level. As mentioned, this does not lead to an infinite regress, because higher level models are more general, more uncertain, and more vague and fuzzy.

Let us note that in the hierarchical structure (Fig. 2) concept-models at the bottom level of the hierarchy correspond to objects directly perceived in the world. Perception mechanisms to a significant extent are determined by sensor organs which evolved over billions of years. Models at this level are to a large extent the result of evolution and to a lesser extent the result of cultural constructions. These models are "grounded" in "real" objects existing in the surrounding world. For example, "food" objects are perceived not only by the human mind, but also by all pre-human animals.

This is not true for concept-models at higher levels of the hierarchy. These more abstract and more general models are cultural constructs (to some extent). They cannot be perceived directly in the surrounding world (e.g., concept-models of "rationality," or "meaning and purpose of life"). These concepts cannot just emerge in the mind on their own as some useful combination of simpler concepts. Because there are a huge number of combinations of simpler concepts, an individual human being does not have enough time in his or her life to accumulate enough experiential evidence to verify the usefulness of these combinations. These higher level concepts accumulate in cultures due to languages. An individual mind is assured about the usefulness of certain high-level concept-models because he can talk about them with other members of the society (with a degree of mutual understanding). Concepts acquired from language are not automatically related to events or combinations of objects in the surrounding world. For example, every five-year-old knows about "good guys" and "bad guys." Yet, still at 40 or 70 nobody could claim the he or she can perfectly use these models to understand the surrounding world. Philosophers and theologians have argued about the meaning of good and evil for thousands of years, and these arguments are likely to continue forever. The study of mechanisms relating language concepts to concept-models of cognition have just begun [10, 12, 34, 37, 38].

The hierarchical structure of the mind is not a separate mechanism, independent from the knowledge instinct. Detailed neural and mathematical mechanisms connecting these two are still a matter of ongoing and future research [10, 12, 34, 37]. Here we outline some basic principles of the knowledge instinct operation in the mind hierarchy. Previous sections described the mechanism of differentiation, creating diverse and detailed models, acting at a single level of hierarchy. At a single level, the meaning of each model is to satisfy the knowledge instinct by finding patterns in the input data, bottom-up signals, and adapting to these patterns. There are also meanings and purposes related to bodily instincts: for example, food objects can be used to satisfy needs for food and desires for eating. In this chapter we limit our discussion to spiritual needs, to the knowledge instinct.

We have discussed that models acquired deeper meanings and purposes at higher hierarchical levels. The knowledge instinct acting at higher levels and aesthetic emotions at higher levels are perceived more consciously than at lower levels. The pure aesthetic feeling of harmony between our knowledge and the surrounding world at lower levels is below threshold of conscious registration in our minds. We do not feel much joy from the understanding of simple objects around us. But we do enjoy solving complex problems that required a lot of time and effort. This emotional feel of harmony from improving-creating high level concept-models is related to the fact that high level concepts unify many lower level concepts and increase the overall meaning and purpose of our diverse knowledge. Jung called this synthesis, which he emphasized is essential for psychological well being.

Synthesis, the feel of overall meaning and purpose of knowledge, is related to the meaning and purpose of life, which we perceive at the highest levels of the hierarchy of the mind. At those high levels models are intrinsically vague and undifferentiated, not only in terms of their conceptual content, but also in terms of differentiation of conceptual and emotional. At the highest levels of the mind the two are not quite separable. This inseparability, which we sometimes feel as a meaning and purpose of our existence, is important for evolution and survival. If the hierarchy of knowledge does not support this feel, the entire hierarchy would crumble, which was an important (or possibly the most important) mechanism of destruction of old civilizations. The knowledge instinct demands satisfaction at the lowest levels of understanding concrete objects around, and also at the highest levels of the mind hierarchy, understanding of the entire knowledge in its unity, which we feel as meaning and purpose of our existence. This is the other side of the knowledge instinct, a mechanism of *synthesis* [29].

8 Evolutionary Dynamics of Consciousness and Cultures

8.1 Neurodynamics of Differentiation and Synthesis

Every individual mind has limited experience over the lifetime. Therefore, a finite number of concept-models are sufficient to satisfy the knowledge instinct. It is well appreciated in many engineering applications, that estimating a large number of models from limited data is difficult and unreliable; many different solutions are possible, one no better than the other. Psychologically, the average emotional investment in each concept decreases with an increase in the number of concepts, and a drive for differentiation and creating more concepts subsides. Emotional investment in a concept is a measure of the meaning and purpose of this concept within the mind system, that is, a measure of synthesis. Thus, the drive for differentiation requires synthesis. More synthesis leads to faster differentiation, whereas more differentiation decreases synthesis.

In a hierarchical mind system, at each level some concepts are used more often than others, they acquire multiple meanings, leading to a process opposite to differentiation. These more general concepts "move" to a higher hierarchical levels. These more general, higher-level concepts are invested with more emotion. This is a process of synthesis increase.

Another aspect of synthesis is related to language. Most concepts within individual minds are acquired with the help of language. Interaction between language and cognition is an active field of study (see [12] for neurodynamics of this interaction and for more references). Here we do not go into the details of this interaction, we just emphasize the following. First, creation of new concepts by differentiation of inborn archetypes is a slow process, taking millennia; results of this process, new concepts, are stored in language, which transmits them from generation to generation. Second, a newborn mind receives this wealth of highly differentiated concepts "ready-made," that is without real-life experience, without understanding and differentiating cognitive concepts characterizing the world; a child at 5 or 7 can speak about much of existing cultural content, but it would take the rest of life to understand, how to use this knowledge. This is directly related to the third aspect of language-cognition interaction: language model-concepts are not equivalent to cognitive model-concepts. Language models serve to understand language, not the world around. Cognitive models that serve to understand the world are developed in individual minds with the help of language. This development of cognitive models from language models, connection of language and cognition is an important aspect of synthesis.

Let us dwell a bit more on this aspect of synthesis. Learning language is driven by the language instinct [39]; it involves aesthetic emotions; a child likes to learn language. However, this drive and related emotions subside after about 7, after language is mostly learned. During the rest of life, the knowledge instinct drives the mind to create and improve cognitive models on the basis of language models [12]. This process involves aesthetic emotions related to learning cognitive concepts. Again, synthesis involves emotions.

People are different in their ability to connect language and cognition. Many people are good at talking, without fully understanding how their language concepts are related to real life. On any subject, they can talk one way or another without much emotional investment. Synthesis of language and cognition involves synthesis of emotional and conceptual contents of psyche.

Synthesis of emotional and conceptual is also related to hierarchy. Higher level concepts are more general and vaguer. They are less differentiated not only in their conceptual precision, but also their conceptual and emotional contents are less differentiated. Important high-level concepts are more emotional than low-level, mundane, everyday concepts. They are also less conscious (remind, more differentiation leads to more conscious content). Therefore, synthesis connects language and cognition, concepts and emotions, conscious and unconscious. This is opposite of differentiation; we all have high-value concepts (related to family life, or to political cause, or to religion) which are so important to us and so emotional, that we cannot "coldly analyze," cannot differentiate them. "Too high" level of synthesis invests concepts with "too much" emotional-value contents, so that differentiation is stifled.

To summarize, differentiation and synthesis are in complex relationships, at once symbiotic and antagonistic. Synthesis leads to spiritual inspiration, to active creative behavior leading to fast differentiation, to creation of knowledge, to science and technology. At the same time, "too" high level of synthesis stifles differentiation. Synthesis is related to hierarchical structure of knowledge and values. At the same time, high level of differentiation discounts psychological emotional values of individual concepts, and destroys synthesis, which was the basis for differentiation. In Sects. 3, 4 and 5 we presented a NMF / DL mathematical model of neurodynamics of differentiation. We lack at present same detail level of neurodynamics of synthesis. In this section we make first steps toward developing mathematical evolutionary model of interacting differentiation and synthesis. Both mechanisms act in the minds of individual people. Future detailed models will develop neural mechanisms of synthesis, will account for mechanisms of cognition, emotion, and language, and will study multi-agent systems, in which each agent possesses complex neurodynamics of interaction between differentiation and synthesis. We call such an approach neural micro-dynamics. Lacking these micro-dynamics models, in this section we develop simpler models averaged over population.

8.2 Macro-dynamics

As a first step here we develop simplified evolutionary dynamic models similar to mean field theories in physics. These models describe the neural mechanisms of differentiation, synthesis, and hierarchy using measures averaged over population of interacting agents, abstracting from details of emotional and language mechanisms. A future challenge would be to relate these models to nonlinear dynamic models discussed in this book in chapters by Walter Freeman and Robert Kozma. We call this averaging method "neural macro-dynamics." We start with simplest dynamic equations inspired by neurodynamics of differentiation and synthesis, discuss their properties, and evaluate needed modification toward developing a "minimal" realistic model. Results of this analysis can be used in sociological cultural studies to understand past, present, and future of cultures, emerging cultural phenomena, and to improve current and future models.

We characterize accumulated knowledge, or differentiation, by a "mean field" averaged quantity, D, which represents the average number of concept-models used in a population. When considered alone, separate from other mechanisms driving neurodynamics, the simplest dynamical equation is

$$dD/dt = a. \tag{9}$$

This equation describes linear growth in the complexity of culture as measured by accumulated knowledge, or differentiation, D. The next step toward more realistic modeling accounts for the fact that differentiation involves developing new, more detailed models from the old ones. Therefore the speed of differentiation is proportional to accumulated knowledge, i.e.,

$$dD/dt = aD. \tag{10}$$

Here, a is a constant. The solution of this equation describes an exponential growth of knowledge, i.e.,

$$D(t) = D_0 \exp(at). \tag{11}$$

Both of the above equations could be considered "minimally realistic" in the short term. In the long term, however, they are too optimistic, too simple, and not realistic.

We know that continuous growth in knowledge may exist in some cultures over limited time periods, however occasionally the growth in knowledge and conceptual diversity is interrupted and culture disintegrates or stagnates. This is true in all known cultures, e.g., Western culture disintegrated and stagnated during the Middle Ages. Whereas some researchers have attributed the disintegration of Roman Empire to barbarians or to lead poisoning [40], here we would like to search for possible intrinsic spiritual, neurodynamic mechanisms.

According to our previous discussions, and following Jung analysis [29], a more complicated dynamic of knowledge accumulation involves synthesis, S. Synthesis characterizes the relationship between knowledge and its instinctive, emotional, value in the society. For example, a ratio of the similarity measure and differentiation, LL/D, measures a degree of satisfaction of the knowledge instinct (2) per concept-model. A closely related, but more instrumental, measure available for sociological research [41] is an average measure of emotional investment per concept in a society. With the growth of differentiation, the emotional value of every individual concept diminishes, and therefore the simplest neurodynamic equation for synthesis is (below, b is a constant)

$$\mathrm{dS/dt} = -\mathrm{bD}. \tag{12}$$

According to the previous analysis, synthesis inspires creativity and stimulates differentiation. The simplest modification of (9), accounting for influence of synthesis is

$$\mathrm{dD/dt} = \mathrm{aS}. \tag{13}$$

These equations are similar to the brain dynamic equations for KII-sets discussed in R. Kozma chapter in this book. Together, (12) and (13) lead to oscillating solutions with frequency ω and phase ϕ_0, i.e.,

$$\begin{aligned} \mathrm{D(t)} &= D_0\, cos(\omega t + \phi_0),\ \omega = sqrt(ab) \\ \mathrm{S(t)} &= -(D_0\, \omega/a) sin(\omega t + \phi_0). \end{aligned} \tag{14}$$

These solutions are unsatisfactory, however, because here D and S can assume negative values, whereas differentiation and synthesis cannot become negative by their very definition.

A more realistic equation for differentiation would account for the following. The speed of differentiation is proportional to accumulated knowledge, D, and is enhanced by synthesis, S, and is therefore proportional to D*S. We have to take into account that, psychologically, synthesis is a measure of the meaning and purpose in knowledge and culture, it is a necessary condition for human existence, and it has to remain positive. When synthesis falls below certain positive value, $\mathrm{S_0}$, knowledge loses any value, culture disintegrates, and differentiation reverses its course, i.e.,

$$\mathrm{dD/dt} = \mathrm{aD(S - S_0)}. \tag{15}$$

Still, (12) is unsatisfactory since it always leads to decrease in synthesis, so that any cultural revival and long term accumulation of knowledge is impossible. The long-term joint solution of (15) and (12) is $D \approx 0$, $S \approx S_0$.

From the previous analysis, we know that synthesis is created in hierarchies. Diverse, differentiated, knowledge at particular level in a hierarchy acquires meaning and purpose at the next level. The simplest measure of hierarchy, H, is the number of hierarchical levels, on average, in the minds of the population. A useful measure would have to account for conceptual hierarchy and the hierarchy of values. Accounting for hierarchical synthesis, (12) can be re-written as

$$dS/dt = -bD + dH. \tag{16}$$

Here, d is a constant. If the hierarchy, H, is genetically or culturally fixed to a constant value, (16) and (15) have several joint solutions. Let us explore them. First, there is a long-term solution with constant knowledge and synthesis:

$$\begin{aligned} D &= (b/d)H \\ S &= S_0 u \end{aligned} \tag{17}$$

Here, differentiation and synthesis reach constant values and do not change with time. The hierarchy of concepts (and values) is rigidly fixed. This could be a reasonable solution, describing highly conservative, traditional societies in a state of cultural stagnation. The conceptual hierarchy, H, reaches a certain level, then remains unchanged, and this level forever determines the amount of accumulated knowledge or conceptual differentiation. Synthesis is at a low level S_0. All cultural energy is devoted to maintaining this synthesis, and further accumulation of knowledge or differentiation is not possible. Nevertheless, such a society might be stable for a long time. Some Polynesian and New Guinean cultures, lacking writing or complex religion and practicing cannibalism, still maintained stability and survived for millennia [42]. Chinese culture had stagnated since early BCE until recent times, although at much higher level of the hierarchy. It would be up to cultural historians and social scientists to evaluate whether such cultures are described by the above mathematical solution and, if so, what particular values of model parameters are appropriate.

Alternatively, if evolution starts with $S > S_0$, differentiation first grows exponentially $\sim \exp(a(S-S_0)t)$. This eventually leads to the term $-bD$ in (16) overtaking dH, so that synthesis diminishes and the differentiation growth exponent is reduced. When $S < S_0$, differentiation falls until $bD = dH$, at which point differentiation grows again, and the cycle continue. This type solution is illustrated in Fig. 3

Here, the solid line indicates the cycles of differentiation. When comparing this line with historical cultural data, one should remember that the time scale here has not been determined. Cycles that peak when cultures flourish and end with devastation and loss of knowledge take centuries. However, we should not disregard much shorter cultural cycles, for example, fascism in Germany or communism in Soviet Union, which have occurred in the 20th century. Figure 3 indicates the loss of about

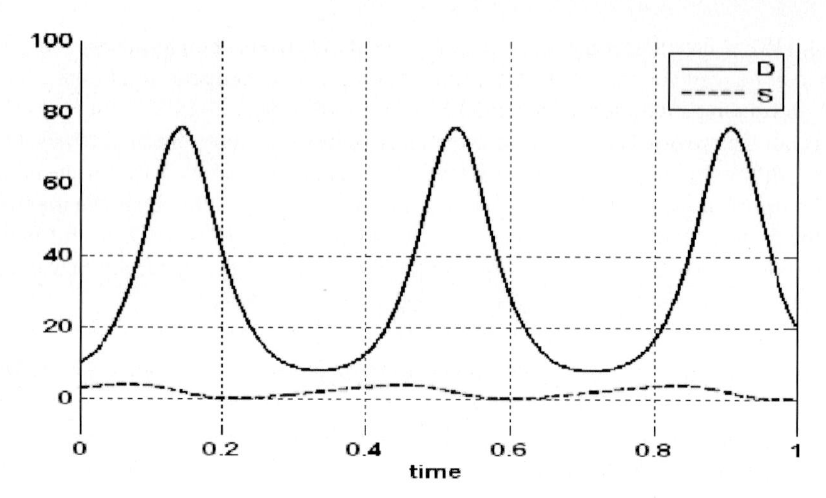

Fig. 3 Evolution of differentiation and synthesis described by (15, 16) with parameter values $a = 10$, $b = 1$, $d = 10$, $S_0=2$, $H_0 = 3$, and initial values $D(t=0) = 10$, $S(t=0) =3$

85% of knowledge (D) within a cycle; is this reasonable? Before answering this question, we should emphasize that the frequency of oscillations and top-to-bottom amplitude depend upon the values of the parameters used. We had no data with which we could select scientifically correct values for the parameters. It will be up to sociologists and cultural historians to decide upon appropriate parameter values. Another topic for future studies will be the appropriate measure of D. Possibly, the proper measure of D is an average knowledge per person, not over the entire population, but only over the part of population actively involved in running states. In "well managed" societies, educated people are actively involved in society management. In "badly managed" societies, like Soviet Union, educated people were excluded from voicing their opinion, and a few poorly educated people made the decisions. Therefore, 85% loss of knowledge during fast oscillations may represent the loss of knowledge and synthesis in the "ruling class," but not in the entire society.

Notwithstanding these arguments, the wild oscillations in differentiation and synthesis shown in Fig. 3 may not be reasonable. It might be an indication that (15, 16) are simplified and may be missing some important mechanisms creating synthesis. Roles of mechanisms such as religion, art, music are discussed in the last section; their mathematical modeling is beyond the scope of this chapter.

Oscillating solutions similar to Fig. 3 are also possible if evolution starts with $S < S_0$. First, differentiation will fall, but then dH will exceed bD (in 16), synthesis will grow and thus the oscillating solutions ensue. These oscillating solutions describe many civilizations over extended periods of time, e.g. Western civilization over millennia. Again, it would be up to cultural historians and social scientists to evaluate which cultures are described by this solution, and what particular values of model parameters are appropriate.

The dashed line in Fig. 3 indicates the cycles of synthesis. In this example synthesis falls to 0, which is probably not realistic. We could have keep synthesis strictly positive by selecting different values of parameters, but these kinds of detailed studies are not our purpose here. We would like to emphasize that there is presently no scientific data that can be used to select reasonable parameter values for various societies; this is a subject of future research. Similarly, the many cycles exactly repeated in this figure indicate the simplistic nature of this model.

8.3 Expanding Hierarchy

Expanding knowledge in the long term requires expanding hierarchical levels. As discussed, differentiation proceeds at each hierarchical level, including the highest levels. In this process, knowledge accumulating at a particular level in the hierarchy may lead to certain concept-models being used more often than others. These concepts used by many agents in a population in slightly different ways acquire more general meanings and give rise to concepts at a higher level. Thus, increasing differentiation may induce more complex hierarchy, and the hierarchy expands, i.e.,

$$\mathrm{dH/dt = e\ dD/dt.} \tag{18}$$

Equation (18), (16), and (15) describe a culture expanding in its knowledge content and in its hierarchical complexity. For example, a solution with fixed high level of synthesis can be described by

$$\begin{aligned}
\mathrm{S} &= \mathrm{const} > \mathrm{S_0},\\
\mathrm{D(t)} &= \mathrm{D_0 \exp(a(S - S_0)t)},\\
\mathrm{H(t)} &= \mathrm{H_0 + e_c\ D_0 \exp(a(S - S_0)t)}.
\end{aligned} \tag{19}$$

This solution implies the following "critical" value for parameter e,

$$\mathrm{e_c = b/d.} \tag{20}$$

Figure 4 illustrates this expanding-culture solution with constant synthesis. If $e > e_c$, than synthesis, differentiation, and hierarchy grow indefinitely, as shown in Fig. 5.

These solutions with unbounded growth in knowledge, its hierarchical organization and, in Fig. 5, growing stability (synthesis) are too optimistic compared to the actual evolution of human societies.

If $e < e_c$, then synthesis and knowledge hierarchy collapse when differentiation destructs synthesis. However, when differentiation falls, $H_0 > e_c\ D_0 \exp(\ a(S - S_0)t\ )$, synthesis again starts growing, leading to the growth of differentiation. After a fast flourishing period, synthesis again is destructed by differentiation when its influence on synthesis overtakes that of the hierarchy, and culture collapses. These periods of collapse and growth alternate, as shown in Fig. 6

This assumption of the hierarchy growing in sync with differentiation (18) is too optimistic. The growth of hierarchy involves the differentiation of models at the highest level, which involve concepts of the meaning and purpose of life. These

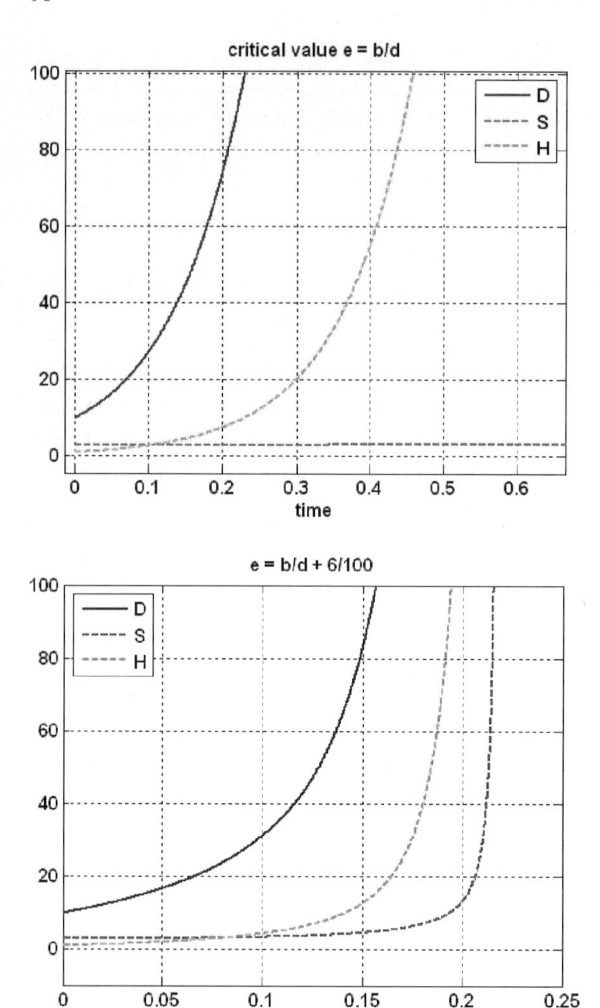

Figs. 4-5 Exponentially expanding solutions. Evolution of differentiation, synthesis, and hierarchy is described by (15, 16, 19) with parameter values a = 10, b = 1, d = 10, S_0=2, H_0 = 1, and initial values D(t=0) = 10, S(t=0) = 3. In Fig. 4 e = b/d = 0.1 (20); in Fig. 5 e = 1.06

concepts cannot be made fully conscious, and in many societies they involve theological and religious concepts of the Highest. Changes in these concepts involve changes of religion, such as from Catholicism to Reformation, they involve national upheavals and wars, and they do not always proceed smoothly as in (18). Currently we do not have theory adequate to describe these changes; therefore we proceed within a single fixed religious paradigm. This can be approximately described as constant hierarchy H, as in the previous section. Alternatively we can consider slowly expanding hierarchy,

$$H(t) = H_0 + e * t. \tag{21}$$

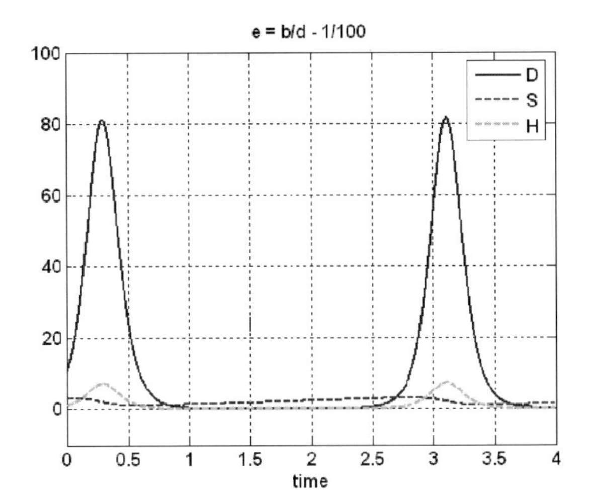

Fig. 6 Alternating periods of cultural growth and stagnation, same parameter values as above, except e = 0.99 < b/d

The solution of (15, 16, 21) is illustrated in Fig. 7

This growing and oscillating solution might describe Judeo-Christian culture over the long period of its cultural evolution. Whereas highly ordered structure is a consequence of the simplicity of equations, this solution does not repeat exactly the same pattern, rather the growth of hierarchy leads to growth of differentiation, and to faster oscillations. Note, the evolution and recoveries from periods of stagnation in Western culture were sustained by the growing hierarchy of knowledge and values. This stable, slow growing hierarchy was supported by religion. However, science

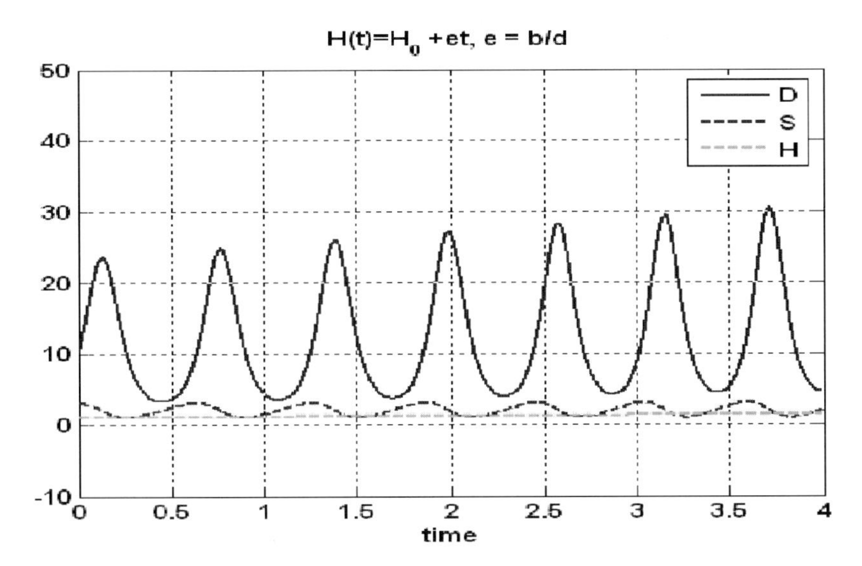

Fig. 7 Oscillating and growing differentiation and synthesis (15, 16, 21); slow growth corresponds to slowly growing hierarchy, e = 0.1. Note, increase in differentiation leads to somewhat faster oscillations

has been replacing religion in many people's minds (in Europe more so than in the US) approximately since the Enlightenment (the 18th c.). The current cultural neurodynamics in Western culture are characterized by the predominance of scientific differentiation and the lack of synthesis. More and more people have difficulty connecting scientific highly-differentiated concepts to their instinctual needs. Many turn to psychiatrists and take medications to compensate for a lack of synthesis. The stability of Western hierarchical values is precarious, and during the next downswing of synthesis hierarchy may begin to disintegrate, leading to cultural collapse. Many think that this process is already happening, more so in Europe than in the US.

8.4 Dual Role of Synthesis

The previous section considered only the inspirational role of synthesis. The effect of synthesis, as discussed previously, is more complex: high investment of emotional value in every concept makes concepts "stable" and difficult to modify or differentiate [12]. Therefore, a high level of synthesis leads to stable and stagnating culture. We account for this by changing the effect of synthesis on differentiation as follows:

$$\mathrm{dD/dt = aDG(S),\ \ G(S) = (S - S_0)\exp(-(S - S_0)/S_1)} \tag{22}$$

$$\mathrm{dS/dt = -bD + dH} \tag{23}$$

$$\mathrm{H(t) = H_0,\ \ or\ H(t) = H_0 + e^{*}t.} \tag{24}$$

Solutions similar to those previously considered are possible: a solution with a constant value of synthesis similar to (17), as well as oscillating and oscillating-growing solutions.

A new type solution possible here involves a high level of synthesis with stagnating differentiation. If dH > bD, then according to (23) synthesis grows exponentially, whereas differentiation levels off, and synthesis continues growing. This leads to a more and more stable society with high synthesis, with high emotional values attached to every concept, while knowledge accumulation stops, as shown in Fig. 8.

Cultural historians might find examples of stagnating internally stable societies. Candidates are Ancient Egypt and contemporary Arab Moslem societies. Of course, these are only suggestions for future studies. Levels of differentiation, synthesis, and hierarchy can be measured by scientific means, and these data should be compared to the model. This would lead to model improvement, as well as to developing more detailed micro-neurodynamic models, simulating large societies of interacting agents, involving the mind subsystems of cognition and language [43]. And we hope that understanding of the processes of cultural stagnation will lead to overcoming these predicaments and to improvement of human condition.

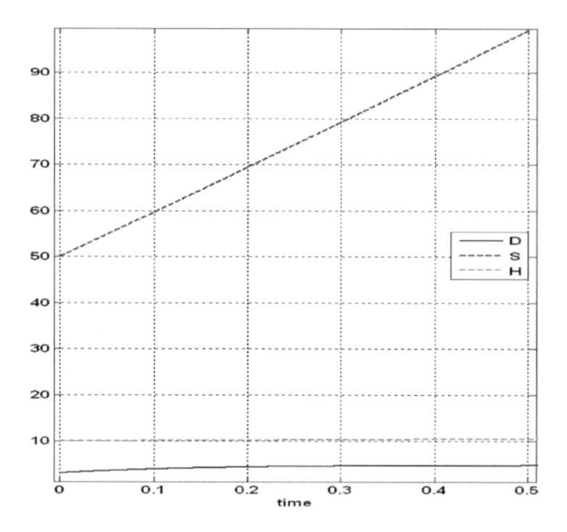

Fig. 8 Highly stable society with growing synthesis, high emotional values attached to every concept, while knowledge accumulation stops; parameter values: $D(t=0)=3$, $H_0 = 10$, $S(t=0) = 50$, $S_0 = 1$, $S_1 = 10$, $a = 10$, $b = 1$, $d = 10$, $e=1$

8.5 Interacting Cultures

Let us now study the interaction of cultures having different levels of differentiation and synthesis. Both are populations of agents characterized by NMF-minds and evolutionary (22, 23, 24). Culture $k=1$ is characterized by parameters leading to oscillating, potentially fast growing differentiation and a medium oscillating level of synthesis ("dynamic" culture). Culture $k=2$ is characterized by slow growing, or stagnating differentiation, and high synthesis ("traditional" culture). In addition, there is a slow exchange by differentiation and synthesis among these two cultures (examples: the US and Mexico (or in general, immigrants to the US from more traditional societies); or academic-media culture within the US and "the rest" of the population). Evolutionary equations modified to account for the inflow and outflow of differentiation and synthesis can be written as

$$dD_k/dt = a_k D_k G(S_k) + x_k D_{\underline{k}} \tag{25}$$

$$dS_k/dt = -b_k D_k + d_k H_k + y_k S_{\underline{k}} \tag{26}$$

$$H_k = H_{0k} + e_k * t \tag{27}$$

Here, the index $\underline{k}$ denotes the opposite culture, i.e., for k=1, $\underline{k} = 2$, and v.v; parameters $x_k$and $y_k$determine the interaction or coupling between the two cultures. Figure 9 illustrates sample solutions to these equations.

In Fig. 9 the evolution starts with two interacting cultures, one traditional and another dynamic. Due to the exchange of differentiation and synthesis among the cultures, traditional culture acquires differentiation, looses much of its synthesis, and becomes a dynamic culture. Let us emphasize that although we tried to find parameter values leading to less oscillations in differentiation and more stability, we did not find such solutions. Although parameters determining the exchange of differentiation and synthesis are symmetrical in two directions among cultures, it is interesting to note that traditional culture does not "stabilize" the dynamic one, the effect is mainly one-directional, that is, traditional culture acquires differentiated

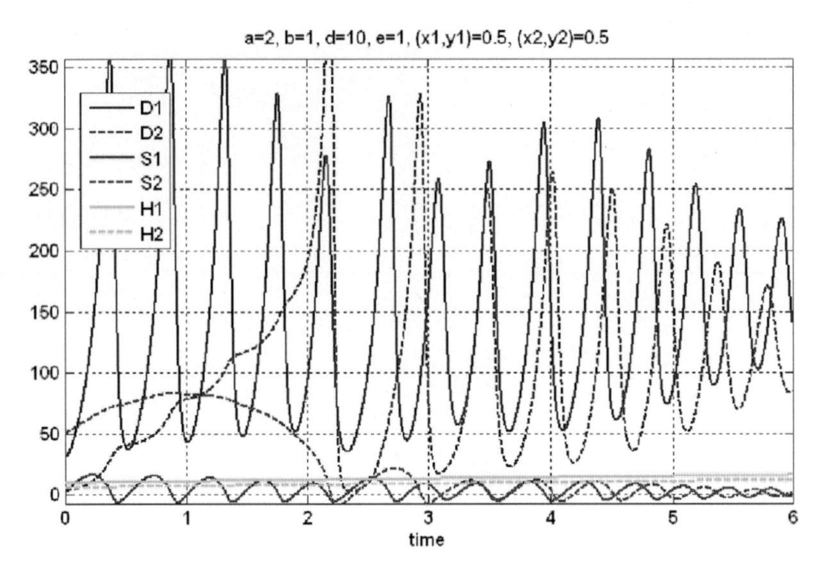

Fig. 9 Effects of cultural exchange (k=1, solid lines: D(t=0)= 30, H_0= 12, S(t=0) = 2, S_0= 1, S_1= 10, a = 2, b = 1, d = 10, e=1, x = 0.5, y = 0.5; k=2, dotted lines: D(t=0)= 3, H_0= 10, S(t=0) = 50, S_0= 1, S_1= 10, a = 2, b = 1, d = 10, e=1, x = 0.5, y = 0.5). Transfer of differentiated knowledge to less-differentiated culture dominates exchange during t < 2 (dashed curve). In long run (t > 6) cultures stabilize each other and swings of differentiation and synthesis subside (note however, that in this example hierarchies were maintained at different levels; exchange of hierarchical structure would lead to the two cultures becoming identical)

knowledge and dynamics. Wild swings of differentiation and synthesis subside a bit only after t > 5, when both cultures acquire a similar level of differentiated knowledge; then oscillations can partly counterweigh and stabilize each other at relatively high level of differentiation. It would be up to cultural historians and social psychologists, to judge if the beginning of this plot represents contemporary influence of American culture on the traditional societies. And if this figure explains why the influence of differentiation-knowledge and not highly-emotional stability-synthesis dominates cultural exchanges (unless "emotional-traditionalists" physically eliminate "knowledge-acquiring ones" during one of their period of weakness). Does partial stabilization beyond t > 5 represent the effect of multiculturalism and explain the vigor of contemporary American society?

This question is addressed in Fig. 10, which extends Fig. 9 to longer time scale. In long run (t > 5) cultures stabilize each other and swings of differentiation and synthesis subside. Note, that in this example hierarchies were maintained at different levels. Is this representative of Catholic and Protestant communities coexisting with approximately equal levels of differentiation and synthesis, but different hierarchies? This is a question for social psychologists. We would like to emphasize that co-existence of different cultures is beneficial in long run: both communities evolve with more stability.

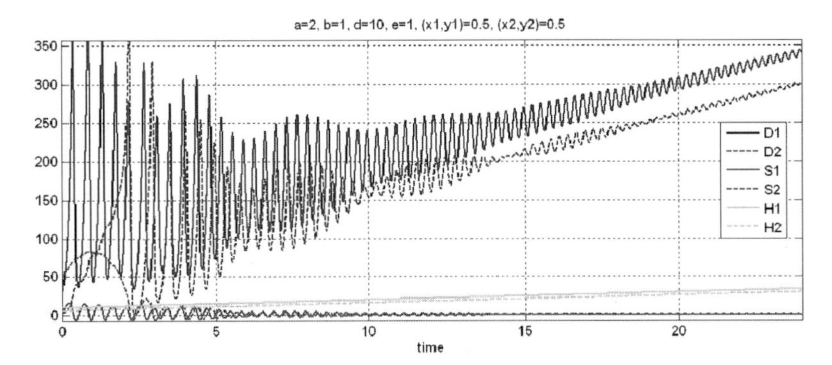

Fig. 10 Effects of cultural exchange, same as Fig. 5 at longer time scale. In long run ($t > 5$) cultures stabilize each other and swings of differentiation and synthesis subside. Note, that in this example hierarchies were maintained at different levels (exchange of hierarchical structures would lead to the two cultures becoming identical)

9 Future Directions

9.1 Neurodynamics of Music: Synthesis of Differentiated Psyche

High levels of differentiation, according to models in the previous section, are not stable. By destroying synthesis, differentiation undermines the very basis for knowledge accumulation. This led in the previous section to wild oscillations in differentiation and synthesis. Here we analyze an important mechanism of preserving synthesis along with high level of differentiation, which will have to be accounted for in future models.

Let us repeat that synthesis, which is a feeling of meaning and purpose, is a necessary condition of human existence. Synthesis is threatened by the differentiation of knowledge. It is difficult to maintain synthesis in societies, like contemporary Western societies, which possess high differentiation and much abstract knowledge. In contrast, it is easier to maintain synthesis in traditional societies, where much of knowledge is directly related to the immediate needs of life. Since time immemorial, art and religion have connected conceptual knowledge with emotions and values, and these provided cultural means for maintaining synthesis along with differentiation. A particularly important role in this process belongs to music, since music directly appeals to emotions [44, 45].

Music appeared from the sounds of voice, i.e., from singing. The prosody or melody of voice sounds, rhythm, accent, tone, and pitch are governed by neural mechanisms in the brain. Images of neural activity (obtained by magnetic resonance imaging, MRI) show that the human brain has *two centers controlling melody of speech*; an ancient center located in the limbic system, and a recent one in the cerebral cortex. The ancient center is connected with direct uncontrollable emotions, whereas the recent center is connected with concepts and consciously controlled emotions. This fact was learned from medical cases in which patients with a damaged cortex lost the ability for speaking and understanding complex phrases, while still were able to comprehend sharply emotional speech [46].

Prosody of speech in primates is governed by a single ancient emotional center in the limbic system. Conceptual and emotional systems in animals are less differentiated than in humans. Sounds of animal cries engage the entire psyche, rather than concepts and emotions separately. An ape or bird seeing danger does not think about what to say to its fellows. A cry of danger is *inseparably* fused with recognition of a dangerous situation, and with a command to oneself and to the entire flock: "Fly!" An evaluation (emotion of fear), understanding (concept of danger), and behavior (cry and wing sweep)–are not differentiated. The conscious and unconscious are not separated: recognizing danger, crying, and flying away is a fused concept-emotion-behavioral *synthetic* form of thought-action. Birds and apes can not control their larynx muscles *voluntarily*.

Emotions-evaluations in humans have separated from concepts-representations and from behavior. For example, when sitting around the table and discussing snakes, we do not jump on the table uncontrollably in fear every time "snakes" are mentioned. This *differentiation* of concepts and emotions is driven by language. Prosody or melody of speech is related to cognition and emotions through aesthetic emotions. This connection of concepts with emotions, conscious models with unconscious archetypes, is *synthesis. The human voice engages concepts and emotions. Melody of voice is perceived by ancient neural centers involved with archetypes, whereas conceptual contents of language involves conscious concepts. Human voice, therefore, involves both concepts and emotions; its melody is perceived by both conscious and unconscious; it maintains synthesis and creates wholeness in psyche.* [47]

Over thousands of years of cultural evolution, music perfected this inborn ability. *Musical sound engages the human being as a whole*—such is the nature of archetypes, ancient, vague, undifferentiated emotions-concepts of the mind. Archetypes are non-differentiated, their emotional and conceptual contents, their high and low are fused and exist only as possibilities. By turning to archetypes, music gets to the most ancient unconscious depths as well as to the loftiest ideas of the meaning of existence. This is why folk songs, popular songs, or opera airs might affect a person more strongly than words or music separately. The synthetic impact of a song, connecting the conscious and unconscious, explains the fact that sometimes mediocre lyrics combined with second-rate music might still impact listeners. When music and poetry truly correspond with each other and reach high artistic levels, a powerful psychological effect occurs. This effect uncovers mechanisms of the mysterious co-belonging of music and poetry. *High forms* of art effect synthesis of the most important models touching the meaning of human existence. *Popular songs*, through interaction of words and sounds, connect the usual words of everyday life with the depths of the unconscious. This explains why in contemporary culture, with its tremendous number of differentiated concepts and lack of meaning, such an important role is taken by popular songs. [9, 34, 48].

Whereas language evolved as the main mechanism for the differentiation of concepts, music evolved as the main mechanism for the differentiation of emotions (conscious emotions in the cortex). This differentiation of emotions is necessary for unifying differentiated consciousness: synthesis of differentiated knowledge entails emotional interactions among concepts [49]. This mechanism may remedy a disturbing aspect of the oscillating solutions considered in the previous section, i.e., the wild oscillations of differentiation and synthesis. During every period of cultural slowdown about 85% of knowledge collapsed. In previous sections we defined the knowledge instinct as the maximization of similarity, and we defined aesthetic emotions as changes in similarity. Future research will have to make the next step, which will define the mechanism by which differentiated aesthetic emotions unify contradictory aspects of knowledge. We will model neural processes, in which diverse emotions created by music unify contradictory concepts in their manifold relations to our cognition as a whole. We will have to understand processes in which the knowledge instinct differentiates itself and the synthesis of differentiated knowledge is achieved.

9.2 Experimental Evidence

The knowledge instinct is clearly a part of mind operations [1, 2, 3]. Can we prove its ubiquitous nature and connection to emotional satisfaction or dissatisfaction? Can we measure aesthetic emotions during perception (when it is usually subliminal)? Can we measure aesthetic emotions during more complex cognition (when it is more conscious)? Does brain compute similarity measures, and if so, how is it done? Does it relate to aesthetic emotions as predicted by the knowledge instinct theory? Does it operate in a similar way at higher levels in the hierarchy of the mind? Operations of the differentiated knowledge instinct, and the emotional influence of concepts on cognition of other concepts, are virtually obvious experimental facts. However, detailed quantitative studies of this phenomenon are missing. For example, can we prove that emotionally sophisticated people can better tolerate cognitive dissonances (that is, conceptual contradictions) than people less sophisticated emotionally (it would be important to control other variables, say IQ).

Dan Levine studies emotional effects on learning [50]. In his experiments normal subjects gradually accumulated cognitive knowledge, whereas emotionally impaired patients could not properly accumulate cognitive knowledge. Subject emotions in his experiments were not related to any bodily need, rather these were aesthetic emotions. Are these aesthetic emotions limited to the cortex, or are ancient emotional mechanisms also involved?

Mechanisms of conceptual differentiation at a single level in a hierarchy described in Sect. 4 correspond to psychological and neurophysiological experimental evidence. These include the interaction between bottom-up and top-down signals, and resonant matching between them as a foundation for perception [6, 51]. Experimental evidence is less certain for these mechanisms being repeated at each hierarchical level. Experimental evidence for dynamic logic is limited to the fact that imagination (concept-models voluntary recollected from memory with closed eyes) are vague and fuzzy relative to actual perceptions with open eyes. Dynamic logic makes a specific suggestion that top-down (model) signals form a vague-fuzzy image that gradually becomes more specific until it matches the perceived object. This prediction might be amenable to direct verification in psychological experiments.

Norman Weinberger studied the detection of a specific acoustic tone, using an electrode to measure the response from the cellular receptive fields for acoustic frequencies in the brain [52]. Dynamic logic predicts that the initial response will be fuzzy and vague. During learning, the neural response will gradually become more specific, more "tuned." This trend was actually experimentally observed. As expected according to dynamic logic, the frequency receptive field became more "tuned," in the auditory cortex. The auditory thalamus, however, an evolutionarily older brain region, did not exhibit dynamic-logic learning. It would be more difficult to confirm or disprove this mechanism at higher levels in the hierarchy.

9.3 Problems for Future Research

Future experimental research will need to examine, in detail, the nature of hierarchical interactions, including the mechanisms of learning hierarchy. This research may reveal to what extent the hierarchy is inborn vs. adaptively learned. Studies of the neurodynamics of interacting language and cognition have already begun [10, 12, 53]. Future research will need to model the differentiated nature of the knowledge instinct. Unsolved problems include: neural mechanisms of emerging hierarchy, interactions between cognitive hierarchy and language hierarchy [11, 12]; differentiated forms of the knowledge instinct accounting for emotional interactions among concepts in processes of cognition, the infinite variety of aesthetic emotions perceived in music, their relationships to mechanisms of synthesis [34, 47, 48]; neural mechanisms of interactions of differentiation and synthesis, and evolution of these mechanisms in the development of the mind during cultural evolution.

Cultural historians can use the results of this chapter as a tool for understanding the psychological mechanisms of cultural evolution. The results may explain how differentiation and synthesis have interacted with language, religion, art, and especially music, and how these interactions have shaped the evolution of various cultures. Social psychologists can use the results of this chapter as a tool for understanding the psychological mechanisms governing present conditions. It is possible to measure the levels of differentiation and synthesis in various societies, and to use this knowledge for improving human conditions around the world. It will also be possible to predict future cultural developments, and to use this knowledge for preventing strife and stagnation, and for stimulating wellbeing.

Acknowledgment I am thankful to D. Levine, R. Deming, B. Weijers, and R. Kozma for discussions, help and advice, and to AFOSR for supporting part of this research under the Lab. Task 05SN02COR, PM Dr. Jon Sjogren.

References

1. Harlow, H.F., & Mears, C. (1979). The Human Model: Primate Perspectives, Washington, DC: V. H. Winston and Sons.
2. Berlyne, D. E. (1960). Conflict, Arousal, And Curiosity, McGraw-Hill, New York, NY; Berlyne, D. E. (1973). Pleasure, Reward, Preference: Their Nature, Determinants, And Role In Behavior, Academic Press, New York, NY.
3. Festinger, L. (1957). A Theory of Cognitive Dissonance, Stanford, CA: Stanford University Press.
4. Grossberg, S. (1988). *Neural Networks and Natural Intelligence.* MIT Press, Cambridge, MA.
5. Zeki, S. (1993). *A Vision of the Brain* Blackwell, Oxford, England.
6. Ganis, G., and Kosslyn, S. M. (2007). Multiple mechanisms of top-down processing in vision.
7. Kant, I. (1790). *Critique of Judgment,* tr. J.H.Bernard, Macmillan & Co., London, 1914.
8. Damasio, A.R. (1995). Descartes' Error: Emotion, Reason, and the Human Brain. Avon, NY, NY.
9. Perlovsky, L.I. 2001. Neural Networks and Intellect: using model based concepts. New York: Oxford University Press.
10. Perlovsky, L.I. (2006). Toward Physics of the Mind: Concepts, Emotions, Consciousness, and Symbols. Phys. Life Rev. 3(1), pp. 22–55.
11. Perlovsky, L.I. (2004). Integrating Language and Cognition. *IEEE Connections*, Feature Article, **2**(2), pp. 8–12.
12. Perlovsky, L.I. (2006). Symbols: Integrated Cognition and Language. Chapter in A. Loula, R. Gudwin, J. Queiroz, eds. Semiotics and Intelligent Systems Development. Idea Group, Hershey, PA, pp. 121–151.
13. Bellman, R.E. (1961). Adaptive Control Processes. Princeton University Press, Princeton, NJ.
14. Minsky, M.L. (1968). Semantic Information Processing. The MIT Press, Cambridge, MA.
15. Brooks, R.A. (1983). Model-based three-dimensional interpretation of two-dimensional images. IEEE Trans. Pattern Anal. Machine Intell., **5**(2), 140–150.
16. Perlovsky, L.I., Webb, V.H., Bradley, S.R. & Hansen, C.A. (1998). *Improved ROTHR Detection and Tracking Using MLANS* . AGU Radio Science, **33**(4), pp. 1034–44.
17. Singer, R.A., Sea, R.G. and Housewright, R.B. (1974). Derivation and Evaluation of Improved Tracking Filters for Use in Dense Multitarget Environments, IEEE Transactions on Information Theory, IT-20, pp. 423–432.
18. Perlovsky, L.I. (1996). *Gödel Theorem and Semiotics.* Proceedings of the Conference on Intelligent Systems and Semiotics '96. Gaithersburg, MD, v.2, pp. 14–18.
19. Perlovsky, L.I. (1998). *Conundrum of Combinatorial Complexity.* IEEE Trans. PAMI, **20**(6) p. 666-70.
20. Kecman, V. (2001). Learning and Soft Computing: Support Vector Machines, Neural Networks, and Fuzzy Logic Models (Complex Adaptive Systems). The MIT Press, Cambridge, MA.
21. Marchal, B. (2005). Theoretical Computer Science & the Natural Sciences, Physics of Life Reviews, **2**(3), pp. 1–38.
22. Perlovsky, L.I. (1997). *Physical Concepts of Intellect.* Proc. Russian Academy of Sciences, **354**(3), pp. 320–323.
23. Perlovsky, L.I. (2006). Fuzzy Dynamic Logic. New Math. and Natural Computation, **2**(1), pp.43-55.
24. Perlovsky, L.I. (1996). *Mathematical Concepts of Intellect.* Proc. World Congress on Neural Networks, San Diego, CA; Lawrence Erlbaum Associates, NJ, pp. 1013–16
25. Perlovsky, L.I. (2002). Physical Theory of Information Processing in the Mind: concepts and emotions. SEED On Line Journal, 2002 2(2), pp. 36–54.
26. Cramer, H. (1946). *Mathematical Methods of Statistics,* Princeton University Press, Princeton NJ.
27. Perlovsky, L.I. and Deming, R.W. (2006). Neural Networks for Improved Tracking. IEEE Transactions on Neural Networks (in press).

28. Singer, R.A., Sea, R.G. and Housewright, R.B. (1974). Derivation and Evaluation of Improved Tracking Filters for Use in Dense Multitarget Environments, *IEEE Transactions on Information Theory*, **IT-20**, pp. 423–432.
29. Jung, C.G., 1921, *Psychological Types.* In the Collected Works, v.6, Bollingen Series XX, 1971, Princeton University Press, Princeton, NJ.
30. Jung, C.G. (1934). *Archetypes of the Collective Unconscious.* In the Collected Works, V. 9,II, Bollingen Series XX, 1969, Princeton University Press, Princeton, NJ.
31. Taylor, J. G. (2005). Mind And Consciousness: Towards A Final Answer? Physics of Life Reviews, **2**(1), p. 57.
32. James, W. (1890). In "The Principles of Psychology", 1950, Dover Books.
33. Jaynes, J. (1976). The Origin of Consciousness in the Breakdown of the Bicameral mind. Houghton Mifflin Co., Boston, MA; 2nd edition 2000.
34. Perlovsky, L.I. (2007). *The Knowledge Instinct.* Basic Books. New York, NY.
35. Kant, I. (1798) Anthropologie in pragmatischer Hinsicht, see Anthropology from a Pragmatic Point of View, Cambridge University Press (2006), Cambridge, England.
36. Carpenter, G.A. & Grossberg, S. (1987). A massively parallel architecture for a self-organizing neural pattern recognition machine, Computer Vision, Graphics and Image Processing, 37, 54–115; (2003). Adaptive Resonance Theory. In M.A. Arbib (Ed.), The Handbook of Brain Theory and Neural Networks, 2nd Ed., Cambridge, MA: MIT Press, 87-90.
37. Perlovsky, L.I. (2006). *Modeling Field Theory of Higher Cognitive Functions.* Chapter in A. Loula, R. Gudwin, J. Queiroz, eds., Artificial Cognition Systems. Idea Group, Hershey, PA, pp. 64–105.
38. Perlovsky, L.I. (2006). *Neural Networks, Fuzzy Models and Dynamic Logic.* Chapter in R. Köhler and A. Mehler, eds., Aspects of Automatic Text Analysis (Festschrift in Honor of Burghard Rieger), Springer, Germany, pp. 363–386.
39. Pinker, S. (2000). The Language Instinct: How the Mind Creates Language. Harper Perennial.
40. Demandt, A. (2003). 210 Theories, from Crooked Timber weblog entry August 25; http://www.crookedtimber.org/
41. Harris, C. L., Ayçiçegi, A., and Gleason, J. B. (2003). Taboo words and reprimands elicit greater autonomic reactivity in a first language than in a second language, Applied Psycholinguistics, 24, pp. 561–579
42. Diamond, J. (2004). Collapse: How Societies Choose to Fail or Succeed, Viking, New York, NY.
43. Perlovsky, L.I. (2005). Evolving Agents: Communication and Cognition. Chapter in Autonomous Intelligent Systems, Eds: V. Gorodetsky, J. Liu, V.A. Skormin. Lecture Notes in Computer Science, 3505 / 2005. Springer-Verlag GmbH.
44. Crystal, D. (1997). The Cambridge encyclopedia of language, second edition. Cambridge: Cambridge University Press.
45. Perlovsky, L.I. (2005). Evolution of Consciousness and Music, Zvezda, 2005, 8, pp. 192–223 (Russian); http://magazines.russ.ru/zvezda/2005/8/pe13.html
46. Damasio, A.R. (1994). Descartes' Error: Emotion, Reason, and the Human Brain. Grosset/Putnam, New York, NY.
47. Perlovsky, L.I. (2006). Co-evolution of Consciousness, Cognition, Language, and Music. Tutorial lecture course at Biannual Cognitive Science Conference, St. Petersburg, Russia.
48. Perlovsky, L.I. (2005). Music–The First Principles, 2005, http://www.ceo.spb.ru/libretto/kon_lan/ogl.shtml.
49. Perlovsky, L.I. (2006). Joint Evolution of Cognition, Consciousness, and Music. Lectures in Musicology, School of Music, University of Ohio, Columbus.
50. Levine, D. (2007). Seek Simplicity and Distrust It: Knowledge Maximization versus Effort Minimization. IEEE International Conference on Knowledge Intensive Multi-Agent System, KIMAS'07, Waltham, MA.
51. Freeman, W.J. (1975). *Mass action in the nervous system.* Academic Press, New York, NY.
52. Weinberg, N. http://www.dbc.uci.edu/neurobio/Faculty/Weinberger/weinberger.htm.
53. Fontanari, J.F. and Perlovsky, L.I. (2006). Meaning creation and communication in a community of agents. World Congress on Computational Intelligence (WCCI). Vancouver, Canada.

Using ADP to Understand and Replicate Brain Intelligence: The Next Level Design?

Paul J. Werbos

Abstract Since the 1960's I proposed that we could understand and replicate the highest level of intelligence seen in the brain, by building ever more capable and general systems for adaptive dynamic programming (ADP) – like "reinforcement learning" but based on approximating the Bellman equation and allowing the controller to know its utility function. Growing empirical evidence on the brain supports this approach. Adaptive critic systems now meet tough engineering challenges and provide a kind of first-generation model of the brain. Lewis, Prokhorov and I have done some work on second-generation designs. I now argue that mammal brains possess three core capabilities – creativity/imagination and ways to manage spatial and temporal complexity – even beyond the second generation. This chapter reviews previous progress, and describes new tools and approaches to overcome the spatial complexity gap. The Appendices discuss what we can learn about higher functions of the human mind from this kind of mathematical approaches.

1 Introduction

No one on earth today can write down a complete set of equations, or software system, capable of learning to perform the complex range of tasks that the mammal brain can learn to perform. From an engineering viewpoint, this chapter will provide an updated roadmap for how to reach that point. From a neuroscience viewpoint, it will provide a series of *qualitative* but *quantifiable* theories of how intelligence works in the mammal brain. The main text of the chapter was written for an engineering workshop; it explains the basic mathematical principles, and how they relate to some of the most important gross features of the brain. A new appendix, written for arxiv.org, discusses more of the implications for comparative neuroscience and for our subjective experience as humans.

The main text of this chapter will not address the human mind as such. In nature, we see a series of *levels* of intelligence or consciousness [19]; for example, within the vertebrates, M. E. Bitterman [2] has shown that there are major qualitative jumps from the fish to the amphibian, from the amphibian to the reptile, and from the reptile to even the simplest mammal. Ninety-nine percent of the higher parts of the human brain consist of structures, like the six-layer neocortex, which exist in the

L. I. Perlovsky and R. Kozma, *Neurodynamics of Cognition and Consciousness.*

smallest mouse, and show similar general-purpose-learning abilities in the mouse; therefore, the scientific goal of understanding the kind of learning and intelligence that we see in the smallest mouse is an important step towards understanding the human mind, but it is certainly not the whole thing.

Section 2 of this chapter will briefly review the general concepts of optimization and ADP. It will give a few highlights from the long literature on why these offer a central organizing principle both for understanding the brain and for improving what we can do in engineering. Section 3 will review the first and second generation ADP designs, and their relevance to brain-style intelligence. Section 4 will discuss how to move from second-generation designs to the level of intelligence we see in the brain of the smallest mouse – with a special emphasis on how to handle spatial complexity, by learning symmetry groups, and to incorporate that into an ADP design or larger brain. Appendices A-D go on to discuss *comparative* cognitive neuroscience (other classes of vertebrate brain), and the significance of this work to understanding higher capabilities of the human brain and subjective experience. Appendix 5 gives a brief explanation of how intelligence at the level of a bird brain could do far better than today's systems for managing electric power grids.

2 Why Optimality and Why ADP?

2.1 Optimality as an Organizing Principle for Understanding Brain Intelligence

For centuries, people have debated whether the idea of optimization can help us understand the human mind. Long ago, Aristotle proposed that all human efforts and thought are ultimately based on the pursuit of (maximizing) happiness – a kind of inborn "telos" or ultimate value. Utilitarians like John Stuart Mill and Jeremy Bentham carried this further. A more updated version of this debate can be found in [7] and in the chapter by Perlovsky; here I will only review a few basic concepts.

To begin with [14], animal behavior is ultimately about choices as depicted in Fig. 1.

The simplest types of animals may be born with fixed rules about what actions to take, as a function of the state of their environment as they see it. More advanced animals, instead, have an ability to select actions which are somehow based

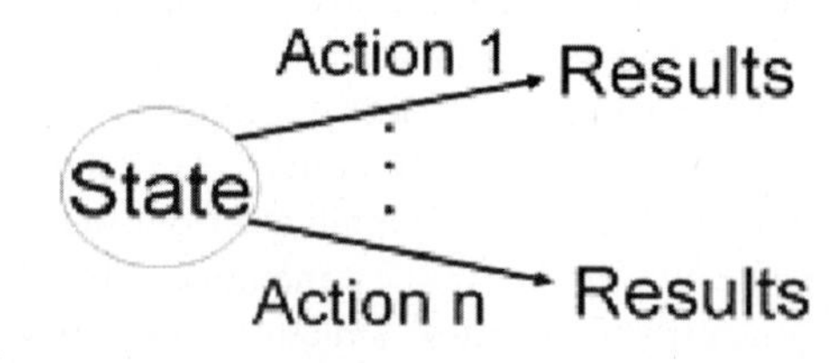

Fig. 1 Schematic views of simple behavior on a given set of possible actions

on the *results* that the actions might have. *Functionality* of the brain is about making choices which yield better results. *Intelligence* is about *learning* how to make better choices.

To put all this into mathematics, we must have a way to evaluate which results are "better" than which other results. Von Neumann's concept of Cardinal Utility function [9] provides that measure; it is the foundation of decision theory [10], risk analysis, modern investment analysis, and dynamic programming, among others. Before Von Neumann, most economists believed that (consumer) values or preferences are strictly an *ordinal* relationship; for example, you can prove by experiment that most people prefer steak over hamburger, and that they prefer hamburger over nothing, but it seemed meaningless to argue that "the gap between hamburger and nothing is bigger than the gap between steak and hamburger." Von Neumann showed that we *can* make these kinds of numerical comparisons, once we introduce probabilities into our way of thinking. For example, given a 50-50 coin flip between steak or nothing, versus a certainty of hamburger, many people prefer hamburger. Of course, a more complete utility function for human consumers would account for other variables as well.

Usually, when we talk about discrete "goals" or "intentions," we are not talking about the long-term values of the organism. Rather, we are talking about subgoals or tactical values, which are intended to yield better results or outcomes. The utility function which defines what is "better" is the foundation of the system as a whole.

Next consider the analogy to physics.

In 1971–1972, when I proposed a first generation model of intelligence, based on ADP, to a famous neuroscientist, he objected: "The problem with this kind of model is that it creates an anthropomorphic view of how the brain works. I have spent my entire life teaching people how to overcome a bachtriomorphic view of the frog. Thinking about people by using empathy could be a disaster for science. Besides, even in physics, we know that the universe is maximizing a kind of utility function, and we don't think of the universe in anthropomorphic terms."

From a strictly objective viewpoint, his argument actually supports the idea of trying to use optimization as a central organizing principle in neuroscience. After all, if it works in physics, in a highly rigorous and concrete way, why not here? If we can unify our functional understanding of the brain not only with engineering, but with subjective experience and empathy, isn't this a source of strength, so long as we keep track of which is which? In fact, my real goal here has been to develop the kind of mathematical understanding which really does help us to unify our cybernetic understanding, our objective understanding of brain and behavior, and our subjective understanding of ourselves and others.

But does it really work that way in physics? Partly so. According to classical physics, the universe really does solve the optimization problem depicted here:

The universe has a kind of "utility function," L ($\underline{\mathbf{x}}$, t). It "chooses" the states ö of all particles and fields at all times t by choosing states which maximize the total sum of L across all of space time, between time t_- and time t_+, subject to the requirement that they provide a continuous path from the fixed state at some initial time t_- and some final time t_+. This elegant formalism, due to Lagrange, provides a very simple parsimonious description of the laws of physics; instead of specifying

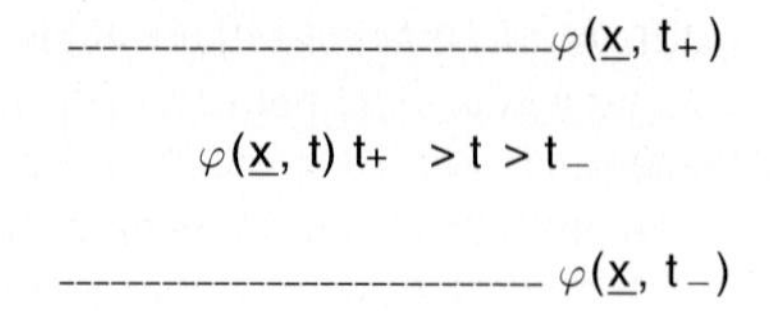

Fig. 2 Illustration of the state vector $\varphi(\underline{\mathbf{x}}, t)$ evolving between time instances t_- and t_+

n dynamic laws for n types of particle or field, we can specify the "Lagrangian function L," and derive all the predictions of physics from there. In order to perform that calculation, we can use an equation from classical physics, the Hamilton-Jacobi equation, which tells us how to solve deterministic optimization problems across time or space-time.

But that is not the whole story. Hamilton and Lagrange had many debates about whether the universe really maximizes L – or does it minimize it or find a minmax solution? Does the physical universe find something that looks like the outcome of a two-person zerosum game? By the time of Einstein, it appeared so. Modern quantum theory gets rid of the deterministic assumption, but adds random disturbance in a very odd way. It actually turns out that we can recover something like Lagrange's original idea, which fits the tested predictions of modern quantum theory, by introducing a stochastic term whose statistics are symmetric both in space and in time; however, the details are beyond the scope of this chapter. (See `http://www.werbos.com/reality.htm`.)

To describe the brain, it is not enough to use the old optimization rule of Hamilton and Jacobi. We need to consider the stochastic case, because animals, like us, cannot predict our environment in a deterministic way. The foundation for optimization over time in the stochastic case is the Bellman equation, a great breakthrough developed by Bellman in 1953, made possible by Von Neumann's concept of Cardinal Utility function.

The principles of optimality are important to fundamental physics – but also to thermo-dynamics, and to the physics of emergent phenomena in general. Those details are beyond the scope of this chapter.

Finally, let me address two of the most common questions which people tend to ask when I talk about the brain as an "optimization machine."

First: If brains are so optimal, why do humans do so many stupid things? Answers: Brains are designed to *learn* approximate optimal policy, as effectively as possible with *bounded computational resources* (networks of neurons), starting from a less optimal start. They never learn to play a perfect game of chess (nor will our computers, nor will any other algorithm that can be implemented on a realistic computer) because of constraints on computational resources. *We just do the best we can.*

Also, when one human (a researcher) criticizes another, we are seeing a comparison between *two* highly intelligent systems. Some brains learn faster than others. In my view, humans themselves are an intermediate state towards an even higher/faster intelligence, as discussed in the Appendices of this paper.

Second question: if this optimization theory of the brain is correct, wouldn't brains get stuck in local minima, just like artificial optimization programs when confronted with a complex, nonlinear environment? Answers: they do indeed. Every person on earth is caught in a "local minimum," or rut, to some degree. In other words, we could all do a bit better if we had more creativity. But look at those hairy guys (chimpanzees) in the jungle, and the rut they are in!

The optimization theory of the brain implies that our brains *combine* an incremental learning ability with an ability to learn to be more creative – to do better and better "stochastic search" of the options available to us. There are a few researchers in evolutionary computing or stochastic search who tell us that their algorithms are guaranteed to find the global optimum, eventually; however, those kinds of guarantees are not very realistic because, for a system of realistic complexity, they require astronomical time to actually get to the optimum.

2.2 Optimality and ADP in Technology

The benefits of adaptive dynamic programming (ADP) to technology have been discussed by many other authors in the past [11], with specific examples. Here I will review only a few highlights. (See also the example of the electric power grid, discussed at the end of the Appendix.)

Many control engineers ask: "Why try to find the optimal controller out of all possible controllers? It is hard enough just to keep things from blowing up – to stabilize them at a fixed point." In fact – the most truly stable controllers now known are nonlinear feedback controllers, based on "solving" the "Hamilton-Jacobi-Bellman" equation. But in order to implement that kind of control, we need mechanisms to "numerically solve" (approximate) the Bellman equation as accurately as possible. ADP is the machinery to do that.

Furthermore – there are times when it is impossible to give a truly honest absolute guarantee of stability, under accurate assumptions. Certainly, a mouse running through the field has no way to guarantee its survival – nor does the human species as a whole, in the face of the challenges now confronting us. (See `www.werbos.com`.) In that kind of real-world situation, the challenge is to *maximize the probability* of survival; that, in turn, is a stochastic optimization problem, suitable for ADP, and not for deterministic methods. (Recent work by Gosavi has explored that family of ADP applications.) Verification and validation for real complex systems in the real world is heavily based on empirical tests and statistics already.

Finally, in order to address nonlinear optimization problems in the general case, we absolutely must use universal nonlinear function approximators. Those could be Taylor series – but Barron showed years ago that the simplest form of neural networks offer more accurate nonlinear approximation that Taylor series or other linear basis function approximators, in the general case, when there is a large number of state variables. Use of more powerful and accurate approximators (compatible with distributed hardware, like emerging multicore chips) is essential to more accurate approximations and better results.

3 First and Second Generation ADP Models of Brain Intelligence

3.1 Origins and Basics of the First Generation Model

Backpropagation and the first true ADP design both originated in my work in the 1970's, as shown in the simplified chart of Fig. 3.

In essence, the founders of artificial intelligence (AI) – Newell, Shaw and Simon, and Minsky [4] – proposed that we could build brain-like intelligent systems by building powerful reinforcement learning systems. However, after a great deal of experimentation and intuition and heuristic thinking, they could not design systems which could optimize more than a few variables. Knowing that the brain can handle many thousands of variables, they simply gave up – just as they gave up on training simplified neural models (multilayer perceptrons). Amari, at about the same time, wrote that perhaps derivatives might be used somehow to train multilayer perceptrons – but suggested that it would be unlikely to work, and did not provide any algorithm for actually calculating the required derivatives in a distributed, local manner.

In 1964, I – like many others – was deeply inspired by Hebb's classic book on intelligence [5]. Inspired by the empirical work on mass action and learning in the brain (by Lashley, Freeman, Pribram and others), he proposed that we would not really need a highly complex model in order to explain or reproduce brain-like intelligence. Perhaps we could generate intelligence as the emergent result of learning; we could simply construct billions of model neurons, each following a kind of universal neuron learning rule, and then intelligence could emerge strictly as a result of learning. I tried very hard to make that work in the 1960's, and failed. The

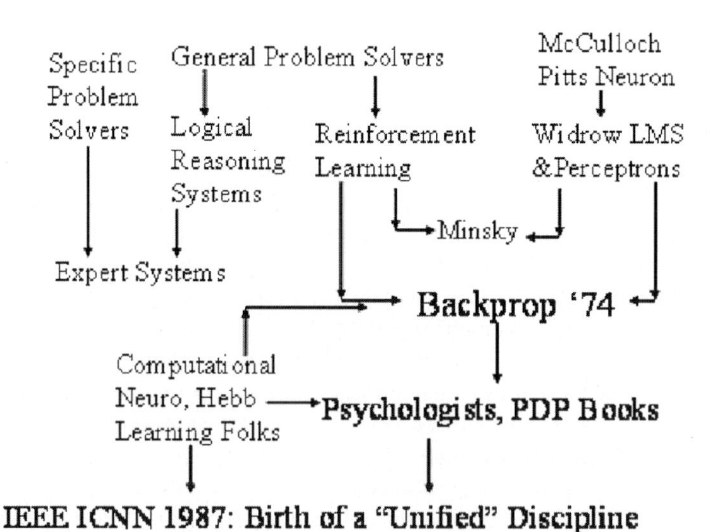

Fig. 3 Historical perspective on the birth of back propagation and ADP

key problem was that Hebb's approach to a universal learning rule is essentially calculating correlation coefficients; those are good enough to construct useful associative memories, as Grossberg showed, but not to make good statistical predictions or optimal control. They are simply not enough by themselves to allow construction of an effective general-purpose reinforcement learning machine.

By 1971–1972, I realized that Hebb's vision could be achieved, if we relax it only very slightly. It is possible to design a general purpose reinforcement learning machine, if we allow just three types of neuron and three general neuron learning rules, instead of just one.

Actually, the key insight here came in 1967. In 1967 (in a paper published in 1968 [14]), I proposed that we could overcome the problems with reinforcement learning by going back to basic mathematical principles – by building systems which learn to approximate the Bellman equation. Use of the Bellman equation is still the only exact and efficient method to compute an optimal strategy or policy of action, for a general nonlinear decision problem over time, subject to noise. The equation is:

$$J(\underline{\mathbf{x}}(t)) = \frac{Max < U(\underline{\mathbf{x}}(t), \underline{\mathbf{u}}(t)) + J(\underline{\mathbf{x}}(t+1)) >}{1+r}, \underline{\mathbf{u}}(t)$$

where $\underline{\mathbf{x}}$(t) is the state of the environment at time t, $\underline{\mathbf{u}}$(t) is the choice of actions, U is the cardinal utility function, r is the interest or discount rate (exactly as defined by economics and by Von Neumann), where the angle brackets denote expectation value, and where J is the function we must solve for in order to derive the optimal strategy of action. In any state $\underline{\mathbf{x}}$, the optimal $\underline{\mathbf{u}}$ is the one which solves the optimization problem in this equation. A learning system can learn to approximate this policy by using a neural network (or other universal approximator) to approximate the J function and other key parts of the Bellman equation, as shown in Fig. 4, from my 1971–1972 thesis proposal to Harvard:

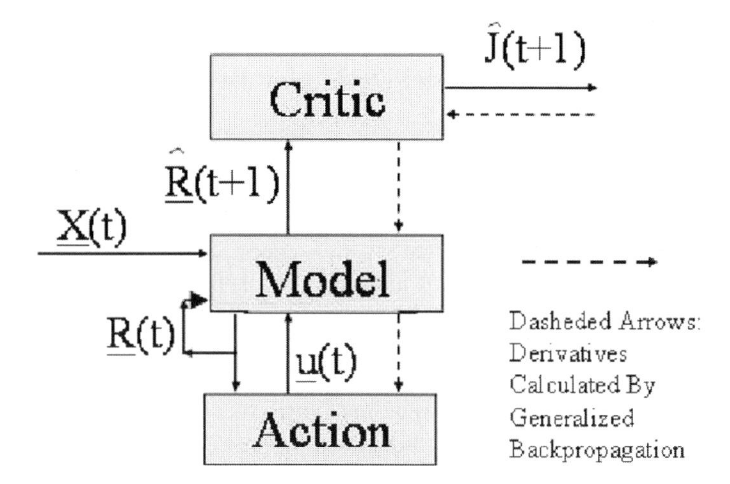

Fig. 4 Learning system based on the approximate solution of the Bellman equation. Emergent intelligence is possible if we allow three types of neurons [16]

In that design, I needed a *generalized* form of backpropagation as a tool to calculate the essential derivatives or sensitivity coefficients needed to allow correct incremental learning of all three parts. I formulated and proved a new chain rule for "ordered derivatives" which makes it possible to compute the required derivatives exactly through any kind of large nonlinear system, not just neural networks.

Intuitively, the "Action'' network here is the one which actually computes or decides the actions to be taken by the organism. The "Model" network learns how to predict changes in the environment of the organism, and it also learns how to estimate the objective state of reality ($\underline{\mathbf{R}}$), which is more than just the current sensory input ($\underline{\mathbf{X}}$) to the organism. The "Critic" network estimates the "J" function, which is a kind of *learned* value function, similar to what is called "secondary reinforcement" in animal learning.

For my PhD thesis (reprinted in entirety in [16]), I included the proof, and many applications of backpropagation to systems other than neural networks. In [11, 12], I described how *generalized backpropagation* can be used in a wide variety of applications, including ADP with components that could be neural networks or *any other* nonlinear differentiable system.

The method which I proposed to adapt the Critic network in 1971–1972 I called "Heuristic Dynamic Programming" (HDP). It is essentially the same as what was later called "the Temporal Difference Method." But I learned very early that the method does not scale very well, when applied to systems of even moderate complexity. It learns too slowly. To solve this problem, I developed the core ideas of two new methods – dual heuristic programming (DHP) and Globalized DHP (GDHP) – published in a series of papers from 1977 to 1981 [13, 14, 15]. To prove convergence, in [3], I made small but important changes in DHP; thus [3] is the definitive source for DHP proper. For more robust extensions, see the final sections of [18]. See [11] and [3] for reviews of practical applications of HDP, DHP, GDHP and related adaptive critic systems.

4 A Second Generation Model/Design for Brain-style Intelligence

By 1987, I realized that the brain has certain capabilities beyond what any of these first-generation designs offer. Thus I proposed [15] a second generation theory. Key details of that theory were worked out in [3] and in papers with Pellionisz, See the flow chart below and the papers posted at `www.werbos.com`.

The second generation design was motivated in part by trying to understand the brain. It was also motivated by an engineering dilemma. The dilemma is that truly powerful foresight, in an ADP system, requires the use of Critic Networks and Model Networks which are far more powerful than feedforward neural networks (or Hebbian networks or Taylor series or linear basis function networks). It requires the use of recurrent networks, including networks which "settle down" over many cycles of an inner loop calculation before emitting a calculation. That, in turn, requires a relatively low sampling rate for calculation; about 4–8 frames per second is the rate observed for the cerebral cortex of the mammal brain, in responding

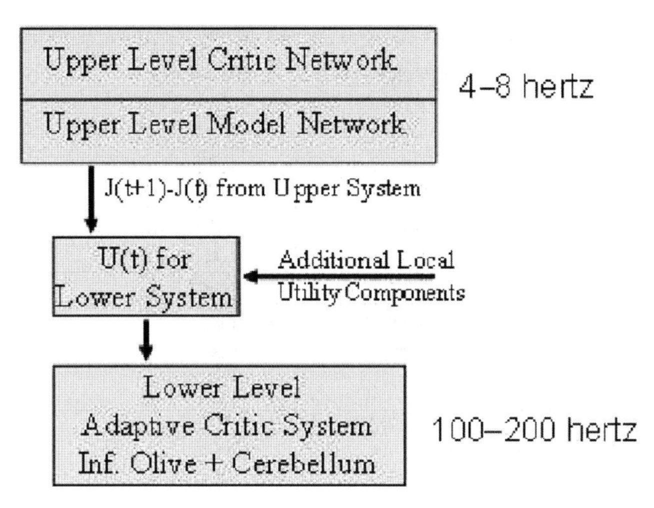

Fig. 5 Second generation design of brain-like intelligence

to new inputs from the thalamus, the "movie screen" watched by the upper brain. However, smooth muscle control requires a much higher bandwidth of control; to achieve that, I proposed that the brain is actually a kind of master-slave system. In chap. 13 of [3], I provided equations for an "Error Critic for motor control" which provide one possible design for a fast model-free "slave" neural network, matching this model. In that design, the "memory neurons" which estimated the vector $\underline{\mathbf{R}}$ are in the Action Network; I proposed that these are simply the Purkinje cells of the cerebellum. They are trained by a kind of distributed DHP-like Critic system, based partly in the inferior olive and partly near the Purkinje cells themselves.

Danil Prokhorov, in various IJCNN papers, showed how that kind of fast design (and some variations he developed) works well, by certain measures, in computational tests. Recent formal work in Frank Lewis's group at the University of Texas (ARRI) has shown strong stability results for continuous-time model free ADP designs which *require* an external value input, exactly like what this master-slave arrangement would provide.

Intuitively... the "master" is like the coach within you, and the "slave" is like the inner football player. The football player has very fast reflexes, and is essential to the game, but he needs to strive to go where the more far-seeing coach sends him. The coach can learn more complex stuff faster than the football player, and responds to a more complex strategic picture. Lower-level stability is mainly provided by the football player.

In 1987, Richard Sutton read [15], and arranged for us to discuss it at great length in Massachusetts. This was the event which injected the idea of ADP into the reinforcement learning school of AI. The paper is cited in Sutton's chapter in [8], which includes an implementation of the idea of "dreaming as simulation" discussed in [15].

4.1 Engineering Roadmap and Neuroscience Evidence for Third Generation Theory/Design

In 1992, I believed that we could probably replicate the level of intelligence we see in the basic mammal brain, simply by refining and filling in these first and second generation theories of how the brain works. In fact, the first and second generation design already offer potential new general-purpose adaptive capabilities far beyond what we now have in engineering. It is still essential that we continue the program of refining and understanding and improving these classes of designs as far as we can go – both for the sake of engineering, and as a prerequisite to set the stage for even more powerful designs.

I have suggested that half of the funding aimed at reverse engineering the brain should still go towards the first and second generation program – half towards the ADP aspects, and half towards the critical subsystems for prediction, memory and so on. (See www.eas.asu.edu/∼nsfadp.) Because those are complicated issues, and I have written about them elsewhere, I will not elaborate here.

More and more evidence has accumulated suggesting that optimization (with a predictive or "Model" component) is the right way to understand the brain. For example, Nicolelis and Chapin, in *Science*, reported that certain cells in the thalamus act as advance predictors of other cells. More important, when they cut the existing connections, the thalamo-cortical system would adapt in exactly the right way to relearn how to predict. This is clear evidence that the thalamo-cortical system – the biggest part of the brain – is in great part an adaptive "Model" network, a general-purpose system for doing adaptive "system identification" (as we say in control theory). Barry Richmond has observed windows of forwards and backwards waves of information in this circuit, fully consistent with our Time-Lagged Recurrent Network (TLRN) model of how such a Model network can be constructed and adapted.

Papez and James Olds senior showed decades ago how cells in the "limbic system" convey "secondary reinforcement signals," exactly as we would predict for an adaptive Critic component of the brain. More recent work on the dopamine system in the basal ganglia suggests even more detailed relations between reinforcement learning and actual learning in neural circuits.

A key prediction of the engineering approach has always been the existence of subcircuits to compute the derivatives – the generalized backpropagation – required here. When we first predicted backwards synapses, to make this possible, many ridiculed the engineering approach. But later, in Science, Bliss et al reported a "mysterious" but strong reverse NMDA synapse flow. Spruston and collaborators have reported backpropagation flows (totally consistent with the mathematics of generalized backpropagation) in cell membranes. The synchronized clock signals implied in these designs are also well-known at present to "wet," empirical neuroscientists.

More details – and the empirical implications which cry out for follow-on work – are discussed in some of the papers on my web page, such as papers for books edited by Pribram.

One interesting recent thought.: From engineering work, we have learned that the complexity of the *learning* system needed to train a simple input-output system or learned policy is far greater than the complexity of the input-output system itself. A simple example comes from Kalman filtering, where the "scaffolding" matrices (P, etc.) needed for consistent filtering are n times as large as the actual state estimates themselves; n is the number of state variables. Another example comes from the neural network field, where Feldkamp and Prokhorov have shown that Time-Lagged Recurrent Networks (TLRN) do an excellent job of predicting engine behavior – even across different kind of engines. It performs far better than the direct use of Extended Kalman Filters (EKF) in state estimation, and performs as well as full particle filter methods, which are far more expensive (especially when there are many variables in the system, as in the brain). To train these TLRNS, they use backpropagation through time *combined with* "Distributed EKF (DEKF) training," a kind of training which requires updating matrices. Could it be that "junk DNA" includes a large system whose purpose is to tune the adaptation of the "coding DNA," which are after all only a small portion of our genetic system? Could it be that individual neurons do contain very complex molecular memories after all – memories invisible to our conscious mind, but essential to more efficient learning (such as the matrices for DEKF learning)? These are important empirical issues to explore.

5 Bridging the Gap to the Mammal-brain Level

AI researchers like Albus [1] have long assumed that brains must have very complex, explicit, hard-wired hierarchies of systems to handle a high degree of complexity in space and in time. By 1997, I became convinced that they are partly right, because I was able to formulate modified Bellman equations which allow much faster learning in cases where a state space can be sensibly partitioned in a (learnable) hierarchical way [20, 21]. Nature would not neglect such an opportunity – and it fit well with emerging new knowledge about the basal ganglia, and ideas from Pribram.

Recent biological data does not support the older hierarchy ideas form AI, but it clearly call out for some kind of specific mechanisms in three core areas: (1) a "creativity/imagination" mechanism, to address the nonconvex nature of complicated optimization problems; (2) a mechanism to exploit modified Bellman equations, in effect; and (3) a mechanism to handle spatial complexity.

The flow chart in Fig. 6 summarizes a strawman model of the creativity mechanism which I proposed in 1997 [12]. I hoped to stimulate broad research into "brain-like stochastic search." (See my web page for a CEC plenary talk on that challenge). Wunsch, Serpen, Thaler, Pelikan and Fu's group at Maryland have all done important early work relevant to this task, but it hasn't really come together. Likewise, work in the last few years on temporal complexity has not done full justice to the modified Bellman equations, and has not shown as much progress as hoped for; part of the problem is that temporal complexity is usually associated with spatial

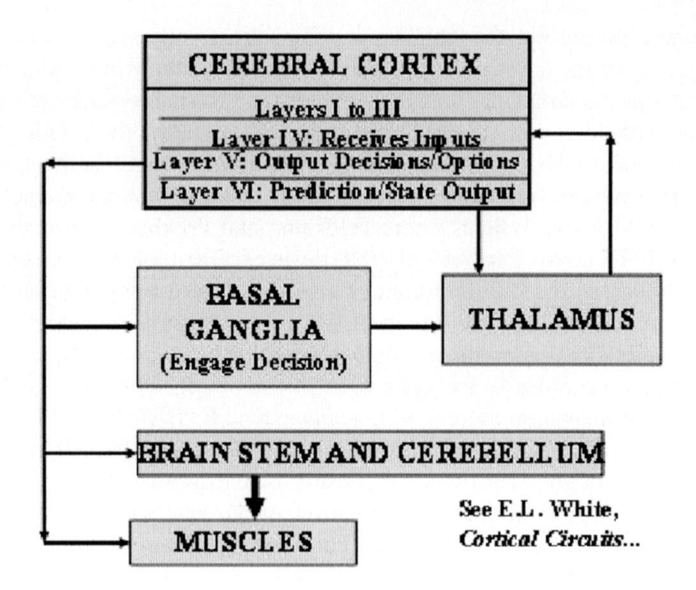

Fig. 6 Third generation view of creativity imagination: Layer V = "Option Networks". See www.werbos.com/WerbosCEC99.htm; important work by Serpen, Pelikan, Wunsch, Thaler, Fu – but still wide open; Widrow testbed

complexity as well. Also, there is new evidence from neuroscience which has not yet been assimilated on the technology side.

The most exciting opportunity before us now is to follow up on more substantial progress and new ideas related to spatial complexity.

An important early clue towards spatial complexity came from the work of Guyon, LeCun, and others at AT&T, illustrated in Fig. 7.

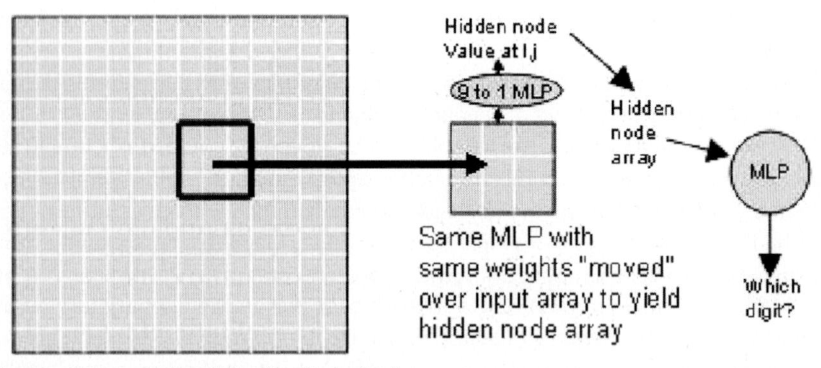

Fig. 7 Moving window net: clue re complexity. Exploiting symmetry of Euclidean translation crucial to reducing number of weight, making large input array learnable

The most accurate ZIP code digit recognizer then came from a simple MLP network, *modified* to exploit symmetry with respect to spatial translation. Instead of independently training hidden neurons to process pieces of an image, they would train a *single* hidden neuron, and *re-use* it in different locations, by moving it around the image. LeCun later called this a "conformal neural network." He had excellent results training it by backpropagation in many image processing tasks. Nevertheless, these feed-forward networks could still not learn the more complex kinds of mappings, like the connectedness mapping described long ago by Minsky [13]; it is not surprising that a network which could not handle connectedness, or learn to emulate Hough relaxation of image data, could not learn how to segment an entire ZIP code.

In 1994, Pang and I demonstrated a network that could solve these problems – a "Cellular SRN," (CSRN), which combines the key capabilities of a Simultaneous Recurrent Network [3] and a "conformal" network. This immediately allows prediction and control and navigation through complex two-dimensional scenes (and images) far more complex than an MLP could ever learn. That did not become immediately popular, in part because the learning was slow and tools were not available to make it easy for people to take advantage of the great brain-like power of such networks. This year, however, Ilin, Kozma and I reported [6] a new learning tool which dramatically speeds up learning, and is available from Kozma as a MatLab tool. This by itself opens the door for neural networks to solve complex problems they could never really handle in the past. (There is great room for research to speed it up even more, but it is ready for practical use already.)

In 1997 [12] and in subsequent tutorials (and a patent), I proposed a more general approach to exploiting symmetry, which I called the ObjectNet. Instead of mapping a complex input field into M rectangular cells, all governed by a common "inner loop" neural network, one may map it into a network of k types of "Objects," with k different types of "inner loop" neural networks. This has great potential in areas like electric power and image processing, for example. A conventional MLP or recurrent network can learn to manage perhaps a few dozen variables in a highly nonlinear system – but how can one design a neural network which inputs the thousands of variables of an entire electric power grid and predict the system as a whole? Object nets provide a way of doing that. The ObjectNet is essentially a functional, explicit way to implement the classic notion from cognitive science of a "cognitive map."

This year, Venayagamoorthy has published preliminary results showing how an ADP system based on a simple, feedforward version of ObjectNet can handle power systems more complex than the earlier first-generation brain-like systems (which already outperformed more conventional control methods). More astonishing – David Fogel used a simple ObjectNet as the Critic in a system adapted to play chess. This was the world's first computer system to achieve master-class performance in chess *without* using a supercomputer and without using detailed clues and advice from a human; it *learned* how to play the game at that level.

But all of this is just a beginning. At `www.face-rec.org`, a series of reviews basically show that two of the three top working systems today rely on neural network concepts (vonderMalsburg and Wechsler). The key to face recognition turns out to be the ability to *learn* new "invariants" or transformations, more complex than simple two-dimensional translation. This offers some easy short-term possibilities: to exploit CSRNs to learn the relevant mappings, which could never be learned before. But it also poses a very fundamental question: how can the *brain* learn such transformations? (The topic of face recognition is also discussed in the chapter by Wong and Cho.)

Here is a new, more formal way to think about what is going on here. The first challenge here is to learn "symmetries of the universe." More concretely, the challenge to the brain is to learn a family of vector maps f_α such that:

$$Pr(f_\alpha(\underline{\mathbf{x}}(t+1)|f_\alpha(\underline{\mathbf{x}}(t)) = Pr(\underline{\mathbf{x}}(t+1)|\underline{\mathbf{x}}(t))$$

for all α and the same conditional probability distribution Pr.

This new concept may be called stochastic invariance. Intuitively, the idea is that the probability of observing $\underline{\mathbf{x}}$(t+1) after observing $\underline{\mathbf{x}}$(t) should be the same as the probability of observing the *transformed* version of $\underline{\mathbf{x}}$(t+1) after observing the *transformed* version of $\underline{\mathbf{x}}$(t). For example, if the probability of observing a red square in the left side of the visual field is 0.25 two seconds after you observe a blue circle in the left side of the visual field, then the probability of observing a red square in the *right* side should also be 0.25 two seconds after you observe a blue circle in the right side, *if* the interchange of left and ride sides is one of the symmetries of nature.

Once a brain learns these symmetries, it may exploit them in one or more of three ways:

1. "Reverberatory generalization": after observing or remembering a pair of data $\{\underline{\mathbf{x}}(t+1), \underline{\mathbf{x}}(t)\}$, also train on $\{f_\alpha(\underline{\mathbf{x}}(t+1)), f_\alpha(\underline{\mathbf{x}}(t))\}$;
2. "Multiple gating": after inputting $\underline{\mathbf{x}}$(t), pick α so as to use f_α to map $\underline{\mathbf{x}}$(t) into some canonical form, and learn a universal predictor form canonical forms. (This is analogous to the Olshausen model, which is *very* different in principle from neuroscience models of spontaneous or affective gating and attention.)
3. "Multimodular gating": like multiple gating, except that multiple parallel copies of the canonical mapping are used in parallel to process more than one subimage at a time in a powerful way.

Human brains seem to rely on the first two, or the second. Perhaps higher levels of intelligence could be designed here. But this begs the question: how could these maps be learned? How could the brain learn to map complex fields into a condensed, canonical form for which prediction is much easier to learn? How can the "Objects" in an ObjectNet be learned? This suggests an immediate and astonishingly simple extension of the ObjectNet theory. In 1992, I proved basic consistency results for a new architecture called the "Stochastic Encoder/Decoder Predictor" (SEDP).[12, Chap. 13]. SEDP *directly* learns condensed mappings. It is an adaptive nonlinear generalization of Kalman filtering, explicit enough to allow the learning of symmetry relations. As with the earlier HDP and CSRN architectures, it will require many specific tricks to improve its learning speed. (e.g., exploitation of nearest neighbor relation in the learning, and salience flows?). It provides a principled way to *learn* the symmetry groups which are the foundation for a principled approach to spatial complexity.

References

1. J. Albus. Outline of intelligence. *IEEE Trans. Systems, Man and Cybernetics*, 21(2), 1991.
2. M. Bitterman. The evolution of intelligence. *Scientific American*, Jan 1965.
3. White & D.Sofge (Eds). Handbook of intelligent control. *Van Nostrand*, 1992.
4. E. A. Feigenbaum and J. Feldman. Computers and thought. *McGraw-Hill*, 1963.
5. D. O. Hebb. Organization of behavior. *Wiley*, 1949.
6. R. Ilin, R. Kozma, and P. J. Werbos. Cellular srn trained by extended kalman filter shows promise for adp. *Proc. Int. Joint Conf. Neural Networks, Vancouver, Canada*, pp. 506–510, Jul 2006.
7. D. Levine and E. Elsberry. Optimality in biological and artificial networks? *Erlbaum*, 1997.
8. W. T. Miller, R. Sutton, and P. Werbos (eds). Neural networks for control. *MIT Press*, 1990.
9. J. Von. Neumann and O. Morgenstern. The theory of games and economic behavior. *Princeton NJ: Princeton U. Press*, 1953.
10. H. Raiffa. Decision analysis. *Addison-Wesley*, 1968.
11. Jennie Si, Andrew G. Barto, Warren Buckler Powell, and Don Wunsch (eds). Handbook of learning and approximate dynamic programming. *Wiley/IEEE*, 2004.
12. Sutton TD and S.Sutton. Learning to predict by the methods of temporal differences,. *Machine Learning*, 3:9–44, 1988.
13. P. Werbos. Multiple models for approximate dynamic programming. *In K. Narendra, ed., Proc. 10th Yale Conf. on Learning and Adaptive Systems. New Haven: K.Narendra, EE Dept., Yale U., 1998. Also posted at www.werbos.com/WhatIsMind.*
14. P. Werbos. The elements of intelligence. *Cybernetica (Namur)*, 3, 1968.
15. P. Werbos. Building & understanding adaptive systems: A statistical/numerical approach to factory automation & brain research. *IEEE Trans. SMC*, Jan/Feb 1987.
16. P. Werbos. The roots of backpropagation: From ordered derivatives to neural networks and political forecasting. *Wiley*, 1994.
17. P. Werbos. Brain-like design to learn optimal decision strategies in complex environments. *In M.Karny et al eds, Dealing with Complexity: A Neural Networks Approach. Springer, London*, 1998.
18. P. Werbos. Stable adaptive control using new critic designs. *Posted at http://arxiv.org as adap-org/9810001*, Oct 1998.
19. P. Werbos. What do neural nets and quantum theory tell us about mind and reality? *In K. Yasue, M. Jibu & T. Della Senta, eds, No Matter, Never Mind : Proc. of Toward a Science of Consciousness. John Benjamins*, 2002.

Appendix: More Links to the Brain and to Human Subjective Experience

Appendix A: The Larger Context

The text of this chapter *mainly* tries to look down on the most basic mammal brain – "the soulless rat" – and understanding it in engineering terms. It tries to unify an optimization approach – and the engineering needed to work out the mathematical and function details – with cognitive or systems neuroscience. In a way, it is providing a pathway for trying to truly unify computational neuroscience with systems and cognitive neuroscience. That specific effort at unification is a great, concrete opportunity for a major new stream of scientific research. My comments about subjective human experience in that main text were really just heuristics to aid understanding. It is important to keep straight what is or is not part of that new stream of science.

At the same time, we as scientists or as humans have a larger agenda. In this appendix, I will talk more about that larger agenda. As humans, we can look at ourselves in the mirror, objectively, using the same scientific tools we would use to look at a mouse or a rat. But we can also look at our own inner experience. The larger challenge here is to arrive at an understanding which can fit *three* sources of empirical data or validation – the data from ordinary experiments on brain and behavior, seen objectively; the data from subjective experience, seen within ourselves; and the data we get from systematically testing the *functional* capabilities of our models/designs when applied to complex tasks, as in engineering. This kind of unified understanding is what we really need in order to better understand ourselves. The chapters by Kozma, Levine and Perlovsky address the higher cognitive functions of the human mind.

But how can we arrive at such a unified understanding, when one side is struggling hard to rise as high as a mouse, and the other is well beyond it?

In parallel to the scientific track, which this chapter has focused on, the new theory opens up a new humanistic track, which is less quantitative for now, but broader in scope. The human mind includes everything I just talked about regarding the "mouse brain," but includes two major extensions or refinements, in my view: (1) a partially evolved new capacity for symbolic reasoning, language and empathy, based in part on "mirror neurons" and new structures in the NRTP (a small but important sheath around the thalamus); and (2) a kind of embryonic collective intelligence effect, which goes far beyond what we easily see in the laboratory. This humanistic track has much to gain from the scientific track, because *everything we see in the mouse brain is also present in us* (even if we do sniff less often).

Again, I will only give a few new highlights here. I have posted a number of papers on various aspects of the scientific and humanistic track at `http://www.werbos.com`. Several of those papers include flowcharts associating components of the brain with components of an integrated ADP design. The chapter by Levine also discusses relations between cognitive models and modules in the brain.

Appendix B: Comparative Cognitive Neuroscience

At the start of this chapter, I briefly referred to the classic, important work of M.E. Bitterman, who demonstrated major *qualitative* differences in intelligence between the reptile, the bird, the mammal, and so on. The "Rockefeller" series on the brain, *The Neurosciences* edited by Schmitt and Worden, also contains a beautiful summary of the differences in the "wiring diagrams" between these different levels of brain.

The key unique feature of the mammal brain is the *six-layer* type of cortex, the "neocortex," which accounts for more than half the weight of a human brain. This new structure basically evolved as a *merger* of two different three-layer cortices from lower organisms. This has important connections to the story of this chapter.

Even the bird brain has an ability to handle spatial complexity far beyond what our engineering systems can do today. After all, complex tasks in image processing and spatial "reasoning" are essential to the life of a bird. Thus I would claim that the Stochastic Encoder/Decoder Predictor (SEDP) structure I mentioned in Sect. 4 exists even within the three-layer cortex of the bird which performs such tasks.

Like Kalman filtering, SEDP tries to estimate or impute a representation of the true state of the external variable, by estimating (or "filtering") its value R_i in real time. For each such estimated variable, SEDP has to consider three different numerical values at each time – a value *predicted* from the past, a value *estimated* by accounting for new sensory data, and a value *simulated* in order to reflect the uncertainty of the estimate. How could a brain structure maintain a coordination of three different numerical values, which fits the required calculations? The obvious way would be to evolve a very large kind of neuron which first computes the prediction in its first section, and then tests the prediction and updates it in a connected large second section. And then adds some random noise at a final, third stage. Strange as this may sound, it fits beautifully with the actual structure of giant pyramid neurons, which are found even in the three-layer cortex of the bird. This also requires a kind of clock pulse to synchronize the calculations, controlling the interface where the prediction enters the estimation part of the neuron; in fact, Scheibel and Scheibel long ago reported such a nonspecific modulation input touching the middle part of the giant pyramid cells in the human six-layer cortex.

This chapter has argued that we can build ADP systems which are very effective in coping with spatial complexity, even before we do all we can with imagination/creativity and temporal complexity. Could it be that nature has already done the same thing, with the bird or even with vertebrates lower than the bird?

Figure on creativity in Sect. 4 illustrates a further idea from [17]. On that slide, you can see an arrow going from Layer V of the neocortex down to the basal ganglia. The idea here is as follows. When the two early three-layer cortices became fused, in the mammal, the *stochastic* aspect of the SEDP design became harnessed to solve an additional task: the task of stochastic search or creativity in making decisions. In effect, layer V of the mammal brain provides a new capability to *suggest options* for decisions. Is it a case where Layer V proposes and the basal ganglia disposes? Perhaps. And perhaps a powerful imagination or creativity is what mammal brains have that lower brains do not. (Lower vertebrates may of course possess simpler, less powerful stochastic exploration mechanisms, similar to the earlier work by Barto, Sutton and Anderson on reinforcement learning, or to the stochastic stabilization mechanisms discussed in the later sections of [18].)

But before that kind of creativity or imagination can evolve, there must first be some wiring to implement the concept of "decision." That is where complexity in time comes into it. Complexity in time is the main focus of [17] and [13]. Some tutorial flow charts and slides about this are also posted, in the later sections of the tutorials posted on my web page. Here again, I will only present some new thoughts.

At the IJCNN05 conference in Montreal, the neuroscientist Petrides presented some important and persuasive new results on the "advanced" parts of the human neocortex. Here is a simplified version of his conclusions: "I have thoroughly reviewed and re-assessed the experiments of my illustrious predecessors, like Goldman-Rakic, and discovered that their interpretations were not quite right. Today, I will speak about the two highest parts of the human cerebral cortex, the dorsolateral prefrontal cortex and the orbitofrontal. It turns out that these two regions are there to allow us to answer the two greatest challenges to the human mind: 'Where did I leave my car this time in the parking lot? And what was I trying to do anyway?' "

More concretely – these parts of the cerebral cortex of the human connect directly to the basal ganglia, where decisions are "engaged." Based on classic artificial intelligence, some neuroscientists once speculated that the brain contains a kind of explicit hierarchy of decisions within decisions, used to structure time into larger and larger intervals. They speculated that different connections to different parts of the basal ganglia might refer to different levels of the temporal hierarchy. But recent studies of the basal ganglia do not fit that idea. Instead, they suggest that the "hierarchy in time" is more implicit and deliberately fuzzy. (I received incredible hate mail from people claiming to represent the American Association for Artificial Intelligence when Lotfi Zadeh and I suggested a need for a mathematically well-defined fuzzy relationship here!) Petrides' work suggests that the connection from dorsolateral cortex to the basal ganglia proposes the "verb" or "choice of discrete decision block type" to the basal ganglia. It also suggests that the "hierarchy" is implicit and fuzzy, based on how one decision may engage others – but may in fact be forgotten at times because of limited computational resources in the brain. (Indeed, in the real world of human life, I have seen many, many examples of people producing poor results, because they seemed to forget what they were really trying to do, or why. People do crash cars and they do crash entire nations.) Perhaps, in the human, the orbitofrontal or another region tends to propose the "object" or parameters of the decision to be made. Yet I would guess that even birds can make decisions (albeit unimaginative decisions), and possess similar connections to their own well-developed basal ganglia from more primitive cerebral cortex.

A discussion of brain evolution would not be complete without some reference to the great evolutionary biologist George Gaylord Simpson. Simpson pointed out that we should not assume that all the attributes of modern mammals evolved at the same time. Looking at the indentations of early mammal skulls, he postulated that the modern mammal brain evolved much later than the other attributes. The evolutionary niche of the fast breeding, teeming mice living under the dinosaurs allowed relatively rapid evolution – but it still required many millions of years, because of the complexity of the wiring that had to be changed. Could it be that the breakthrough in the creativity and imagination of the mouse was what forced the later dinosaurs (like the Maiasaurus) to put more energy into protecting their eggs – and weakened them enough to allow their easy demise when confronted with environmental stress? Humans today, like that early mouse, also represent the early stages of the kind of "quantum break-through" that Simpson discussed.

Appendix C: Links to Subjective Experience

Scientists need to know how to give up empathy, on a temporary and tentative basis, in order to map out and improve what they can learn from laboratory experiments in neuroscience and psychology and engineering as clearly as possible. But if human decision-makers did that all the time, they would be well on track towards blowing up the world. Furthermore, by unifying our subjective experience and our scientific understanding, we can arrive at important new testable insights in all of these realms. When we totally give up empathy, we are basically renouncing the use of our own mirror neurons, and lowering the level of intelligence that we express in our work.

The well-known neuroscientist and psychologist Karl Pribram has often made this point in more graphic terms. "Animal behavior is really all about the four f's – food, fear, ... Any so-called model of human intelligence that doesn't give a central place to them is totally crazy." A "theory of intelligence" that does not address emotions is actually more useful as a specimen of human immaturity and neurosis than as a scientific theory. Isaac Newton – a true scientist – was willing to overcome his squeamishness, and insert a needle into his own eye and observe what he saw, in order to test his new ideas about color and light. To better understand the mind, we need to overcome our own squeamishness and look more forcefully into our own subjective experience. The chapters by Levine and Perlovsky also discuss the central role of emotions as a part of human intelligence.

To understand the full database of human subjective experience, we ultimately need to consider, especially, the work of Sigmund Freud, Carl Jung and the larger streams of "folk psychology" and diverse civilizations found all across the streams of human culture. My web page does make those connections, though of course they are not so concise and compartmentalized as an engineering design can be. The chapter by Perlovsky discusses Jung's concepts of differentiation and synthesis as mechanisms of cultural volition.

The work of Freud is especially important to the theory/model presented here, because it was one of the sources which inspired it in the first place. In my earliest paper on the idea of ADP[14], I proposed that we approximate the Bellman equation somehow to develop a working mathematical design/theory – but I did not know how. I also suggested that Freud's theory of "psychodynamics" might turn out to be right, and to work, if only we could translate it into mathematics. The workable 1971–1972 design shown above – the first true adaptive critic design – was the result of my going ahead and doing what I proposed in that paper. (I hope that some of you, dear reader, will be able to do the same for what I propose here, since I do not have as many years left to me, and have other life-or-death responsibilities as well.) Instead of letting myself be overcome by squeamishness at Freud's fuzzy language, I looked hard and figured out how to translate it into mathematics. That is what backpropagation really is – a translation into mathematics (and generalization) of Freud's concepts of psychodynamics.

Freud had many other insights which are important to the program of research proposed here. For example, when he discusses the "struggle" between the "ego" and the "id," what is he really talking about? In that discussion, his "id" is really just the part of the brain that uses associative memory or nearest-neighbor associations to make predictions. (For example, if someone wearing a blue shirt hits you, when you are a child, you may feel full of fear years later whenever you see a person wearing a blue shirt. That's an id-iotic response.) The "ego" includes a more coherent kind of global cause-and-effect style of prediction, like what we get in engineering when we train Time-Lagged Recurrent Network using the kind of tools that Ford has used, combining back-propagation and DEKF methods. Even when we try to build a "second generation brain" – a brain far simpler than the brain of a bird, let alone a mouse – it is essential that we learn how to combine these "id" and "ego" capabilities together. This is the basis for my proposal for "syncretism," discussed in various papers in my web page. The work to implement and understand syncretism is part of the research described in Sect. 3 – even before we reach so high as the brain of the bird.

Neural network designs used in engineering today are very different from most of the models in computational neuroscience (CNS). The reason is that most people in the CNS niche today rely heavily on Hebbian models, which is really powerful only for creating memory. Some CNS people have suggested that the entire brain is nothing but a memory machine – a suggestion just as absurd as the AI idea of a brain without emotions; it is another example of a theory motivated by a lack of powerful enough mathematical tools to cope with the basic empirical reality which stares us in the face. Engineering today mainly focuses on the larger-scale systems tools which address the "ego" aspect. To build the best possible "second generation" brain, we need to unify these two aspects, exactly as Freud and I have proposed. Those aspects also exist in the highest levels of intelligence in the human mind and human social existence. The chapters

Appendix D: A Humorous Postscript on an Engineering Application

As I finish this chapter, I can imagine someone asking: "Dr. Werbos, are you really trying to tell us that some kind of automated bird brain really could improve on what we are doing today to control the East Coast electric power grid? That is a huge insult to the thousands of brilliant engineers, economists, and lawyers who have devised that system."

Yes, I am saying that a kind of automated bird brain could do better than a lot of what we are doing today, but it is not intended as an insult. Here are some reasons why it is possible.

First, in controlling the power grid, we rely heavily on *automated* systems, in part because events happen too fast for humans to keep up with at the lowest levels. It also helps to be able to communicate at the speed of light. A distributed, automated bird brain would be far more advanced than what we have today at the high-bandwidth aspect of control – if it is properly interfaced with the other parts.

Second, the best system we have today to manage most aspects of the grid is called Optimal Power Flow (OPF). By optimizing the entire system as one large system, it provides integrated results much better than we could get from separately managed independent components all working on their own. OPF is the most widely used university-developed algorithm in use in electric utilities today. But today's OPF lacks the ability to *anticipate* future possible developments. Electric power grids have large and immediate needs to inject more foresight into the automated systems, for many reasons. (e.g. anticipating and avoiding possible "traffic jams," efficient "time of day pricing," shifting sources and loads between day and night, especially as solar power and plug-in hybrid cars become more popular.) Traditional second-generation ADP can provide the necessary foresight, but cannot handle all the complexity. This is why spatial complexity – including ObjectNets – is an essential think to add to ADP. Of course, the nodes in such a distributed network might well need to use new chips, like the Cellular Neural Network chips now being manufactured for use in image processing, in order to make fullest use of the new mathematical designs.

This extension of OPF – called "Dynamic Stochastic OPF" – is discussed at length in chapters by myself and by James Momoh in [11]. (Momoh developed the original version of OPF distributed to electric utilities by the Electric Power Research Institute, which cosponsored the workshop on DSOPF which led to those chapters.)

Neurodynamics of Intentional Behavior Generation

Robert Kozma

Abstract The chapter reviews mechanisms of generation and utilization of knowledge in human cognitive activity and in artificial intelligence systems. First we explore experience-based methods, including top-down symbolic approaches, which address knowledge processing in humans. Symbolic theories of intelligence fall short of explaining and implementing strategies routinely produced by human intelligence. Connectionist methods are rooted in our understanding of the operation of brains and nervous systems, and they gain popularity in constructing intelligent devices. Contrary to top-down symbolic methods, connectionism uses bottom-up emergence to generate intelligent behaviors. Recently, computational intelligence, cognitive science and neuroscience have achieved a level of maturity that allows integration of top-down and bottom-up approaches, in modeling the brain.

We present a dynamical approach to higher cognition and intelligence based on the model of intentional action-perception cycle. In this model, meaningful knowledge is continuously created, processed, and dissipated in the form of sequences of oscillatory patterns of neural activity distributed across space and time, rather than via manipulation of certain symbol system. Oscillatory patterns can be viewed as intermittent representations of generalized symbol systems, with which brains compute. These dynamical symbols are not rigid but flexible and they disappear soon after they have been generated through spatio-temporal phase transitions, at the rate of 4–5 patterns per second in human brains. Human cognition performs a granulation of the seemingly homogeneous temporal sequences of perceptual experiences into meaningful and comprehendible chunks of concepts and complex behavioral schemas. They are accessed during future action selection and decisions. This biologically-motivated computing using dynamic patterns provides an alternative to the notoriously difficult symbol grounding problem and it has been implemented in computational and robotic environments.

1 Prologue: On the Language of the Brain

Invention of digital computers over half a century ago fascinated scientists with enormous opportunities created by this new research tool, which potentially paralleled the capabilities of brains. Von Neumann has been one of the pioneers of this

L. I. Perlovsky, R. Kozma, *Neurodynamics of Cognition and Consciousness.*

new digital computing era. While appreciating potential of computers, he warned about a mechanistic parallel between brains and computers. In his last work about the relationship between computers and brains, he pointed out that the operation of brains can not obey the potentially very high precision of algorithms postulated by Turing machines [1], and thus it is absolutely implausible that brains would use such algorithms in their operations. At higher levels of abstraction, in the last pages of his final work, Von Neumann contends that the language of the brain can not be mathematics [2]. He continues:

> It is only proper to realize that language is a largely historical accident. The basic human languages are traditionally transmitted to us in various forms, but their very multiplicity proves that there is nothing absolute and necessary about them. Just as languages like Greek and Sanskrit are historical facts and not absolute logical necessities, it is only reasonable to assume that logics and mathematics are similarly historical, accidental forms of expression. They may have essential variants, i.e. they may exist in other forms than the ones to which we are accustomed. Indeed, the nature of the central nervous system and of the message systems that it transmits, indicate positively that this is so. We have now accumulated sufficient evidence to see that whatever language the central nervous system is using, it is characterized by less logical and arithmetic depth than what we are normally used to. (Von Neumann, 1958)

If the language of the brain is not mathematics, if it is not a precise sequence of well-defined logical statements such as used in mathematics, then what is it? Von Neumann was unable to elaborate on this question due to his early tragic death. Half a century of research involving artificially intelligent computer designs could not give the answer either. This is partly due to the fact that Von Neumann's warning about principal limitations of the early designs of digital computers, called today "Von Neumann computer architectures" fell on deaf ears. Biological and human intelligence uses different ways of operations from the one implemented in symbol-manipulating digital computers. Nevertheless, researchers excited by seemingly unlimited power of digital computers embarked on projects of building increasingly complex computer systems to imitate and surpass human intelligence, without imitating natural intelligence. These projects gave impressive results, but notoriously fell short of producing systems approaching human intelligence.

The last half a century produced crucial advances in brain research, in part due to advances in experimental techniques. These advances are related to the focus of the present work: spatio-temporal dynamics of neural assemblies in the brain. The major challenge has been to reconcile the apparent contradiction between the absence of symbolic representations in brains as evidenced by neurodynamics and the symbolic nature of higher-level cognition and consciousness. In philosophy of artificial intelligence this is addressed as the symbol grounding problem. The dynamical approach to cognition considers brain as a dynamic system moving along a complex non-convergent trajectory influenced by the subject's past and present experiences and anticipated future events. The trajectory may rest intermittently, for a fraction of a second, at a given spatio-temporal pattern. This pattern may have some meaning to the individual based on previous experiences. In this sense one may call this pattern a representation of the meaning of the given sensory influence in the context of the present internal state. However, the spatio-temporal pattern is unstable. A swift phase transition destroys it and moves the system along the trajectory. In other words, the quasi-stable spatio-temporal patterns can be considered as the words, and the phase transitions among patterns as the grammar of the language of the brain during its never ending cognitive processing cycles. The rest of this chapter describes the mathematical and computational model of the brain neurodynamics implementing the above phenomenological approach.

2 Dynamic Approaches in Artificial Intelligence (AI) Research

2.1 Traditional AI Models

Symbolic approaches to knowledge and cognition proved to be powerful concepts dominating the field from the 60's through the 80's. The physical symbol system hypothesis illustrates key components of the symbolic approach [3, 4, 5]. According to this hypothesis, a physical symbol system has the necessary and sufficient means for intelligent action. In practical terms this means that the types of syntactic manipulation of symbols found in formal logic and formal linguistic systems typify this view of cognition. In this viewpoint, external events and perceptions are transformed into inner symbols to represent the state of the world. This inner symbolic code stores and represents all of the system's long-term knowledge. Actions take place through the logical manipulation of these symbols. This way, solutions are found for current problems presented by the environment. Problem solving takes the form of a search through a problem space of symbols, and the search is performed by logical manipulation of symbols through predefined operations (copying, conjoining, etc.). These solutions are implemented by forming plans and sending commands to the motor system to execute the plans to solve the problem. For an overview, see [6].

According to symbolic viewpoint, intelligence is typified by and resides at the level of deliberative thought. Modern examples of systems that fall within this paradigm include SOAR [7] and ACT-R [8]. The symbolic approach models certain aspects of cognition, and is capable of providing many examples of intelligent behavior. However, challenges to this viewpoint of cognition have appeared, both as practical criticisms of the performance of such systems and more philosophical challenges to the physical-symbol system hypothesis. On the practical side, symbolic models are notoriously inflexible and difficult to scale up from small and constrained environments to real world problems.

If symbolic systems are necessary and sufficient for intelligent behavior, why do we have such problems in producing the flexibility of behavior exhibited by biological organisms? On the philosophical side, Dreyfus' situated intelligence approach is a prominent example of a criticism of symbolism. Dreyfus ascertains, following Heidegger's and Merleau-Ponty's traditions, that intelligence is defined in the context of the environment, therefore, a preset and fixed symbol system can not grasp the essence of intelligence [9]. Pragmatic implementations of situated intelligence find their successful implementations in the field of embodied intelligence and robotics [10].

2.2 Connectionism

The connectionist view of cognition provides an alternative theory of the mind to the symbolic approach. Connectionist models emphasize parallel-distributed processing, while symbolic systems tend to process information in a serial fashion. Connectionist approaches represent adaptive and distributed structures, while symbols are static localized structures. Connectionist models offer many attractive features when compared with standard symbolic approaches. They have a level of biological plausibility absent in symbolic models that allows for easier visualization of how brains might process information. Parallel-distributed representations are robust, and flexible. They allow for pattern completion and generalization performance. They are capable of adaptive learning. In short, connectionist models provide a useful model of cognition, which are in many ways complementary to symbolic approaches. Clark [11] categorizes modern connectionism into three generations, as listed below. We also add the fourth generation as the newest development in the field:

- First-generation connectionism: It began with the perceptron and the work of the cyberneticists in the 50's. It involves simple neural structures with limited capabilities. Their limitations draw criticism by representatives of the symbolist AI school in the 60's, which resulted in abandonment of connectionist principles by mainstream research establishment for decades. The resistance to connectivist ideas is understandable; it is in fact a repetition of millennia-old philosophical shift from nominalism to realism [12]. Connectionism has been revived in the mid 80's, thanks to the activities of the PDP research groups work (among others) on parallel distributed processing [13].
- Second-generation connectionism: It gained momentum since the 80's. It extends first-generation networks to deal effectively with complex dynamics of spatio-temporal events. It involves advanced recurrent neural network architectures and a range of advanced adaptation and learning algorithms. For an overview, see [14, 15].
- Third-generation connectionism: It is typified by even more complex dynamic and time involving properties [16]. These models use biologically inspired modular architectures, along with various recurrent and hard-coded connections. Because of the increasing emphasis on dynamic and time properties, third-generation connectionism has also been called dynamic connectionism. Third generation connectionist models include DARWIN [17, 18], and the Distributed Adaptive Control (DAC) models [19, 20, 21].
- Fourth generation connectionism: The newest development of neural modeling, representing an additional step going beyond Clark's original categorization schema. It involves nonconvergent/ chaotic sequences of spatio-temporal oscillations. It is based on advances in EEG analysis, which gave spatiotemporal amplitude modulation (AM) patterns of unprecedented clarity. The K (Katchalsky) models are prominent examples of this category, which are rooted in intuitive ideas from the 70's [22] and gained prominence since the turn of the century.

This chapter elaborates several models related to fourth generation connectionism. It utilizes repeated phase transitions that re-initialize the beta-gamma carrier oscillations by which the firings of innumerable microscopic neurons in brains are organized into macroscopic spatio-temporal patterns that control behavior. A key to these models is the mesoscopic-intermediate-range paradigm [23]. Accordingly, intelligence in brains is rooted in the delicate balance between local fragmentation of individual components at the cellular level, and overall dominance of a unified global component at the brain/hemisphere level. This balance is manifested through metastable dynamic brain states undergoing frequent state transitions. Phase transitions are crucial components of the new generation of connectionist models as they provide vehicles to produce the seamless sequence of spatio-temporal oscillation patterns punctuated by cortical phase transitions. This is the main focus of the present review.

2.3 Dynamic Logic and Integration of Symbolic and Subsymbolic Methods

Dynamic approach to intelligence creates the opportunity of integrating bottom-up and top-down methods of intelligence. It goes beyond bottom-up connectionist approaches by extracting symbolic knowledge from sub-symbolic structures manifested in the form of spatio-temporal fluctuations of brain activity. Traditional approaches to knowledge extraction and ontology generation in machine learning and trained systems have focused on static systems, extracting grammatical rules from dynamical systems or creating ontologies primarily from text or numerical data databases [16]. These approaches are inherently limited by nature of extracted knowledge representations, which are static structures with no dynamics.

Dynamic Logic [12] is a cognitively motivated mechanism of knowledge emergence from initially vague data. Methods for learning knowledge and ontologies from spatio-temporal neural oscillations (biological or artificial) using Dynamic Logic has the promise to give us the tool to study the language of the brain, and thus move the field of computational intelligence and cognitive science out of its entrapment in local representationalism. Corresponding chapters in this volume authored by Perlovsky and by Freeman elaborate different aspects of this promising interrelationship and its potential for future research. In this chapter, cortical phase transitions are described and their roles in higher cognition and consciousness in biological and artificial systems are outlined.

3 Phase Transitions and Cognition

3.1 Cortical Dynamics and Transitions

Dynamical approach to brain functions gained popularity in the past few decades. The theory of nonlinear systems and chaos provides a mathematical tool to analyze complex neural processes in brains. Katchalsky [24] described spatio-temporal oscillations and sudden transitions in neural systems. Emergence of small number of macroscopic activity patterns as the result of the interaction of huge amount of microscopic components has been attributed to the "slaving principle" Haken's synergetics [25]. Accordingly, the relatively slow macroscopic order parameters "enslave" the faster microscopic elements and produce large-scale spatio-temporal cooperative patterns in a system at the edge of stability. This theory has been developed by Kelso into the concept of metastable brains [26, 27]. Metastability is manifested as rapid transitions between coherent states as the result of the complementary effects of overall coordination and disintegration of individual components. The Haken-Kelso-Bunz (HKB) model gives a coherent phenomenology and mathematical formulation of metastability.

Sudden changes in cortical neural activity have been identified in EEG experiments in EEG data. These transitions may signify temporary stable, quasi-stationary spatial segments of activities [22, 28, 29]. Experiments over the gamma frequency band (20 Hz–80 Hz) indicated sustained quasi-stable patterns of activities for several 100 ms, and over spatial scales comparable to the size of the hemisphere. Some authors consider these as atoms of thought, or mental object [30]. Freeman interpreted these finding using dynamic systems theory. Accordingly, the brain's basal state is a high-dimensional/chaotic attractor. Under the influence of external stimuli, the dynamics is constrained to a lower-dimensional attractor wing. The system stays in this wing intermittently and produces the amplitude modulation (AM) activity pattern. After a few 100 ms, the system ultimately jumps to another wing as it explores the complex attractor landscape. Chaotic itinerancy [31] is a mathematical theory that describes the trajectory of a dynamical system as the one intermittently visiting "attractor ruins" as it traverses across the landscape. Chaotic itinerancy has been successfully employed to interpret key features of EEG measurements.

During the past years, a consensus emerges in the literature about the existence of sudden jumps in measured cortical activities. Lehmann identifies "micro-sates" in brain activity and jumps between them [32]. Rapid switches in EEG activity have been described in [33, 34]. Synchronization of neural electrical activity while completing cognitive tasks is studied in various animals, e.g., in cats, rabbits, gerbils, macaque monkeys [28, 35, 36, 37, 38]. Behavioral correlates of jumps between metastable cortical states have been identified [39]. See also chapters by Kelso and Tognoli, and by Bressler in this volume for details. Comprehensive overview of stability, metastability, and transitions in brain activity is given by [40, 41].

For the purposes of the present work we should address several concepts of potential significance to cognitive phase transitions. One of the most influential theories of consciousness is the global workspace theory [42]. There is striking similarity between the cognitive content of phase transitions identified in EEGs [43] and the act of conscious broadcast in Baars' global workspace theory. It can be hypothesized that cortical phase transitions are in fact manifestations of such conscious broadcast events. Brains can be viewed as systems far from equilibrium [25, 44]. In that approach, state transitions occur in systems far from equilibrium as they approach criticality by changing an order parameter. Critical state transition is manifested by slowing the rate of change in the order parameter and by increased amplitude of oscillations in the output.

3.2 Example of Transitions in EEG

The experimental procedures by which the electrode arrays were surgically implanted, the rabbits were trained, and the EEG signals were recorded and stored are documented by [35]. EEG recordings from an array of $8x8$ electrodes implanted on the visual, auditory or somatosensory cortices of trained rabbits have been band-pass filtered over the gamma range (20–80 Hz). Each subject was trained in an aversive classical conditioning paradigm with 20 trials using a reinforced conditioned stimulus (CS+) and 20 trials of an unreinforced conditioned stimulus (CS-) in each session, all with correct conditioned responses.

The amplitudes of the gamma oscillations formed spatially distributed amplitude-modulation (AM) patterns, which showed correlation with the presence of conditioned or unconditioned stimuli. This indicated that the AM patterns carry important information on the discrimination and categorization task of the rabbits. In other words, the categories are manifested in the form of spatially synchronous AM patterns. The synchrony between pairs of EEG records can be measured by various methods. FFT-based analysis is widely used for this purpose. When calculating FFT, a relatively long window function must be applied in order to achieve proper statistical accuracy over the gamma band. Such a broad window, often 50 ms or longer, may smooth over brief transient effects, leading to difficulties in detecting them.

Hilbert transform has the potential to avoid these difficulties, and it has been applied successfully to EEG analysis in the past few years [40, 45, 46]. Hilbert analysis introduces the analytic phase $P_j(t)$ and amplitude $A_j(t)$ measures, and phase differences at the given frequency band. Here t is the time and j is the channel index, distributed across space, $j = 1, \ldots, 64$. For details of the Hilbert transform method as applied to EEG signals, see chapter by Freeman in this volume.

Analytic phase differences between successive temporal instances are shown in Fig. 1, as the function of time (y axis) and channel number (x axis). In the figure the channels are rearranged into a linear array, for the sake of simplicity, although the experiments have been performed on an 8×8 array. Figure 1 shows that for certain time segments, the phase increase has a low, relatively constant level over the entire array. For example, between 200 to 500 steps the phase differences are uniformly low. This is an example of uniformly low phase plateau. During the plateaus, a sustained and synchronized AM pattern is maintained, indicating the emergence of category formation by the subject. At some other time intervals, e.g., at time step 100 and 600, the phase differences are highly variable across the array and cover the range between 0 and $2 * \pi$. The plateaus of constancy in phase difference were separated by jumps and dips with variable phase values. Our interpretation was that the jumps and dips manifested discontinuities in the phase demarcating phase transitions in the cortical dynamics.

3.3 Modeling Phase Transitions in Physics and Biology

Studying phase transitions gains increasing popularity in various research fields beyond physics, including population dynamics, the spread of infectious diseases,

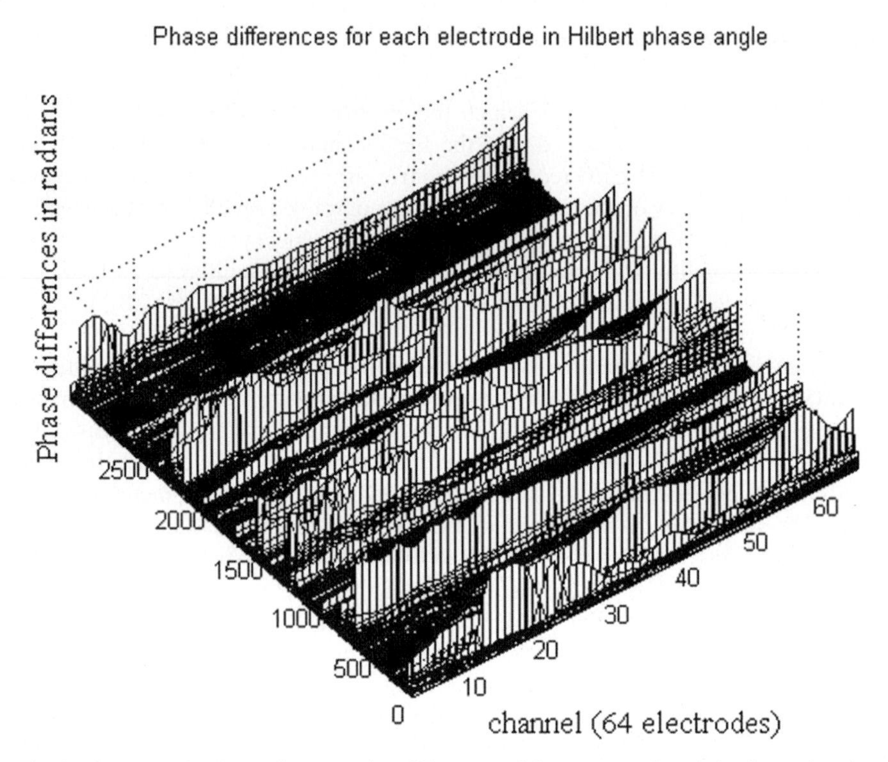

Fig. 1 The raster plot shows the successive differences of the unwrapped analytic phase, changing with channel serial number (right abscissa) and time (left abscissa; the 3000 time steps correspond to 6 s duration of the experiment). This figure is reconstructed from [47]

social interactions, neural systems, and computer networks [48, 49, 50]. Neural signal propagation through axonal effects has velocity in the order of 1–2 m/s, and supports synchronization over large areas of cortex [39, 43, 51]. This creates small-world effects [52, 53] in analogy to the rapid dissemination of information through social contacts. "Small-world" networks can be generated by certain preferential attachment rules between vertices. The number of edges linked to a node is also called its degree. Preferential attachment rule simply states that, during the network evolution, new nodes will be connected to the old nodes in proportion to the degree of the old nodes. As the result, a highly connected node (= a node with high degree) is likely to get even mode connections. This rule can lead to a special hub structure, which makes the network diameter smaller than a pure randomly generated network.

In certain networks, like the www and biological systems, the degree distribution follows a power law, i.e., it is scale-free [54, 55]. Scale-free distribution can be produced by properly specified preferential attachment rule, but a lot of other ways as well. A consistent mathematical theory of scale-free networks is given by [56]. The importance of long-distance correlations has been emphasized by numerous brain theorists [57, 58, 59, 60, 61, 62, 63]. Phase transitions studied by the neuropercolation model offer a new perspective on scale-free behavior in small-world networks relevant to brain functions [64, 65].

Phase transitions between chaotic states constitute the dynamics that explains how brains perform such remarkable feats as abstraction of the essentials of figures from complex, unknown and unpredictable backgrounds, generalization over examples of recurring objects, reliable assignment to classes that lead to appropriate actions, planning future actions based on past experience, and constant updating by way of the learning process [28].

EEG analysis gave spatiotemporal amplitude modulation (AM) patterns of unprecedented clarity [43]. It supported the theory of self-organized criticality (SOC) in neural dynamics [66, 67, 68]. SOC indicates that brains maintain themselves at the edge of global instability by inducing a multitude of small and large adjustments in the form of phase transitions. Phase transitions mean that each adjustment is a sudden and irreversible change in the state of a neural population. SOC provides an internal clock that repeatedly re-initializes the beta-gamma carrier oscillations by which the firings of innumerable microscopic neurons in brains are organized into macroscopic spatiotemporal patterns that control behavior. The meaning of the macroscopic pattern is acquired through the intentional action of the subject, as it will be described in the example of the K (Katchalsky) model based intentional system development.

4 K Architectures and Their Function

4.1 Historical Notes

We propose a hierarchical approach to spatio-temporal neurodynamics, based on K sets. Low-level K sets were introduced by Freeman in the 70's, named in the honor of Aharon Kachalsky, an early pioneer of neural dynamics [22]. K sets are multi-scale models, describing increasing complexity of structure and dynamical behaviors. K sets are mesoscopic models, and represent an intermediate-level between microscopic neurons and macroscopic brain structures. The basic K0 set, for example, describes the dynamics of a cortical microcolums with about 10^4 neurons. K-sets are topological specifications of the hierarchy of connectivity in neuron populations. The dynamics of K-set are modeled using a system of nonlinear ordinary differential equations with distributed parameters. K-dynamics expressed in ODEs predict the oscillatory waveforms that are generated by neural populations. K-sets describe the spatial patterns of phase and amplitude of the oscillations, generated by components at each level. They model observable fields of neural activity comprising EEG, LFP, and MEG.

K sets consist of a hierarchy of components of increasing complexity, including K0, KI, KII, KIII, KIV, and KV systems. They model the hierarchy of the brain starting from the mm scale to the complete hemisphere. Low-level K-sets, up to and including KIII, have been studied intensively in the mid 90's and they model sensory cortices. They have been applied to solve various classification and pattern recognition problems [69]. KIII sets are complex dynamics systems modeling the classification in various cortical areas, having typically hundreds of degrees of freedom. In early applications, KIII sets exhibited extreme sensitivity to model parameters which prevented their broad use in practice [70]. In the past decade systematic analysis has identified regions of robust performance [23], and stability properties of K-sets have been derived [71, 72]. Today, K sets are used in a wide range of applications, including detection of chemicals [73], classification [74, 75], time series prediction [76], and robot navigation [6, 77]. Recent development includes the KIV sets [62] which model intentional behavior. They are applicable to autonomous control [78, 79, 80], and they are the major focus of the present essay.

4.2 Hierarchy of K Sets

4.2.1 KO Set

The KO set represents a non-interacting collection of neurons. It models dendritic integration in average neurons and a sigmoid static nonlinearity for axon transmission. They are described by a state-dependent, linear, 2nd order ordinary differential equation (ODE). The KO set is governed by a point attractor with zero output and stays at equilibrium except when perturbed. KO models a neuron population of about 10^4 neurons. The second order ordinary differential equation describing it is written as:

$$(ab)\frac{d^2P(t)}{dt^2} + (a+b)\frac{dP(t)}{dt} + P(t) = F(t). \tag{1}$$

Here a and b are biologically determined time constants. $P(t)$ denotes the activation of the node as function of time. $F(t)$ includes a nonlinear mapping function $Q(x)$ acting on the weighted sum of activation from neighboring nodes and any external input. The sigmoid function $Q(x)$ is given by the equation:

$$Q(x) = q\{1 - exp(-\frac{1}{q(e^x - 1)})\} \tag{2}$$

In (2), q is the parameter specifying the slope and maximal asymptote of the curve. This sigmoid function is modeled from experiment on biological neural activation [22]. Biological motivation of the $Q(x)$ transfer function, as well its illustration is given in the chapter by Freeman in this volume.

4.2.2 KI Sets

A KI set represents a collection of K0 sets, which can be either excitatory or inhibitory units. At the KI level it is important to have the same type of units in the system, so we have no negative feedback. Accordingly, we can talk about excitatory or inhibitory KI sets, i.e., KI_E and KI_I, respectively. As a result, the dynamics of a KI is described a simple fixed point convergence. If KI has sufficient functional connection density, then it is able to maintain a non-zero state of background activity by mutual excitation (or inhibition). KI typically operates far from thermodynamic equilibrium. The stability of the KI_E set under impulse perturbation is demonstrated using the periglomerular cells in the olfactory bulb [81]. Its critical contribution is the sustained level of excitatory output. Neural interaction by stable mutual excitation (or mutual inhibition) is fundamental to understanding brain dynamics.

4.2.3 KII Sets

A KII set represents a collection of excitatory and inhibitory cells, KI_E and KI_I. KII has four types of interactions: excitatory-excitatory, inhibitory-inhibitory, excitatory-inhibitory, and inhibitory-excitatory. Under sustained excitation from a KI_E set but without the equivalent of sensory input, the KII set is governed by limit cycle dynamics. With simulated sensory input comprising an order parameter, the KII set undergoes a state transition to oscillation at a narrow band frequency in the gamma range.

4.2.4 KIII Sets

The KIII model consists of several interconnected KII sets, and it models a given sensory system in brains, e.g., olfactory, visual, auditory, somatosensory modality. It has been shown that KIII can be used as an associative memory which encodes input data into nonconvergent spatio-temporal oscillations [69, 23]. KIII generates aperiodic, chaotic oscillations with $1/f^2$ spectra. The KIII chaotic memories have several advantages as compared to convergent recurrent networks: (1) they produce robust memories based on relatively few learning examples even in noisy environment; (2) the encoding capacity of a network with a given number of nodes is exponentially larger than their convergent counterparts; (3) they can recall the stored data very quickly, just as humans and animals can recognize a learnt patter within a fraction of a second.

4.2.5 KIV Sets

A KIV set is formed by the interaction of 3 KIII sets. It is used to model the interactions of the primordial vertebrate forebrain in the genesis of simple forms of intentional behavior [82, 62]. KIV models provide a biologically feasible platform to study cognitive behavior associated with learning and action-perception cycle, and as such will be the focus of this review. Figure 2 illustrates the hierarchy of K sets.

4.3 Basic Principles of Neurodynamics

The hierarchical K model-based approach is summarized in the 10 "Building Blocks" of neurodynamics [22, 29]:

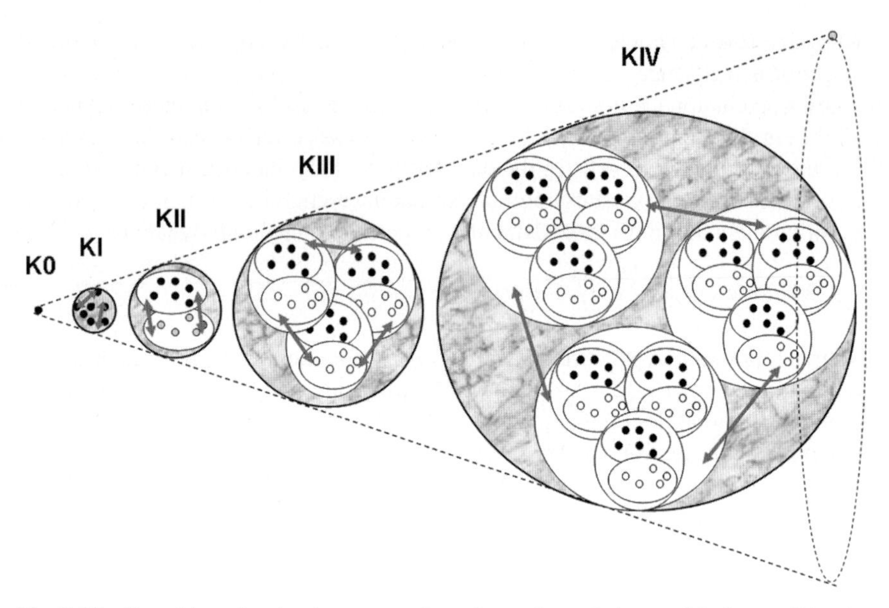

Fig. 2 The K-set hierarchy showing progression of neural population models from cell level to hemisphere wide simulation, K0 through KIV. The progression of models of increasing complexity follows the organizational levels of brains. K0: is a non-interacting collection of neurons described by a state-dependent, linear ODE for dendritic integration and a sigmoid static nonlinearity for axon transmission. K0 is governed by a point attractor with zero output and stays at equilibrium except when perturbed. KI corresponds to a cortical column. It represents a collection of excitatory or inhibitory neurons that has sufficient functional connection density to maintain a state of background activity by mutual excitation or mutual inhibition. KII represents a collection of excitatory and inhibitory populations, which have positive and negative feedback connections. KII can exhibit limit cycle periodic oscillations at a narrow band frequency in the gamma range. KIII is formed by the interaction of several KII sets through long axonal pathways with distributed delays. It simulates the known dynamics of sensory areas that generate aperiodic, chaotic oscillations with $1/f^2$ spectra. KIV is formed by the interaction of 3 KIII sets. It models the hemisphere with multiple sensory areas and the genesis of simple forms of intentional behaviors [82]

1. Non-zero point attractor generated by a state transition of an excitatory population starting from a point attractor with zero activity. This is the function of the KI set.
2. Emergence of damped oscillation through negative feedback between excitatory and inhibitory neural populations. This is the feature that controls the beta-gamma carrier frequency range and it is achieved by KII having low feedback gain.
3. State transition from a point attractor to a limit cycle attractor that regulates steady state oscillation of a mixed E-I KII cortical population. It is achieved by KII with sufficiently high feedback gain.
4. The genesis of broad-spectral, aperiodic/chaotic oscillations as background activity by combined negative and positive feedback among several KII populations; achieved by coupling KII oscillators with incommensurate frequencies.

5. The distributed wave of chaotic dendritic activity that carries a spatial pattern of amplitude modulation AM in KIII.
6. The increase in nonlinear feedback gain that is driven by input to a mixed population, which results in the destabilization of the background activity and leads to emergence of an AM pattern in KIII as the first step in perception.
7. The embodiment of meaning in AM patterns of neural activity shaped by synaptic interactions that have been modified through learning in KIII layers.
8. Attenuation of microscopic sensory-driven noise and enhancement of macroscopic AM patterns carrying meaning by divergent-convergent cortical projections in KIV.
9. Gestalt formation and preafference in KIV through the convergence of external and internal sensory signals leading to the activation of the attractor landscapes leading to intentional action.
10. Global integration of frames at the theta rates through neocortical phase transitions representing high-level cognitive activity in the KV model.

Principles 1 through 7 has been implemented in KIII models and applied successfully in various identification and pattern recognition functions. They serve as the basic steps to create the conditions for higher cognition. Principles 8 and 9 reflect the generation of basic intentionality using KIV sets which is the target of the present overview. Principle 10 expresses the route to high-level intentionality and ultimately consciousness, which is not addressed at present.

5 Construction of Intentional Dynamic Systems at KIV Level

5.1 Aspects of Intentionality

Key features of intentionality as the manifestation of high-level intelligent behavior in humans and animals can be summarized as follows. Intelligence is characterized by the flexible and creative pursuit of endogenously defined goals. Humans and animals are not passive receivers of perceptual information and hey actively search for sensory input. In the process they complete the following sequence [83]:

- Forming hypotheses about expected states the individual may face in the future;
- Expressing the hypotheses as meaningful goals to be aimed at;
- Formulating a plan of action to achieve the goals;
- Informing the sensory and perceptual apparatus about the expected future input, which is a process called re-afference;
- Act into the environment in accordance with the action to achieve the formulated goals;
- Manipulating the sensors, adjust their properties and orientations to receive the sensory data;

- Generalize and categorize the sensory data and combine them into multisensory percepts called Gestalts;
- Verify and test the hypotheses, and update and learn the brain model to correspond better to the perceived data.
- Form new/updated hypotheses and continue the whole process again.

The cyclic operation of prediction, testing by action, sensing, perceiving, and assimilation is called intentionality. The significance of the dynamical approach to intelligence is emphasized by our hypothesis that nonlinear dynamics is a key component of intentional behavior in biological systems [84]. Therefore, understanding dynamics of cognition and its relevance to intentionality is a crucial step towards building more intelligent machines [85]. Specifically, nonconvergent dynamics continually creates new information as a source of novel solutions to complex problems. The proposed dynamical hypothesis on intentionality and intelligence goes beyond the basic notion of goal-oriented behavior, or sophisticated manipulations with symbolic representations to achieve given goals. Intentionality is endogenously rooted in the agent and it can not be implanted into it from outside by any external agency. Intentionality is manifested in and evolved through the dynamical change in the state of the agent upon its interaction with the environment.

5.2 Design of the Intentional KIV System

In this section, we describe the structure and operation of the intentional KIV model based on [62, 82]. KIV has 4 major components. Three of the components are the KIII sets as outlined in the previous section, i.e., the cortex, the hippocampal formation, and the midline forebrain. They are connected through the coordinating activity of the amygdala to the brain stem (BS) and the rest of the limbic system. Figure 3 illustrates the connections between components of KIV. The connections are shown as bidirectional, but they are not reciprocal. The output of a node in a KII set is directed to nodes in another KII set, but it does not receive output from the same nodes but other nodes in the same KII set. In Fig. 3 three types of sensory signals can be distinguished. Each of these sensory signals provides stimulus to a given part of the brain, namely the sensory cortices, midline forebrain (MF), and the hippocampal formation (HF), respectively. The corresponding types of sensory signals are listed below:

- Exteroceptors;
- Interoceptors (including proprioception);
- Orientation signals.

The environmental sensory information enters through broad-band exteroceptors (visual, audio, somatosensory, etc.) and processed in the OB and the PC in the form of a spatially distributed sequence of AM patterns. These AM patterns

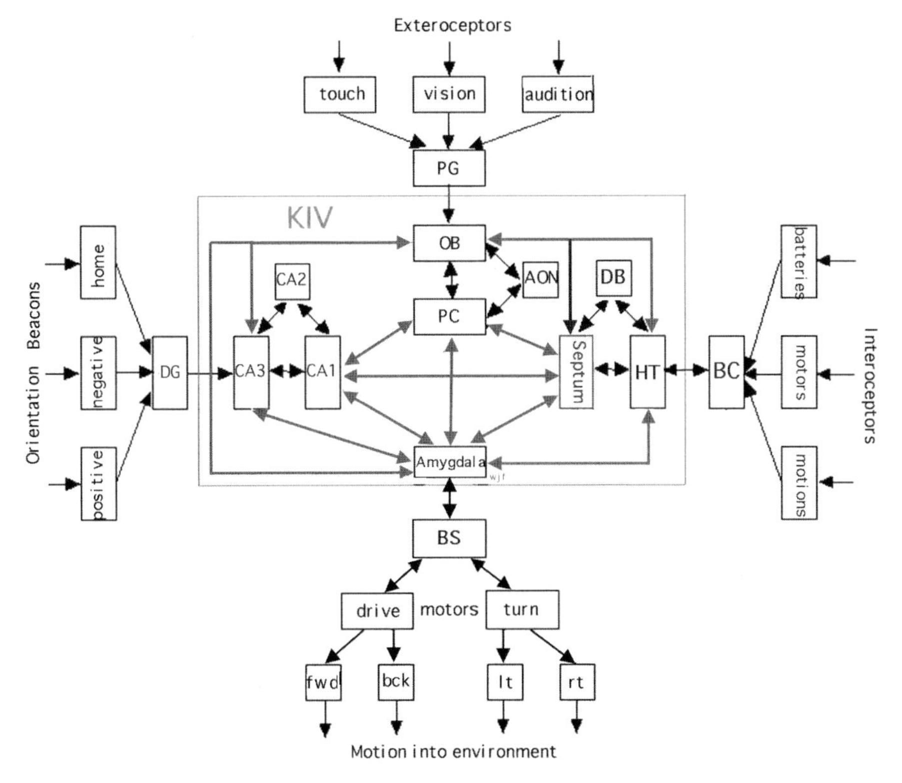

Fig. 3 KIV model of the brain, which consists of three KIII sets (cortex, hippocampal formation, and midline forebrain), the amygdala striatum and BS, brain stem. The amygdala is a KII set, while the brain stem and drives are conventional. The sparse long connections that comprise the KIV set are shown in blue. They are shown as bidirectional, but they are not reciprocal. The convergence location and output are provided by the amygdala, including its corticomedial and basolateral nuclear parts. In this simplified model with no autonomic nervous system, the amygdala provides the goal-oriented direction for the motor system, that is superimposed on local tactile and other protective reflexes [82]

are superimposed on the spatial maps in CA3 that are derived from the orientation beacons. In the present model separate receptor arrays are used for simplicity of receptor specialization and internal connectivity. The cortical KIII system initiates the function of pattern recognition by the agency of sensory input-induced destabilization of high-dimensional dynamics. The input is gated by the septal generator (like a sniff or saccade). This actualizes an attractor landscape formed by previous experience in the OB/PC, which in our model is the common sensorium. The hippocampal KIII system, on the other hand, uses the classification embodied in the outputs of the OB and PC as its content-laden input, to which the DG contributes the temporal and spatial location of environmental events. These events contributed to the previously learned relationships between the expectations and the experience of the system in search of its assigned goal.

A component of the integrated KIV system, the Midline Forebrain formation, receives the interoceptor signals through the basal ganglia, and processes them in the hypothalamus and the septum. MF provides the value system of the KIV, using information on the internal goals and conditions in the animal. It provides the "Why?" stream to the amygdala, which combines this with the "What?" and "Where?" information coming from the cortex and the hippocampus to make a decision about the next step/action to be taken. MF is also a KIII unit, which contributes to the formation of the global KIV coherent state. The coherent KIV state evolves through a sequence of metastable AM patterns, which is also described as the cinematographic principle [43] of brain operation.

5.3 Learning in KIV Models

In order to use the arrays of K sets as novel computational and memory devices, we need to study the effect of learning on the system dynamics. In particular, we need to describe the role of learning and adaptation on phase transitions. The operation of the dynamic memory can be described as follows. In the absence of stimuli the system is in a high dimensional state of spatially coherent basal activity. The basal state is described by an aperiodic (chaotic) global attractor. In response to an external stimulus, the system is kicked out of the basal state into a local memory basin of an attractor wing. This wing is inferred to be of lower dimension than the basal attractor giving spatially patterned amplitude modulation (AM) of the coherent carrier wave. The system resides in this localized wing approximately for the duration of the stimulus and it returns to the basal state after the stimulus ends. This temporal burst process is given a duration of about 200 ms. The system memory is defined as the collection of basins and attractor wings, and a recall is the induction by a state transition of a spatiotemporal gamma oscillation with a spatial AM pattern. Three learning processes are defined [23]:

- Hebbian reinforcement learning of stimulus patterns; this is fast and irreversible;
- Habituation of background activity; slow, cumulative, and reversible;
- Normalization of nodal activities to maintain overall stability; very long-range optimization outside real time.

Various learning processes exist in a subtle balance and their relative importance changes at various stages of the memory process. In the framework of this study, stable operation of the network is studied without using normalization. Habituation is an automatic process in every primary sensory area that serves to screen out stimuli that are irrelevant, confusing, ambiguous or otherwise unwanted. It constitutes an adaptive filter to reduce the impact of environmental noise that is continuous and uninformative. It is continually up-dated in a form of learning, and it can be abruptly canceled ("dis-habituation") by novel stimuli and almost as quickly reinstituted ("re-habituation") if the novel stimuli are not reinforced. It is a central process that does not occur at the level of sensory receptors. It is modeled here by incremental weight decay that decreases the sensitivity of the KIII system to stimuli from the environment that are not designated as desired or significant by accompanying reinforcement. It is expected that learning effect will contribute to the formation of convoluted attractor basins, which facilitate phase transitions in the dynamical model at the edge of chaotic activity.

Learning takes place in the CA1 and CC units of the hippocampus and cortex, respectively. We have two types of learning: Hebbian correlation learning, and habituation. Hebbian learning is paired with reinforcement, reward or punishment; i.e., learning takes place only if the reinforcement signal is present. This is episodic, not continuous, long-term and irreversible. Habituation, on the other hand results in continuous degradation of the response of a cell in proportion of its activity, unless reinforced by long-term memory effects. Note that the KII sets consist of interacting excitatory and inhibitory layers, and the lateral weights between the nodes in the excitatory layers are adapted by the learning effects; see Fig. 4 for illustration.

5.4 Simulating the Operation of the KIV Model

Experiments have been designed and implemented in computer simulation to demonstrate the potential of KIV operating on intentional dynamic principles. In the

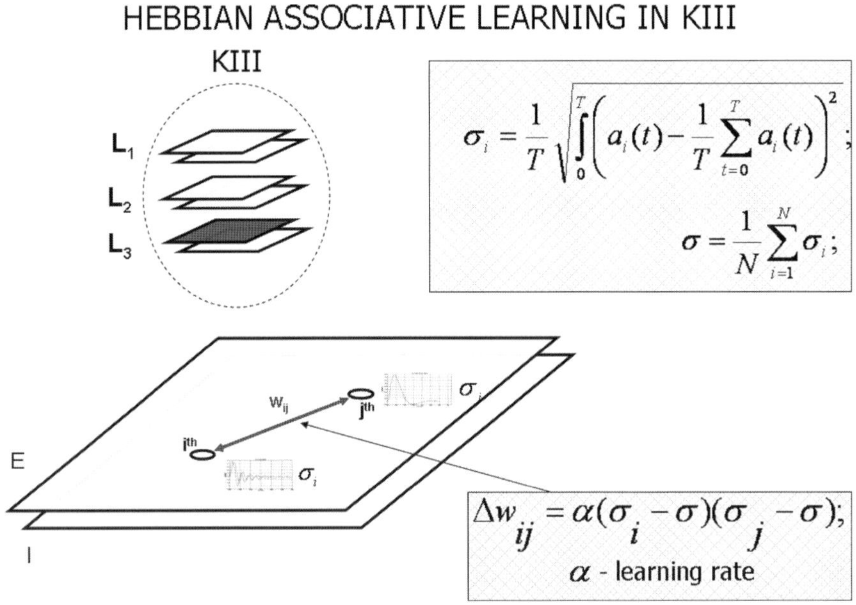

HEBBIAN ASSOCIATIVE LEARNING IN KIII

KIII

L1

L2

L3

$$\sigma_i = \frac{1}{T}\sqrt{\int_0^T \left(a_i(t) - \frac{1}{T}\sum_{t=0}^{T} a_i(t)\right)^2};$$

$$\sigma = \frac{1}{N}\sum_{i=1}^{N}\sigma_i;$$

E

I

W_{ij}

j^{th}

i^{th}

σ_j

σ_i

$$\Delta w_{ij} = \alpha(\sigma_i - \sigma)(\sigma_j - \sigma);$$

α - learning rate

Fig. 4 Illustration of Hebbian correlation learning at the bottom excitatory layer of the KIII set. Due to the nonconvergent nature of typical K sets dynamics, the average error σ_i is calculated for each channel over a time window of T; T is typically 50 to 100 ms in the experiments; based on [76]

experiments we used an autonomous agent moving in a 2-dimensional environment. During its movement, the agent continuously receives two types of sensory data: (1) distance to obstacles; (2) and orientation toward some preset goal location. KIV makes decisions about its actions toward the goal. The sensory-control mechanism of the system is a simple KIV set consisting of 2 KIII sets [79, 86, 87].

The applied KIV set is illustrated in Fig. 5, which has 3 KIII sets: two KIIIs for sensory cortices and a KIII for hippocampus. In the experiment only one sensory cortex is used, which fulfils the functions of the visual cortex. Orientation and visual sensory input are processed in the model hippocampus and cortex, respectively. Each box in Fig. 5 stands for a KI or KII set. For uniformity, we use the terminology of olfaction in cortices for every sensory modality, in conformity with Freeman's original motivation of KIII as the model of the olfactory system. The full KIV model has interoceptors, which are directed to the internal state. In the simplified approach shown in Fig. 5, we do not incorporate interoceptors.

The spatio-temporal dynamics of this system shows sudden changes in the simulated cortical activity, which is in agreement with properties of metastable AM patterns observed in EEG data. An example of the calculated analytical phase difference is shown on Fig. 6, for a simulated period of 4 s and an entorhinal array of consisting of 80 nodes. The intermittent desynchronization is clearly seen at a rate of several times per seconds. These results indicate that the KIV model is indeed a

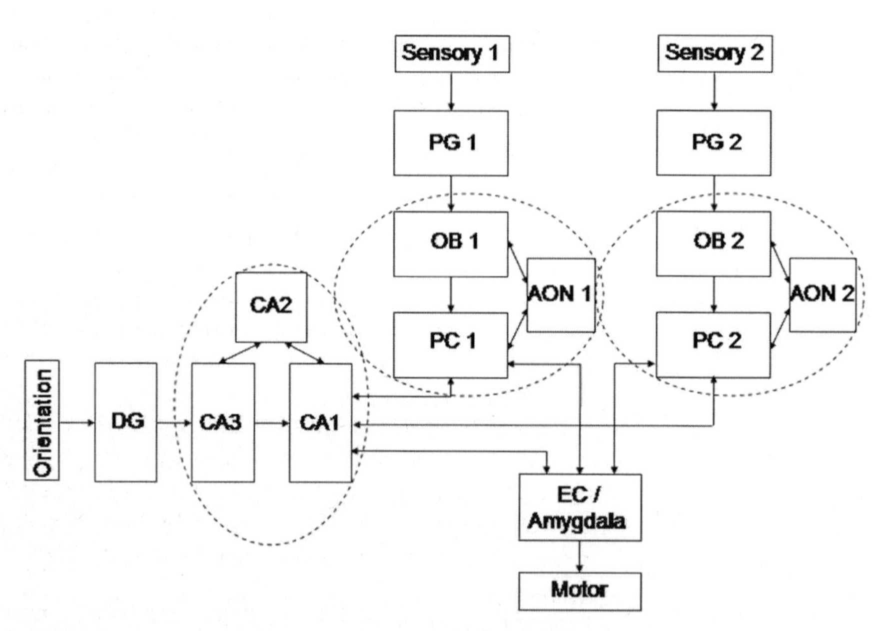

Fig. 5 Schematic view of the KIV model of the cortico-hippocampal formation with subcortical projections for motor actions. KIV consists of KIII sets for the hippocampal formation, 2 sensory cortices, link to motor system coordinated by the entorhynal cortex influenced by amygdala striatum. Notations: PG - periglomerular; OB - olfactory bulb; AON - anterior olfactory nucleus; PC - prepyriform cortex; HF - hippocampal formation; DG - dentate gyrus; CA1, CA2, CA3 - sections of the hippocampus. Dashed circles indicate basic KIII units

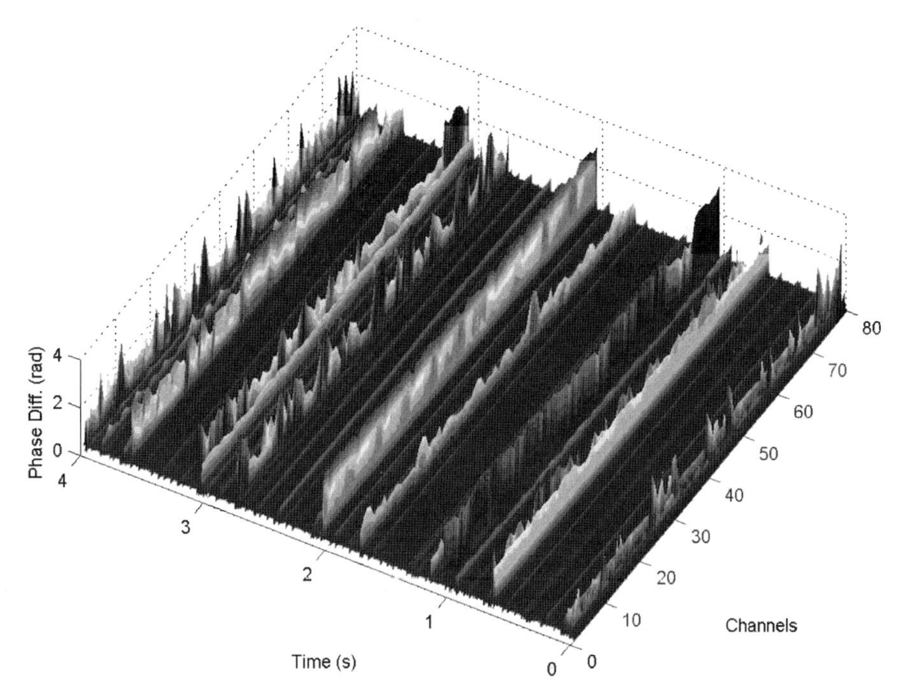

Fig. 6 Illustration of simulations with a K set dynamic memory model of the hemisphere; phase differences in the entorhinal cortex across time and space. The intermittent desynchronization for large part of the array is clearly seen [87]

suitable level of abstraction to grasp essential properties of cortical phase transitions as evidenced in intracranial and scalp EEG and MEG data.

Large-scale sychronization in the cortex, interrupted intermittently by short periods of desynchronization through phase transitions, is an emergent property of the cortex as a unified organ. The intermittent synchronization-desynchronization cycle is a neurophyisiological correlate of intentionality and consciousness. The KIV model is capable of demonstrating this intentional dynamics. KIV is a candidate of implementing intentionality is artificial systems.

Intensive studies are conducted towards the interpretation of the content of the metastable AM patterns between the acts of phase transitions, as building blocks of intentional action. Dynamic logic is an important modeling tool [12] which has been used for system identification in very noisy data with high level of clutter. Dynamic logic, when applied to the analysis of multi-channel EEG data, can provide optimum estimation of AM patterns modeled through the emergence of multiple phase cones [88]. Modeling is done in the context of the given level of knowledge available on the behavior of the autonomous system and its interaction with the environment. Starting with a vague initial model, the description is iteratively improved, leading to more and more precise models. Ultimately a model state is achieved which gives the best possible identification to the problem. Dynamic logic theory shows that such iterative improvements require just a small number of iterations, utilizing a sequence of hierarchical models evolving from vague to precise. Integrating dynamical logic and K sets methodology has the potential to provide a breakthrough in identifying the cognitive meaning of metastable AM patterns, leading toward the identification of the language of the brain.

6 Intentional Dynamics for Robot Control

6.1 Biologically-inspired Approaches to Navigation and Control

Biologically-inspired neural architectures are widely used for control of mobile robots and demonstrated robust navigation capabilities in challenging real life scenarios. These approaches include subsumption methods [89], BISMARC - Biologically Inspired System for Map-based Autonomous Rover Control [90, 91], ethology inspired hierarchical organizations of behavior [92], behavior-based control algorithm using fuzzy logic [93]. Brain-like architectures and modeling brain activity is an increasingly popular area of intelligent control, including learning cognitive maps in the hippocampus [94, 95], the role of place cells in navigation [96], visual mapping and the hippocampus [97, 98, 99], learning in the cortico-hippocampal system [86]. Due to the high level of detail in reproducing the parts of the brain involved in navigation the behavioral analysis includes simple navigational tasks.

The approach using K-sets as dynamical memories is motivated by experimental findings related to activity patterns of population of neurons. Unlike other models, K models are based on experimental findings related to the mesoscopic dynamics of brain activity. The model of navigation includes the activity of the hippocampus and the cortex. As in other approaches, it is assumed that the hippocampus processes information related to global landmarks, and the cortex deals with information received from the vicinity of the agent. An important difference between K model and other biologically-inspired control architectures is the way how information is processed. K models encode information in AM patterns of oscillations generated by the Hebbian synapses in layers of the KIII subsystems, the other models use firing rate and Hebbian synapses [95, 100]. Another important aspect of the models is the learning procedure used for goal finding. Despite these differences, the performance of K model matches or exceeds that of the standard approaches [77, 86].

6.2 SRR2K Robot Testbed

We demonstrate the operation of the KIV system for on-line processing of sensory inputs and onboard dynamic behavior tasking using SRR2K (Sample Return Rover) platform at the Planetary Robotics indoor facility of JPL. The experiments illustrate robust obstacle avoidance combined with goal-oriented navigation by the SRR2K robot. Detailed description of the experiments is given in [80]. Experiments are conducted at the indoor facility of the Planetary Robotics Group, JPL.

It includes an approximately 5×5m irregularly shaped test area covered by sand and rocks imitating natural exploration environments. The terrain layout is variable from smooth surface for easy advance to rough terrain with various hills and slopes posing more challenges to SRR2K traversing through it. The lighting conditions are adjustable at need.

SRR2K is a four-wheeled mobile robot with independently steered wheels and independently controlled shoulder joints of a robot arm; see Fig. 7. Its mass is 7 kg, and the maximum power use during fast movement (30–50 cm/s) is around 35 W can only be sustained for about 6 h without recharging the batteries. In the small experimental environment in this study, no large distances are traveled, so the battery capacity is not an actual limitation for us.

The primary sensing modalities on SRR2K include: (1) a stereo camera pair of 5 cm separation, 15 cm of height and 130 degree field of view (Hazcam); (2) a goal camera mounted on a manipulator arm with 20 degree field of view (Viewcam); (3) internal DMU gyroscope registering along coordinates pitch, roll, and yaw; (4) Crossbow accelerometer in x, y, and z coordinates; (5) a Sun sensor for global positioning information [91]. To simplify measurement conditions and data acquisition, the top-mounted goal camera, the robot arm, and the global positioning sensor are not used in the present experiments. This work is based on measurements by the stereo camera and the DMU unit only. This approach simplifies the technical support and signal monitoring needs, but it also poses a more challenging task for efficient and reliable goal completion.

Fig. 7 Sample Return Rover (SRR2K) situated in the Planetary Robotics indoor facility imitating natural terrain in planetary environments. SRR2K has 4 independently controllable wheels. The figure also shows a robot arm which is not used in the present experiments. The sensory system consists of visual cameras, infra red sensors, accelerometers, and global orientation sensors [80]

At regular time intervals the sensory vectors are written on the on-board computer in a file, which are accessed by the KIV system for further processing. The update of the sensory data arrays happens at about every 30–40 s, which is determined by the speed of SRR2K and the computational time of calculating KIV output. We have the following sensory data: (1) visual data vector consisting of 10 wavelet coefficients determined from the recorded 480×640 pixel image of a Hazcam; (2) IMU recordings using the mean and variance of the gyroscope and accelerometer readings along the 3 spatial coordinates. Statistical moments are determined over the data recording window given by the file refreshment rate as described above; (3) rover heading which is a single angle value determined by comparing the orientation of the rover with respect to the direction of the goal.

6.3 KIV Control Architecture

We apply KIV set for robot navigation. KIV is the brain of an intentional robot that acts into its environment by exploration and learns from the sensory consequences of its actions. KIV operates on the principle of encoding in spatio-temporal oscillations, in analogy with EEG oscillations in the vertebrate brain. By cumulative learning KIV it creates an internal model of its environment, which it uses to guide its actions while avoiding hazards and reaching goals that the human controller defines. We set the simple task of starting from a corner and reach a goal position $GOAL\_XY$ specified at the start of the experiment. The straight road may be not the best when there are some rough areas where difficult to cross, or some hills, which difficult to scale, etc. In this situation we expect that a properly trained SRR2K robot would decide to take a path which avoids the difficult areas, or at least tries to do so. If proper learning and generalization took place, one could change the terrain into a layout which SRR2K never seen before, still it should achieve good performance.

It is important to note that the KIV-based control system will not provide an absolute optimal decision, in general. Or at least this is not likely. Rather it may choose a sub-optimal path. But this in general would be more robust than an optimal path designed by a rule-based method. The rover is at a given state at any instant of its operation, and it transits to a next state based on its present state and available input information. The control algorithm $CRUISE\_XY\_Z$ accepts 2 controlled variables $ANGLE\_TO\_TURN$ and $DISTANCE\_TO\_GO$. These variables are provided by KIV through a control file regularly downloaded to SRR2K. A third variable $DESIRED\_VELOCITY$ can be controlled as well. However, in the limited task for the present project, the velocity was given a value of 10 cm/s which has not been changed for simplicity.

The complete KIV model consists of four major components, namely 3 KIII sets for sensory processing, and a KII for integration, as it has been described previously; see Fig. 3. In the full model, one KIII set models the hippocampus, another one models the cortical region, and the third describes the midline forebrain. The fourth major component is the entorhinal cortex (EC) with amygdala, which is a KII set.

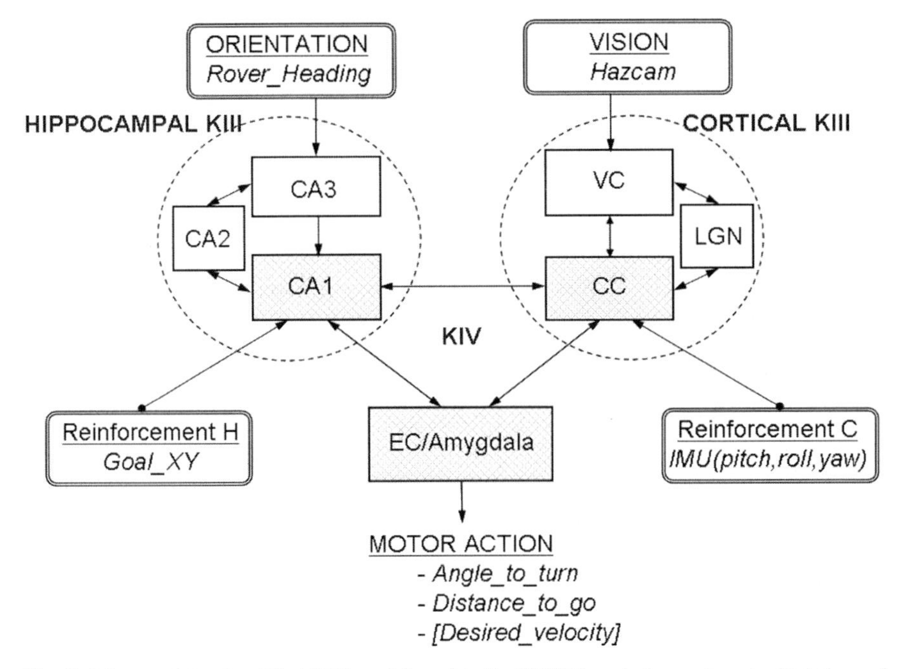

Fig. 8 Schema of the simplified KIV model used in the SRR2K control experiments. Notations of the KII units: CA1, CA2, and CA3 are hippocampal sections; VC and CC are visual cortex and cerebral cortex, respectively, LGN is lateral geniculate nucleus; EC - entorhinal cortex with the amygdala. Specification of orientation and vision sensory signals, and hippocampal and cortical reinforcement signals is given in the text. Shaded boxes indicate locations where learning (CA1 and CC) and recall (EC/Amygdala) take place

EC integrates influences from all parts of the hemisphere, and it provides link to external parts of the limbic system for motor action. In the present work a simplified KIV is used, which has a visual sensory KIII set, a hippocampal KIII set. For simplicity, we have just a reinforcement signal representing the interoceptory unit instead of a full KIII midline forebrain model. Accordingly, the EC integrates the effects of the cortical and hippocampal KIII units. The applied KIV set is depicted in Fig. 8.

6.4 Learning and Control of SRR2K Mobile Robot with Intentional KIV Brain

The KIV-guided robot uses its experience to continuously solve problems in perception and navigation that are imposed by its environment as it pursues autonomously the goals selected by its trainer. There are three phases in the operation of KIV control: (i) learning phase; (ii) labeling phase; and (iii) testing phase. For details see [78].

- At the learning phase, SRR2K explores the environment and builds associations between visual, DMU, and orientation sensory modalities. If it makes a good step during its random exploration, it gets a positive reinforcement signal and a Hebbian learning cycle is completed. Reinforcement signals are given by the DMU values in the case of the visual channel, and by the goal position sensing in the orientation channel. The corresponding learning effects are marked as Reinforcement C and Reinforcement H in Fig. 8.
- At the labeling phase, certain activations from the KIV model are collected as reference patterns. These patterns are activations representing correct motion in a given direction. These activations are then used for finding the right direction of movement for the agent during the testing phase. Using the k-nearest neighbor scheme, the decision is made by selecting the best match between the k predefined AM patterns and the actual activity pattern. Concerning the length of the move, we use for simplicity discrete values 25 cm, 0 cm, or –25 cm, which correspond to move forward, do not move, and backtrack, respectively.
- During tests, SRR2K is placed at a selected position in the environment and is left to move on its own. At each step it captures the sensory data, extracts the main features, and passes on to KIV. Under the influence of the sensory input, the dynamics exhibited by KIV changes and new AM patterns are activated. The activations generated from this test input are matched to the stored reference activations. Using the k-nearest neighbor voting scheme, KIV decides what next step to take.

6.5 Demonstration of Multi-sensory Fusion and Intentional Decision Making in KIV

In the experiments SRR2K used 2 sensory modalities: orientation and short-range vision. We did not use far-field visual information on landmarks and on goal positions, therefore now we do not aim at creating an internal cognitive map based on landmarks [10]. Rather we studied the way KIV builds multisensory associations and how uses those associations for selection of action in the intentional dynamical system.

In a typical scenario, the visual sensing contained information on the terrain in front of SRR2K within at a range of about 0.5 m with 130 degree field of view. This covered the terrain which would be traveled by a forward moving robot in the next one or two steps. Wavelet-based visual processing converted the raw image data into an array of wavelet coefficients which characterized the roughness of the terrain. We divided the visual field into 10 segments each characterized by a single roughness coefficient. An example is shown in Fig. 9, where the 10 coefficients are given along the horizontal axis, while the sequence of time steps progress along the vertical axis. The gray-scale indicates a range of coefficients from black (small coefficient and smooth terrain) to white (large coefficient and rough terrain). Each time step corresponds to one move by SRR2K, which takes about 30 s. One can see bumps within the field of view of the robot on the right at time steps 4 and 16, and

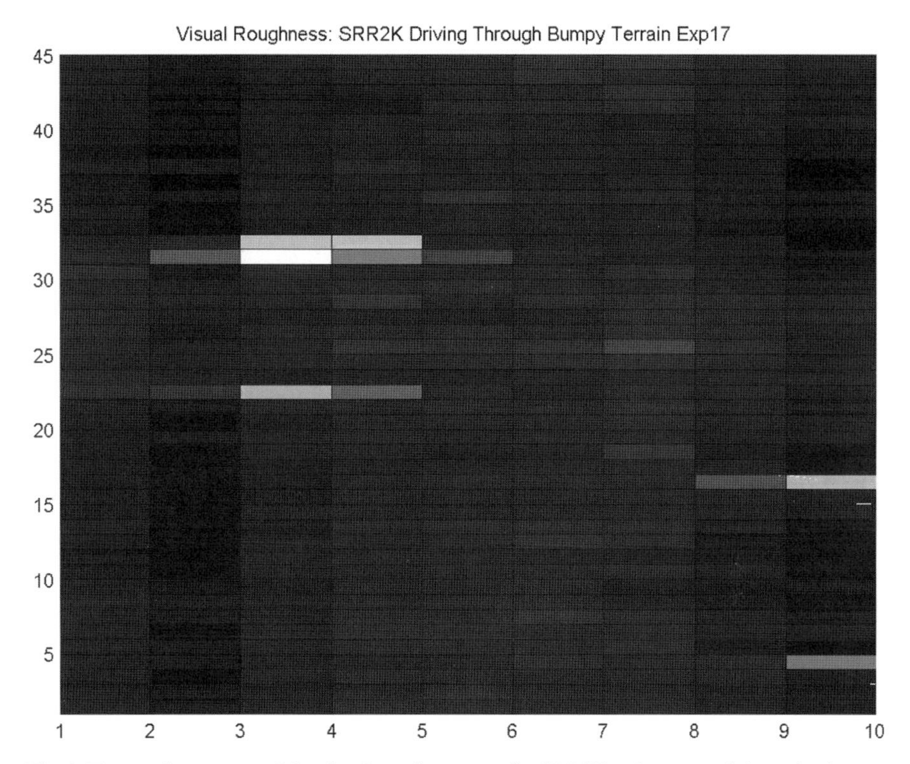

Fig. 9 Temporal sequence of the visual roughness seen by SRR2K as it traversed through a bumpy terrain. Dark tones correspond to smooth surface, and white indicates rough terrain. Bumps are seen within the field of view of the robot on the right at time steps 4 and 16, and on the left at steps 22 and 31–32

on the left at steps 22 and 31–32. In total 45 steps appear sequentially from bottom to top in Fig. 9, but the trajectory has not been straight and the robot took several left and right turns while traversing the bumpy terrain.

Oscillatory components of the DMU gyroscope (root-mean-squared value, RMS) are shown in Fig. 10, upper panel. One can see that the DMU RMS has several peaks which are caused by going over rough areas with rocks. This conclusion is confirmed by the recordings of the accelerometer, which show similar oscillation peaks at the same locations; see Fig. 10, lower panel.

In order to develop an efficient control paradigm, SRR2K should be able to avoid obstacles if it sees them instead of trying to climb trough them even if that means not taking the direct path toward the goal. Once the robot is at the bumps it experiences high oscillations and high RMS values of DMU. It needs to avoid this situation therefore the RMS oscillation is used as a negative reinforcement learning signal. On the other hand, moving toward the goal location is desirable and produces a positive reinforcement learning signal. The essence of the task is to anticipate the undesirable bumpy oscillation a few steps ahead based on the visual signal. SRR2K need to develop this association on its own. A usual AI approach would provide some rule base using the visual signal; for example, turn left if you see a bump on

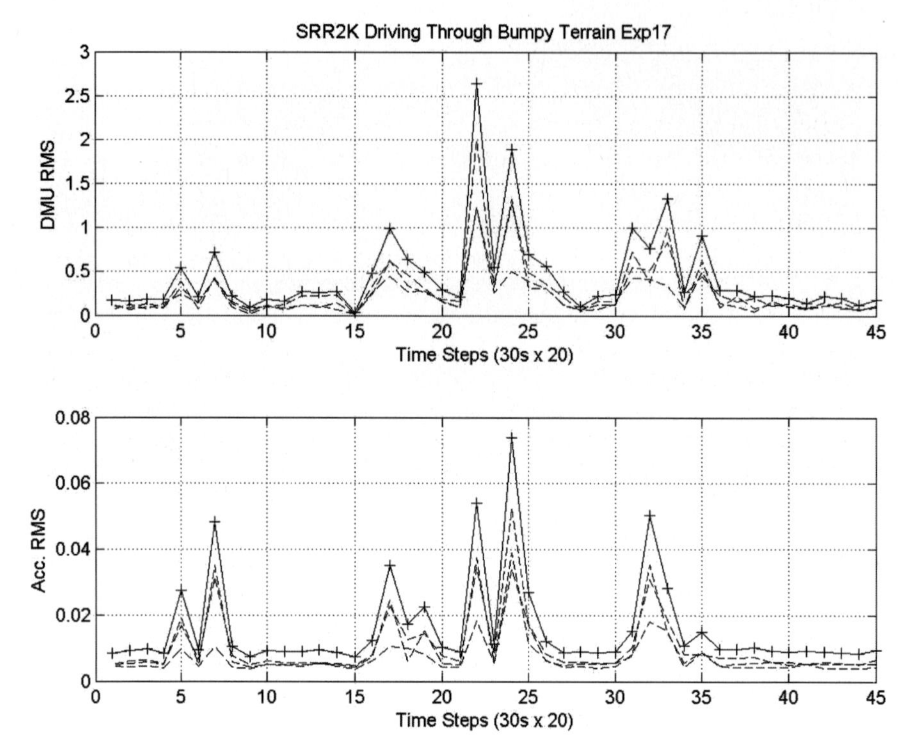

Fig. 10 Oscillatory components of the DMU recordings (upper panel); solid, dash, and dotted line indicate the 3 spatial components as SRR2K traverses the terrain in 45 steps. Oscillatory components of the accelerometer (lower panel); notations are same as above

the right. This may be a successful strategy on relatively simple tasks, but clearly there is no way to develop a rule base which can give instructions in a general traversal with unexpected environmental challenges.

SRR2K has demonstrated that it indeed can learn the association between different sensory modalities, and it traverses successfully a terrain with obstacles, without bumping into them and it reaches the target goal position [80]. It is worth mentioning that KIV based control has the option to take not only instantaneous sensory values, but also data observed several steps earlier. This involves a short-term memory (STM) with a given memory depth. In the given task, a memory could be 3–4 steps deep, or more. One can in fact optimize the learning and system performance based on the memory depth [78]. This has not been used in the SRR2K experiments yet and planned to be completed in the future.

At the next step of the experiments top-mounted camera (Viewcam) can be used to support cognitive map formation. Cognitive maps are generated in a self-organized way rather than supervised learning. This approach is different from existing cognitive map generation techniques [10]. We do not impose rules externally rather we want the system to develop a behavior to embody some meaningful rules. The developed maps can be used to generate rules or update/modify existing rules as needed.

7 Perspectives on Intentional Dynamics

A biologically-motivated pattern-based computation is outlined, which represent a drastic departure from today's digital computer designs, which are based on computation with numbers represented by strings of digits. Brains do not work with digital numbers, rather they operate using a sequence of amplitude-modulated (AM) patterns of activity, which are observed in EEG, MEG, and fMRI measurements. Using neurodynamical principles, we can replace traditional symbol-based computation with pattern-based processing. In our approach, categories are not "a priori" defined. Rather they emerge through the self-organized activity of interacting neural populations in space and time. They emerge and dissolve at a rate of several cycles per second, as dictated by the underlying theta rhythm observed in brains. In the past years, the feasibility of this approach has been demonstrated in various practically relevant problems of classification, pattern recognition, and decision making.

Here we develop a nonlinear dynamic theory of brains for potential deployment in artificial designs. The underlying principle motivated by neurophysiological findings on pattern-based computing in brains through the sequence of cortical phase transitions. There is substantial evidence indicating that cortical phase transitions are manifestations of high-level cognitions and potentially of consciousness, producing switching between metastable AM patterns. KIV models cognitive phase transitions as observed in brains.

Intentional dynamic systems are introduced in the form of the KIV model. KIV has the potential of producing macroscopic phase transitions, which provide the mechanism for fast and robust information processing in KIV, in the style of brains. KIV is a biologically based cognitive brain model capable of learning of and responding to environmental inputs based on its experience robustly and creatively. The KIV-based intentional dynamic brain model addresses the classic symbol grounding problem by means of linking intentional behavior with known mesoscopic neural dynamics. Mesoscopic dynamics is key attribute of our model bridging the gap in our understanding of the mind-body connection.

KIV describes the cognitive action-perception cycle at the core of intentionality, consisting of prediction, planning action, execution, sensing results of action, evaluating prediction, updating new prediction. The cognitive cycle in the dynamical model takes place in real time with frame rates corresponding to ranges observed in humans. KIV has been implemented using robot control test beds in computational simulations and real-life environment. Experiments have been conducted to generate intentional behavior leading to learning, category formation and decision making based on knowledge learned in previous scenarios. Perspectives of future development of intentional systems have been outlined.

Acknowledgment Research presented in this work has been completed while the author has been with NASA Jet Propulsion Laboratory, Pasadena, CA; and with US Air Force Research Laboratory, Hanscom AFB, MA. Financial supports from the following sources are greatly appreciated: JPL Advanced Concepts Office, by Neville Marzwell; National Academies, National Research Council Senior Research Fellowship Office. The author would like to thank to Dr. Walter J Freeman, UC

Berkeley for his continued support. The author would also like to thank to Terry Hunstberger, Hrand Aghazarian, Eddie Tunstel, and to all members of JPL Planetary Robotics; and to Dr. Leonid Perlovsky, AFRL and Harvard University.

References

1. Turing A. M. Computing machinery and intelligence. *Mind*, 59:433–460, 1950.
2. J. V. Neumann. *The computer and the brain*. New Haven, CT, Yale Univ. Press, 1958.
3. A. Newell and H. A. Simon. *Human problem solving*. Englewood Cliffs, NJ: Prentice-Hall, 1972.
4. Newell A. Physical symbol systems. *Cognitive Science*, 4:135–183, 1980.
5. A. Newell. *Unified theories of cognition*. Cambridge, MA: Harvard University Press, 1990.
6. Harter D. and Kozma R. Aperiodic dynamics and the self-organization of cognitive maps in autonomous agents. *Int. J. of Intelligent Systems*, 21:955–971, 2006.
7. Laird J. E., Newell A., and Rosenbloom P. S. Soar: An architecture for general intelligence. *Artificial Intelligence*, 33:1–64, 1987.
8. Anderson J. A., Silverstein J. W., Ritz, S. A. Jones, and R. S. Distinctive features, categorical perception, and probability learning: Some applications of a neural model. *Psychological Review*, 84:413–451, 1977.
9. H. L. Dreyfus. *What Computers Still Can't Do - A Critique of Artificial Reason*. Cambridge, MA, MIT Press, 1992.
10. Mataric M. J. and Brooks R. A. *Cambrian Intelligence*, chapter Learning a distributed map representation based on navigation behaviors, pages 37–58. Cambridge, MA: The MIT Press, 1999.
11. A. Clark. *Mindware: An Introduction to the Philosophy of Cognitive Science*. Oxford University Press, 2001.
12. L. I. Perlovsky. *Neural Networks and Intellect*. Oxford Univ. Press, New York, NY, 2001.
13. D. E. Rumelhart and J. L. McClelland. *Parallel distributed processing: Explorations in the microstructure of cognition*. Cambridge, MA, MIT Press, 1986.
14. C. Bishop. *Neural networks for pattern recognition*. Oxford University Press, 1995.
15. Haykin S. *Neural networks - A comprehensive foundation*. Prentice Hall, NJ, 1998.
16. Towell G. G. and Shavlik J. W. Knowledge-based artificial neural networks. *Artificial Intelligence*, 70:119–165, 1994.
17. Sporns O., Almassy N., and Edelman G. M. Plasticity in value systems and its role in adaptive behavior. *Adaptive Behavior*, 7(3-4), 1999.
18. G. M. Edelman and G. Tononi. *A Universe of Consciousness: How Matter Becomes Imagination*. Basic Books, New York, N.Y., 2000.
19. Vershure P.M., Krose B., and Pfeifer R. Distributed adaptive control: The self-organization of behavior. *Robotics and Autonomous Systems*, 9:181–196, 1992.
20. R. Pfeifer and C. Scheier. *Understanding Intelligence*. Cambridge, MA, MIT Press, 1999.
21. Vershure P.M. and Althaus P. A real-world rational agent: Unifying old and new ai. *Cognitive Science*, 27(4):561–590, 2003.
22. W. J. Freeman. *Mass Action in the Nervous System*. Academic Press, New York, 1975.
23. Kozma. R. Freeman. W.J. Chaotic resonance - methods and applications for robust classification of noisy and variable patterns. *Int. J. Bifurcation and Chaos*, 11:1607–1629, 2001.
24. Katchalsky A., Rowland V., and Huberman B. Dynamic patterns of brain cell assemblies. *Neuroscience Res.Program Bull.*, 12, 1974.
25. H. Haken. *Synergetics: An Introduction*. Berlin: Springer-Verlag, 1983.
26. J. A. S. Kelso. *Dynamic patterns: The self-organization of brain and behavior*. Cambridge, MA, MIT Press, 1995.
27. J.A.S. Kelso and D. A. Engstrom. *The complementary nature*. Cambridge, MA, MIT Press, 2006.

28. Freeman W.J., Burke B.C., and Holmes M.D. Aperiodic phase re-setting in scalp eeg of beta-gamma oscillations by state transitions at alpha-theta rates. *Hum. Brain Mapp.*, 19:248–272, 2003.
29. Freeman W.J. Origin, structure, and role of background eeg activity. part 3. neural frame classification. *Clin. Neurophysiology*, 116:1118–1129, 2005.
30. Changeux J-P. and Dehaene S. Neuronal models of cognitive functions. *Cognition*, 33: 63–109, 1989.
31. Tsuda I. Toward an interpretation of dynamic neural activity in terms of chaotic dynamical systems. *Beh. and Brain Sci.*, 24(5):793–810, 2001.
32. Lehmann D, Strik WK, Henggeler B, Koenig T, and Koukkou M. Brain electric microstates and momentary conscious mind states as building blocks of spontaneous thinking: I. visual imagery and abstract thoughts. *Int. J Psychophysiol.*, 29:1–11, 1998.
33. Fingelkurts A .A. and Fingelkurts A. A. Operational architectonics of the human brain biopotential field: Towards solving the mind-brain problem. *Mind and Brain*, 2:262–296, 2001.
34. Fingelkurts A .A. and Fingelkurts A. A. Making complexity simpler: multivariability and metastability in the brain. *Internat J.Neurosci*, 114:843–862, 2004.
35. Barrie J.M., Freeman W.J., and Lenhart M. Modulation by discriminative training of spatial patterns of gamma eeg amplitude and phase in neocortex of rabbits. *J. Neurophysiol.*, 76: 520–539, 1996.
36. Ohl FW Scheich H and Freeman WJ. Change in pattern of ongoing cortical activity with auditory category learning. *Nature*, 412:733–736, 2001.
37. Ohl FW, Deliano M, Scheich H, and Freeman WJ. Early and late patterns of stimulus-related activity in auditory cortex of trained animals. *Biol. Cybernetics*, 88:374–379, 2003.
38. Bressler S.L. Cortical coordination dynamics and the disorganization syndrome in schizophrenia. *Neuropsychopharmacology*, 28:S35–S39, 2003.
39. Bressler S.L. and Kelso J.A.S. Cortical coordination dynamics and cognition. *Trends in Cognitive Sciences*, 5:26–36, 2001.
40. Le Van Quyen M, Foucher J, Lachaux J-P, Rodriguez E, Lutz A, Martinerie J, and Varela F. Comparison of hilbert transform and wavelet methods for the analysis of neuronal synchrony. *J. Neurosci. Meth.*, 111:83–98, 2001.
41. Werner G. Perspectives on the neuroscience of cognition and consciousness. *BioSystems*, 2006.
42. B. J. Baars. *A cognitive theory of consciousness*. Cambridge Univ. Press, MA, 1988.
43. Freeman W.J. Origin, structure, and role of background eeg activity. part 2. analytic amplitude. *Clin. Neurophysiology*, 115:2077–2088, 2004.
44. I. Prigogine. *From Being to Becoming: Time and Complexity in the Physical Sciences*. WH Freeman, San Francisco, 1980.
45. Lachaux J-P, Rodriquez E, Martinerie J, and Varela FA. Measuring phase synchrony in brain signals. *Hum. Brain Mapp.*, 8:194–208, 1999.
46. Quiroga RQ Kraskov A Kreuz T Grassberger P. Performance of different synchronization measures in real data: A case study on electroencephalographic signals. *Physical Rev E*, 6504(U645-U6 58):art. no. 041903, 2002.
47. Demirer R.M., Kozma R., Caglar M., and Polatoglu Y. Hilbert transform optimization to detect cortical phase transitions in beta-gamma band. *(Submitted)*, 2006.
48. S. Kauffman. *The Origins of Order - Self-Organization and Selection in Evolution*. Oxford Univ. Press, 1993.
49. Crutchfield J.P. The calculi of emergence: Computation, dynamics, and induction. *Physica D*, 75:11–54, 1994.
50. Watts D.J. Strogatz S.H. Collective dynamics of “small-world” networks. *Nature*, 393:440–442, 1998.
51. Bressler S.L. Understanding cognition through large-scale cortical networks. *Current Directions in Psychological Science*, 11:58–61, 2002.
52. D. Watts. *Six Degrees: The Science of a Connected Age*. New York: Norton, 2003.
53. Wang XF and Chen GR. Complex networks: small-world, scale-free and beyond. *IEEE Trans. Circuits Syst.*, 31:6–20, 2003.

54. Albert R. and Barabási A.L. Statistical mechanics of complex networks. *Reviews of Modern Physics*, 74:47, 2002.
55. Barabási A.L. and E. Bonabeau. Scale-free networks. *Scientific American*, 288:60–69, 2003.
56. Bollobas B. and Riordan O. *Handbook of graphs and networks*, chapter Results on scale-free random graphs, pages 1–34. Wiley-VCH, Weinheim, 2003.
57. Ingber L. *Neocortical Dynamics and Human EEG Rhythms*, chapter Statistical mechanics of multiple scales of neocortical interactions, pages 628–681. New York: Oxford U.P., 1995.
58. Hoppensteadt F.C. and Izhkevich E.M. Thalamo-cortical interactions modeled by weakly connected oscillators: could the brain use fm radio principles? *BioSystems*, 48:85–94, 1998.
59. Friston K.J. The labile brain. i. neuronal transients and nonlinear coupling. *Phil Trans R Soc Lond B*, 355:215–236, 2000.
60. Linkenkaer-Hansen K, Nikouline VM, Palva JM, and Iimoniemi R.J. Long-range temporal correlations and scaling behavior in human brain oscillations. *J. Neurosci.*, 15:1370–1377, 2001.
61. K. Kaneko and I. Tsuda. *Complex Systems: Chaos and Beyond. A Constructive Approach with Applications in Life Sciences*. Springer Verlag, 2001.
62. Kozma R., Freeman W.J., and Erdi P. The kiv model - nonlinear spatio-temporal dynamics of the primordial vertebrate forebrain. *Neurocomputing*, 52-54:819–825, 2003.
63. Stam CJ, Breakspear M, van Cappellen van Walsum A.M, and van Dijk BW. Nonlinear synchronization in eeg and whole-head recordings of healthy subjects. *Hum Brain Mapp*, 19:63–78, 2003.
64. Kozma R., Puljic M., Bollobas B., Balister P., and Freeman W.J. Phase transitions in the neuropercolation model of neural populations with mixed local and non-local interactions. *Biol. Cybernetics*, 92:367–379, 2005.
65. Balister P., Bollobas B., and Kozma R. Large deviations for mean field models of probabilistic cellular automata. *Random Structures and Algorithms*, 29:399–415, 2006.
66. Bak P., Tang C., and Wiesenfeld K. *Phys. Rev. Lett.*, 59:381, 1987.
67. P. Bak. *How Nature Works - The Science of Self-Organized Criticality*. Springer Verlag, N.Y., 1996.
68. H. J. Jensen. *Self-organized criticality - Emergent behavior in physical and biological systems*. Cambridge University Press, 1998.
69. Chang H.J. and Freeman W.J. Parameter optimization in models of the olfactory neural system. *Neural Networks*, 9:1–14, 1996.
70. Freeman W.J., Chang H.J., Burke B.C., Rose P.A., and Badler J. Taming chaos: stabilization of aperiodic attractors by noise. *IEEE Trans. Circ. and Syst. I.*, 44:989–996, 1997.
71. Xu D. and Principe J. Dynamical analysis of neural oscillators in an olfactory cortex model. *IEEE Trans. Neur. Netw.*, 15:1053–1062, 2004.
72. Ilin R. and Kozma R. Stability of coupled excitatory-inhibitory neural populations and application to control of multi-stable systems. *Physics Lett A*, 360(1):66–83, 2006.
73. Gutierrez-Galvez A. and Gutierrez-Osuna R. Contrast enhancement of sensor-array patterns through hebbian/antihebbian learning. *Proc. 11th Int. Symp. Olfaction and Elect. Nose, Barcelona, Spain*, 2005.
74. Chang H.J., Freeman W.J., and Burke B.C. Optimization of olfactory model in software to give 1/f power spectra reveals numerical instabilities in solutions governed by aperiodic (chaotic) attractors. *Neural Networks*, 11:449–466, 1998.
75. Freeman W.J., Kozma R., and Werbos P.J. Biocomplexity - adaptive behavior in complex stochastic dynamical systems. *BioSystems*, 59(2):109–123, 2001.
76. Beliaev I. and Kozma R. Time series prediction using chaotic neural networks on the cats benchmark test. *Neurocomputing (in press)*, 2007.
77. Harter D. and Kozma R. Chaotic neurodynamics for autonomous agents. *IEEE Trans. Neural Networks*, 16(4):565–579, 2005.
78. Kozma R. and Muthu S. Implementing reinforcement learning in the chaotic kiv model using mobile robot aibo. *IEEE/RSJ Int. Conf. on Intelligent Robots and Systems IROS'04, Sendai, Japan*, 2004.

79. Kozma R., Wong D., Demirer M., and Freeman W.J. Learrning intentional behavior in the k-model of the amygdala and enthorhinal cortex with the cortico-hippocampal formation. *Neurocomputing*, 65-66:23–30, 2005.
80. Huntsberger T., Tunstel E., and Kozma R. *Intelligence for Space Robotics*, chapter Onboard learning strategies for planetary surface rovers, pages 403–422. TCI Press, San Antonio, TX, 2006.
81. W. J. Freeman. *Neurodynamics: An Exploration of Mesoscopic Brain Dynamics*. London, U.K., Springer-Verlag, 2000.
82. Kozma R. and Freeman W.J. Basic principles of the kiv model and its application to the navigation problem. *J. Integrative Neurosci.*, 2:125–140, 2003.
83. Nunez R.E. and Freeman W.J. Restoring to cognition the forgotten primacy of action, intention, and emotion. *J. Consciousness Studies*, 6(11-12):ix–xx, 1999.
84. Harter D. and Kozma R. Navigation and cognitive map formation using aperiodic neurodynamics. *Proc. of 8th Int. Conf. on Simulation of Adaptive Behavior (SAB'04), LA, CA.*, 8:450–455, 2004.
85. Kozma R. and Fukuda T. Intentional dynamic systems - fundamental concepts and applications (editorial). *Int. J. Intell. Syst.*, 21(9):875–879, 2006.
86. Voicu H., Kozma R., Wong D., and Freeman W.J. Spatial navigation model based on chaotic attractor networks. *Connect. Sci.*, 16(1):1–19, 2004.
87. Kozma R. and Myers M. Analysis of phase transitions in kiv with amygdale during simulated navigation control. *IEEE Inf. Joint Conf. Neur. Netw. IJCNN05, Montreal, Canada*, 2005.
88. Kozma R., R. Deming, and L. Perlovsky. Optimal estimation of parameters of transient mixture processes using dynamic logic approach. *Conference on Knowledge-Intensive Multi-Agent Systems KIMAS'07, Boston, MA*, 2007.
89. Gat E., Desai R., Ivlev R., Loch J., and Miller D.P. Behavior control for robotic exploration of planetary surfaces. *IEEE Trans. Robotics and Autom.*, 10(4):490–503, 1994.
90. Huntsberger T.L. and Rose J. Bismarc- a biologically inspired system for map-based autonmous rover control. *Neural Networks*, 11(7-8):1497–1510, 1998.
91. Huntsberger T., Cheng Y., Baumgartner E. T., Robinson M., and Schenker P. S. Sensory fusion for planetary surface robotic navigation, rendezvous, and manipulation operations. *Proc. Int. Conf. on Advanced Robotics, Lisbon, Portugal*, pages 1417–1424, 2003.
92. Tunstel E. Ethology as an inspiration for adaptive behavior synthesis in autonomous planetary rovers. *Autonomous Robots*, 11:333–339, 2001.
93. Seraji H. and Howard A. Behavior-based robot navigation on challenging terrain: A fuzzy logic approach. *IEEE Trans. Robotics and Autom.*, 18(3):308–321, 2002.
94. O'Keefe J and Recce M.L. Phase relationship between hippocampal place units and the eeg theta rhythm. *Hippocampus*, 3:317–330, 1993.
95. Blum K. I. and Abbott L. F. A model of spatial map formation in the hippocampus of the rat. *Neural Computation*, 8:85–93, 1996.
96. Touretzky D. S. and Redish A. D. Theory of rodent navigation based on interacting representations of space. *Hippocampus*, 6(3):247–270, 1996.
97. Bachelder I.A. and Waxman A.M. Mobile robot visual mapping and localization: A view based neurocomputational architecture that emulates hippocampal place learning. *Neural Networks*, 7:1083–1099, 1994.
98. Arleo A. and W. Gerstner. Spatial cognition and neuro-mimetic navigation: A model of hippocampal place cell activity. *Biological Cybernetics*, 83:287–299, 2000.
99. Hasselmo M.E., Hay J., Ilyn M., and Gorchetchnikov A. Neuromodulation, theta rhythm and rat spatial navigation. *Neural Networks*, 15:689–707, 2002.
100. Berthoz A. Trullier O. Wiener S. and Meyer J.-A. Biologically-based artificial navigation systems: Review and prospects. *Progress in Neurobiology*, 51:483–544, 1997.

How Does the Brain Create, Change, and Selectively Override its Rules of Conduct?

Daniel S. Levine

Abstract How do we know to talk openly to our friends but be guarded with strangers? How do we move between work place, club, and house of worship and fit our behavior to each setting? How do we develop context-dependent rules about what we may eat? We solve such problems readily, but are a long way from designing intelligent systems that can do so. Yet enough is known about cognitive and behavioral functions of three regions of prefrontal cortex, and their subcortical connections, to suggest a neural theory of context-dependent rule formation and learning. Rules that an individual follows can change, either because of personal growth or change in stress. This can be partly explained by the interplay between signals from the hippocampus, signifying task relevance, and the amygdala, signifying emotional salience. Both sets of signals influence the basal ganglia gate that selectively disinhibits behaviors in response to context.

1 Introduction

A complex high-order cognitive system clearly needs to develop criteria for what actions to perform and what actions to refrain from performing. The more complex the system's environment, the more flexible and context-sensitive those criteria need to be.

The term "rules" provokes a certain amount of discomfort in the neural network community. To some readers it has the connotations of strict logical IF-THEN rules from a production system as in symbolic artificial intelligence. In fact, Fodor and Pylyshyn [19] and other symbolic cognitive scientists have argued at times that connectionist neural networks are inherently unsuitable for emulating the human capability of symbolic rule generation.

Yet I mean "rules" not necessarily in this strict sense, although clearly the neural networks in our brain are capable of formal logical operations. I simply mean principles, often consciously describable ones, for organizing one's behavior and conduct across a range of situations. More often than not these are heuristics or rules of thumb; that is, "do's" and "don'ts" that are not absolute but tend to apply, in the philosophers' terminology, *ceteris paribus* (all other things being equal).

L. I. Perlovsky and R. Kozma, *Neurodynamics of Cognition and Consciousness.*

Where do the heuristics we humans employ come from? One of the stock answers is: evolution. That is to say, the answer is that our heuristics are based on behavioral patterns that have been selected for survival and reproductive success. Also, evolutionary psychologists tend to believe that despite some species differences in details, the basics of these behavioral patterns are largely shared with other mammals. Yet as explanations of higher-order cognitive function, the standard evolutionary arguments provide starting points but are incomplete, as I hope to show in this chapter.

There are some behavioral patterns that are based in evolution and present in all of us, such as the patterns of self-interested (or self-protective) behavior and the patterns of social bonding (including altruistic) behavior. Yet we all differ in our criteria for *when* to engage in *which* of those behavioral patterns, and in what social and environmental contexts. Moreover, these decision criteria are not exclusively genetic but heavily influenced by learning and by culture [18, 39].

Yes, the hard-core evolutionist will argue, but all this is dependent on the extensive plasticity of our brains (present in all animals but vastly expanded in humans), and evolution selected us for this trait of plasticity. This is certainly true, but that does not entail evolutionary determinism for each of the behaviors that arise from the plasticity.

Moreover, survival and reproduction do not explain all behaviors. In addition, human beings seek well-being: self-actualization, pleasurable stimulation, aesthetic harmony, mastery, and meaning all can be powerful motivators (see, e.g., [14, 18, 42, 51]). While evolutionary fitness can provide plausible functional accounts of most of these motivators, the behaviors they generate have a life of their own apart from their survival or reproductive value.

We now seek to decompose the brain systems involved in the development of rules at many possible levels of complexity.

2 Key Brain Regions for Rule Development

2.1 Hypothalamus, Midbrain, and Brain Stem

The organism's basic physiological needs are represented in deep subcortical structures that are shared with other animals and that have close connections with visceral and endocrine systems. These regions include several nuclei of the hypothalamus and brain stem.

This part of the brain does not have a neat structure that lends itself readily to modeling; consequently, these areas have been largely neglected by recent modelers despite their functional importance to animals, including humans. Yet these deep subcortical areas played key roles in some of the earliest (developed in the 1960s and 1970s) physiologically based neural networks.

Kilmer, McCulloch, and Blum [30] in perhaps the first computer simulated model of a brain region, placed the organism's gross modes of behavior (e.g., eating, drinking, sex, exploration, etc.) in the midbrain and brainstem reticular formation. The reticular formation was chosen because it is necessary for arousal, capable of producing essential (though not stimulus-driven) behaviors in the absence of the cortex, and anatomically organized into semi-autonomous subregions that have been compared to a stack of poker chips. The S-RETIC model of [30] is divided into modules analogous to these poker chips; the behavioral modes compete within each module until the same mode wins the competition in more than half of the modules. A similar idea appears in the selective attention network theory of Grossberg [23] who placed what he called a *sensory-drive heterarchy* in the hypothalamus. Different drives in the heterarchy compete for activation, influenced both by connections with the viscera (giving advantage in the competition to those drives that most need to be satisfied) and with the cortex (giving advantage to those drives for which related sensory cues are available).

This idea of a competitive-cooperative network of drives or needs has not yet been verified but seems to have functional utility for behavioral and cognitive modeling. Rule making and learning are partly based on computations regarding what actions might lead to the satisfaction of those needs that survive the competition. Yet the competition among needs is not necessarily winner-take-all, and the best decisions in complex situations are those that go some way toward fulfilling a larger number of needs.

But what do we mean by "needs"? There is considerable behavioral evidence that the word should be expanded beyond the purely physiological ones such as hunger, thirst, sex, and protection. The term should also include the needs for social connections, aesthetic and intellectual stimulation, esteem, and self-fulfillment, for example.

The idea of biological drives and purposes beyond those promoting survival and reproduction goes back at least to the psychologist Abraham Maslow's notion of the *hierarchy of needs*, a concept that has been widely misunderstood. The word "hierarchy" has been misinterpreted to mean that satisfaction of the lower-level needs must strictly precede any effort to fulfill higher-level needs, an interpretation Maslow explicitly denied ([42], p. 26). But neural network modeling, based on dynamical systems, allows for a more flexible meaning for the word "hierarchy" (or, as McCulloch and Grossberg have preferred, heterarchy). It is a hierarchy in the sense that there is a competitive-cooperative network with biases (see, e.g., [24]). This means there tends to be more weight toward the lower-level needs if those are unfulfilled, or if there is too much uncertainty about their anticipated fulfillment (see Fig. 1 for a schematic diagram). However, the bias toward lower-level need fulfillment is a form of risk aversion, and there are substantial individual personality differences in risk aversion or risk seeking that can either mitigate or accentuate the hierarchical biases.

But this still leaves some wide-open questions for the biologically based neural modeler or the neuroscientist. We know something (though our knowledge is not yet definitive) about deep subcortical loci for the survival and reproductive oriented needs. But are there also deep subcortical loci for the bonding, aesthetic, knowledge, or esteem needs? If there are, it seems likely that these needs are shared with most other mammals. Data such as those of [8] suggest that the answer might be yes at least for the bonding needs. These researchers found a hypothalamic site, in a

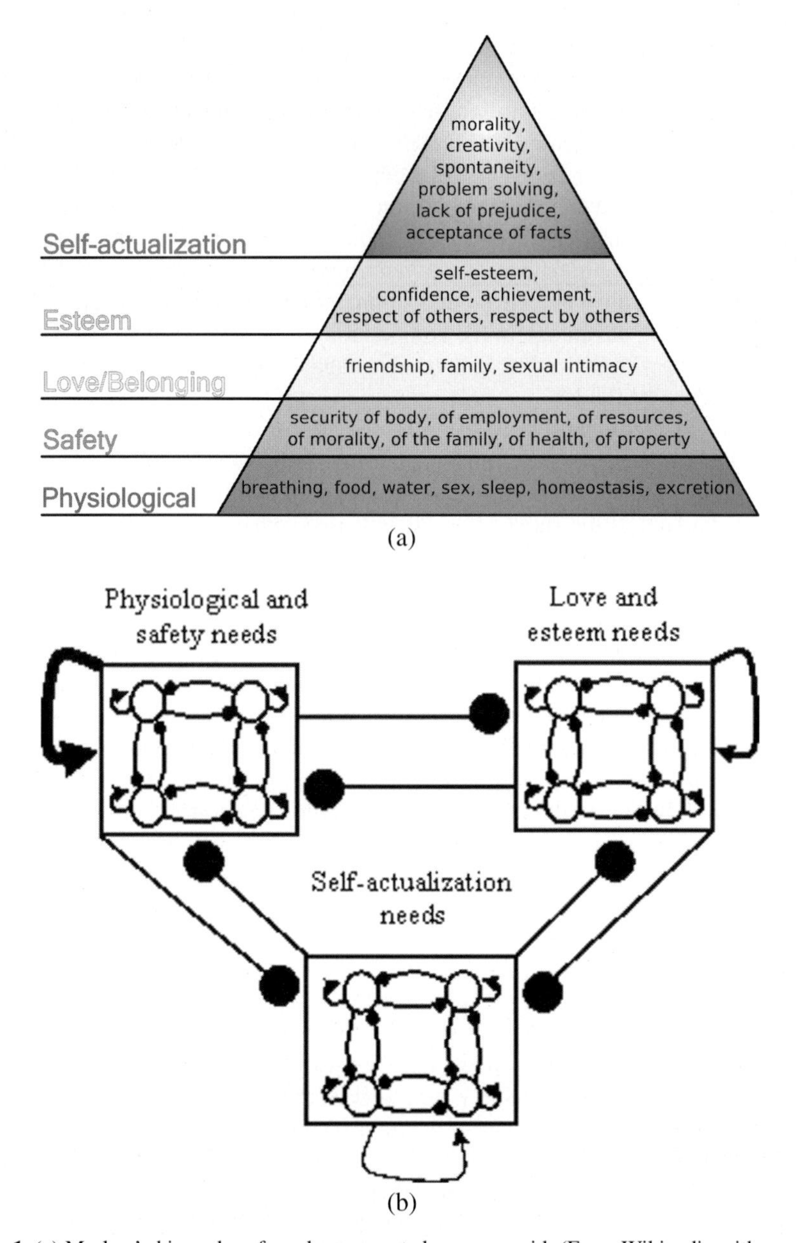

Fig. 1 (**a**) Maslow's hierarchy of needs represented as a pyramid. (From Wikipedia with permission.) (**b**) A neural network rendition of Maslow's hierarchy of needs. Arrows represent excitation, filled circles inhibition. All these needs excite themselves and inhibit each other. But the hierarchy is biased toward the physiological and safety needs which have the strongest self-excitation, represented by the darkest self-arrow; the love and esteem needs have the next darkest self-arrow, and the self-actualization needs the lightest

region called the *paraventricular nucleus*, for the production of oxytocin, a key hormone for social bonding and for pleasurable aspects of interpersonal interactions (including orgasm, grooming, and possibly massage). Also, recent work points to a strong role of the *locus coeruleus*, a midbrain noradrenergic nucleus that is part of the reticular formation, in promoting exploratory behavior [43].

The deeper subcortical structures of the brain played a strong role in qualitative theories by pioneering behavioral neuroscientists who studied the interplay of instinct, emotion, and reason (e.g., [41, 45, 48]), and continue to play a strong role in clinical psychiatric observations. Yet these phylogenetically old areas of the subcortex are often neglected by neural modelers, except for some who model classical conditioning (e.g., [6, 32]). The time is ripe now for a more comprehensive theory of human conduct that will reconnect with some of these pioneering theories from the 1960s and 1970s.

2.2 Amygdala

The amygdala is closely connected with hypothalamic and midbrain motivational areas but "one step up" from these areas in phylogenetic development. The amygdala appears to be the prime region for attaching positive or negative emotional valence to specific sensory events (e.g., [22]). Hence the amygdala is involved in all emotional responses, from the most primitive to the most cognitively driven. In animals, bilateral amygdalectomy disrupts acquisition and maintenance of conditioned responses, and amygdala neurons learn to fire in response to conditioned stimuli.

The amygdala has particularly well studied in relation to fear conditioning [35, 36]. Armony, Servan-Schreiber, Cohen, and LeDoux [1, 2] have developed computational models of fear conditioning in which the amygdala is prominent. In these models, there are parallel cortical and subcortical pathways that reach the primary emotional processing areas of the amygdala. The subcortical pathway (from the thalamus) is faster than the cortical, but the cortex performs finer stimulus discrimination than does the thalamus. This suggests that the two pathways perform complementary functions: the subcortical pathway being the primary herald of the presence of potentially dangerous stimuli, and the cortical pathway performing more detailed evaluations of those stimuli.

Through connections between the amygdala and different parts of the prefrontal cortex, emotionally significant stimuli have a selective processing advantage over nonemotional stimuli. Yet that does not mean that emotional processing automatically overrides selective attention to nonemotional stimuli that are relevant for whatever task the organism is currently performing. If human subjects are involved in an attentionally demanding cognitive task and emotional faces (e.g., faces that showed a fearful expression) are presented at a task-irrelevant location, amygdalar activation is significantly reduced compared to a situation where the emotional faces are task-relevant [52, 53]. This complex interplay of attention and emotion has been captured in various network models involving amygdala as well as both orbital and dorsolateral prefrontal cortex (e.g., [25, 57]).

2.3 Basal Ganglia and Thalamus

A higher-order rule-encoding system requires a mechanism for translating positive and negative emotional linkages into action tendencies or avoidances. This fits with the popular idea of a *gating* system: a brain network that selects sensory stimuli for potential processing, and motor actions for potential performance. Most neuroscientists place the gating system in pathways between the prefrontal cortex, basal ganglia, and thalamus (Fig. 2a); for a theoretical review see [21]. The link from basal ganglia to thalamus in Fig. 2a plays the role of disinhibition; that is, allowing (based on contextual signals) performance of actions whose representations are usually suppressed.

The most important gating area within the basal ganglia is the *nucleus accumbens*. Many neuroscientists identify the nucleus accumbens as a link between motivational and motor systems, and therefore a site of action of rewards — whether these rewards come from naturally reinforcing stimuli such as food or sex, learned reinforcers such as money, or addictive drugs [44].

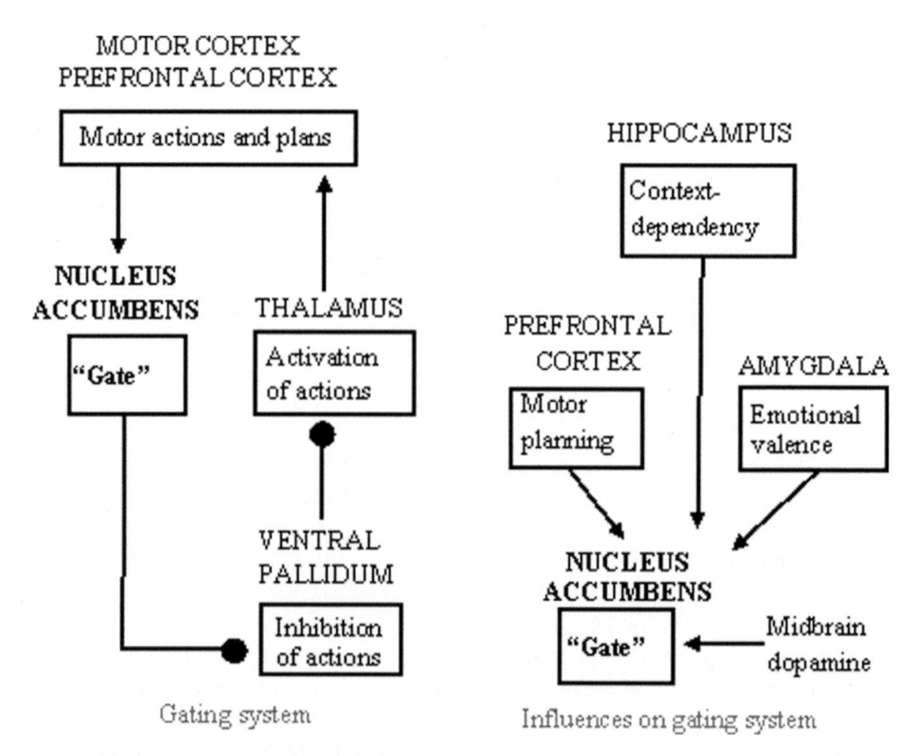

Fig. 2 **Left side**: Loops between nucleus accumbens, thalamus, and cortex for gating behaviors. Nucleus accumbens selectively disinhibits action representations by inhibition of the ventral pallidum, thereby releasing signals from thalamus to cortex. **Right side**: influences on the selective gating from hippocampus (context), amygdala (emotion), and prefrontal cortex (planning). Adapted from [46] with permission from Elsevier science

Clearly, then, influences on the nucleus accumbens from other brain areas are key to choices about which stimulus or action representations are allowed through the gate. Newman and Grace [46] identify three major influences on that area: the first dealing with context, the second with emotion, and the third with plans.

Figure 2b shows some influences on the nucleus accumbens gating system. The influences from the hippocampus are particularly strong: active hippocampal connections can change single accumbens neurons from an inactive to an active state [47]. Since the hippocampus encodes contextual associations for working memory, this can be a vehicle biasing the gates in favor of contextually relevant stimuli.

As Newman and Grace [46] note, there is also a competing bias in favor of emotionally salient stimuli, regardless of the context. This is mediated by connections to the accumbens from the amygdala (Fig. 2b). The hippocampal inputs, associated with "cold" cognition, operate on a slow time scale and promote selective sensitivity to a long-standing task. The amygdalar inputs, associated with "hot" cognition, promote sensitivity to strong and short-duration emotional demands.

Finally, the influences from the frontal cortex are related to planning. We now turn to detailed discussion of the relevant areas of prefrontal cortex.

2.4 Prefrontal Cortex

Many complex movement paradigms involve both reactive and planned movements. The interplay between basal ganglia reactive functions and prefrontal planning functions has been modeled in the case of saccadic eye movements [7].

Three subregions of prefrontal cortex play particularly important, and complementary, roles in the planning and organization of high-level behavior. These regions are the orbitofrontal cortex (OFC), dorsolateral prefrontal cortex (DLPFC), and anterior cingulate cortex (ACC).

The OFC is the area that was damaged in the famous 19th century patient Phineas Gage, and in other patients with deficiencies in decision making and in socially appropriate behavior [15]. Based on such clinical observations and from animal lesion studies, neuroscientists believe the orbitofrontal cortex forms and sustains mental linkages between specific sensory events in the environment and positive or negative affective states. This region creates such linkages via connections between neural activity patterns in the sensory cortex that reflect past sensory events, and other neural activity patterns in subcortical regions (in both the amygdala and hypothalamus) that reflect emotional states.

Longer-term storage of affective valences is likely to be at connections from orbitofrontal cortex to amygdala ([55]; see [20] and [40] for models). Changes that affect behavior ("do" and "don't" instructions, approach toward or avoidance of an object) are likely to be at connections from amygdala to medial prefrontal cortex (incentive motivation) and from orbitofrontal to nucleus accumbens (habit).

Eisler and Levine [18] conjectured that by extension, the OFC might mediate activation of large classes of affectively significant responses, such as the bonding response versus the fight-or-flight response. They suggested a mechanism for such response selection involving reciprocal connections between OFC and the *paraventricular nucleus (PVN)* of the hypothalamus. Different parts of PVN contain various hormones including oxytocin and vasopressin, two hormones that are important for bonding responses (e.g., [11]); and CRF, the biochemical precursor of the stress hormone cortisol which is important for fight-or-flight [34]. The OFC synapses onto another area called the *dorsomedial hypothalamus* that sends inhibitory neurons to PVN that are mediated by the inhibitory transmitter *GABA* (*gamma-amino butyric acid*). This influences selective activation of one or another PVN hormone-producing subregion (see Fig. 3).

The DLPFC is a working memory region, and is more closely connected with the hippocampus rather than the amygdala. It is involved with information processing at a higher level of abstraction than the OFC. For example, in monkeys, OFC lesions impair learning of changes in reward value within a stimulus dimension, whereas DLPFC lesions impair learning of changes in which dimension is relevant [17]. Dehaene and Changeux [16] considered dorsolateral prefrontal as a *generator of*

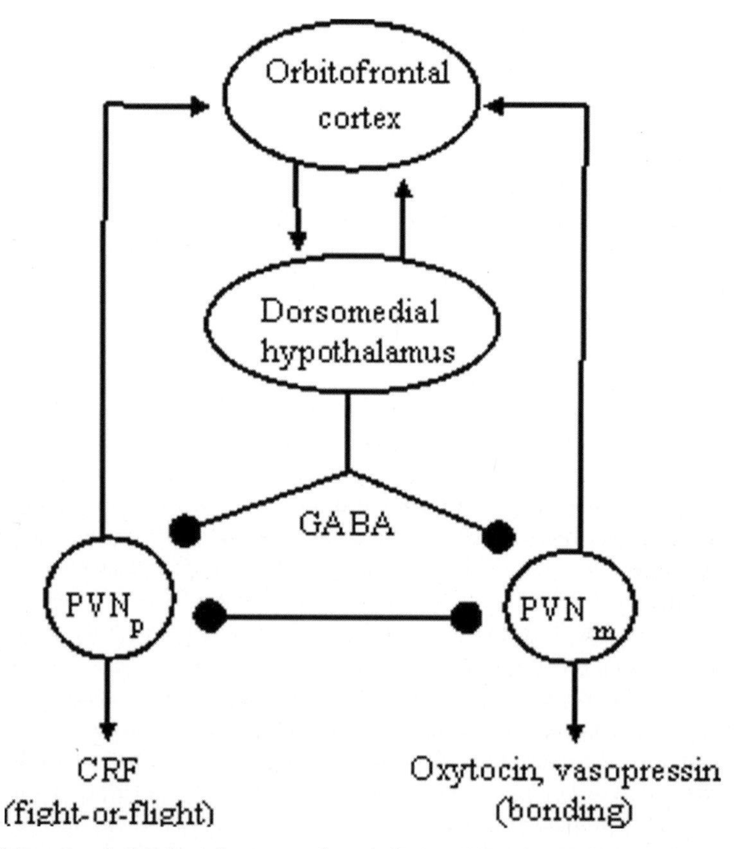

Fig. 3 Influences of orbitofrontal cortex on hypothalamus. GABA is an inhibitory neurotransmitter. The subscripts "p" and "m" stand for "parvicellular" and "magnocellular." Adapted from [18] with the permission of Springer-Verlag

diversity. That is, the DLPFC creates different possible decision rules, whereas OFC affective circuits filter possible rules based on rewards and punishments received from following these rules [15, 45].

The difference between the functions of OFC and DLPFC mirrors the previously discussed difference between "hot" amygdalar and "cold" hippocampal cognition. Yet the "emotional" orbital prefrontal and "task-relevant" dorsolateral prefrontal regions cannot remain entirely separate for effective cognitive functioning. Rather, emotional information and task-relevance information need to be integrated within the brain's overall executive system, so the network can decide when to continue fulfilling a task and when to allow urgent task-external considerations to override task fulfillment. The extensive connections between these two regions (e.g., [3, 49]) constitute one mechanism for integrating these two levels of function.

Yet brain imaging studies also suggest that a third prefrontal area, the ACC, plays an important role in this type of executive integration. Posner and his colleagues [9, 54] found that the ACC is activated when a subject must select or switch among different interpretations or aspects of a stimulus. Also, Yamasaki, LaBar, and McCarthy [58] gave subjects an attentional task with emotional distractors and measured responses of different brain regions to target (task-relevant) stimuli and to distracting stimuli. Yamasaki et al. found that while the areas with increased activation to the two types of stimuli were largely separate, the ACC was the unique area whose activation increased to both target and distracting stimuli.

The ACC is perhaps the most mysterious and subtle area of the whole brain. It was originally considered part of the limbic system rather than the cortex [50] because of the strong roles it plays in emotion related functions. Recent theories of anterior cingulate function have emphasized its role in detection either of potential response error or of conflict between signals promoting competing responses [4, 5, 28].

These three regions of the prefrontal cortex all subserve different aspects of what has come to known clinically as executive function. But in the United States system of government, "executive" has a more specialized meaning, describing one of the three coequal (at least in theory) branches of government, the other two branches being the legislative and the judicial. Let me suggest a fanciful yet plausible analogy between the three large regions of prefrontal cortex and the three branches of our tripartite government. OFC is legislative: as the integrator of social and affective information, and the judge of appropriateness of actions, it is the part of the system which has the "pulse of the people" (the "people" being the subcortical brain and its needs). DLPFC is executive in that sense: it has the level of specialized expertise that allows it to create and try out significant policies. And ACC is judicial: as an error-detector it acts as a brake of sorts on possible activities of the other two areas that stray too far from underlying principles of effective behavior.

Now that we have reviewed widely recognized functions for several subcortical and prefrontal regions, we want to fit them together into a bigger picture. In an earlier article [39], I introduced the ideas of neurobehavioral *angels* and *devils*. The word *angel* was used to mean a stored pattern of neural activities that markedly increases the probability that some specific behavior will be performed in some class of contexts. The pathways representing angels, so defined, include both excitation and inhibition at different loci. These pathways also link brain regions for perception of sensory or social contexts with other regions for planning and performing motor actions, under the influence of yet other regions involved both with affective valuation and with high-level cognition. The term *devil* was used in this article to mean the opposite of an angel; that is, a stored pattern of neural activities that markedly decreases the probability that a specific behavior will be performed in some class of contexts. The devils involve many of the same perceptual, motor, cognitive, and affective brain regions as do the angels, but with some differences in patterns of selective excitation and inhibition.

Finally, based on the usage of Nauta [45], whose work anticipated Damasio's on somatic markers, I used the term *censor* to mean an abstract behavioral filter that encompasses and functionally unifies large classes of what we call angels and devils. This is the level at which rules develop. Individual differences in the structure of these neural censors are tentatively related to some psychiatric classifications of personality and character, as well as to more versus less optimal patterns of decision making.

First we will look at how brain circuits can be put together to subserve the interactions among these angels, devils, and censors. Then we will consider how censors can change over a person's lifetime or with contextual factors, such as the degree that an environment is stressful or supportive.

2.5 Interactions Among These Brain Regions

What is the relationship between the neural representations of these censors and the neural representations of the specific angels and devils the censors comprise? The tentative "Big Picture" appears in Fig. 4, which shows one possible (not unique) network for linking angels and devils with censors. This provides a mechanism whereby activating a specific censor strengthens the angels and devils it is associated with. Conversely, activating a specific angel or devil strengthens the censors it is associated with. (Note: the *same* angel or devil can be associated simultaneously with two conflicting censors.)

Adaptive resonance theory or *ART* [10] has often been utilized to model learnable feedback interactions between two related concepts at different levels of abstraction. Since censors as defined in [39] are abstract collections of angels and devils, it seems natural to model their interconnection with an ART network. Figure 4 includes a possible schema for ART-like connections between angel/devil and censor subnetworks, showing the relevant brain areas.

In each of the two ART subnetworks of Fig. 4 – one subnetwork for individual actions and the other for rules, or classes of actions – the strengths of connections between the two levels can change in at least two ways. First, the interlevel connections undergo long-term learning such as is typical of ART. This is the method by which categories self-organize from vectors whose components are values of features or attributes. In this case actions are vectors of individual movements, whereas censors are vectors of angels and devils. The weight transport from the action module to the

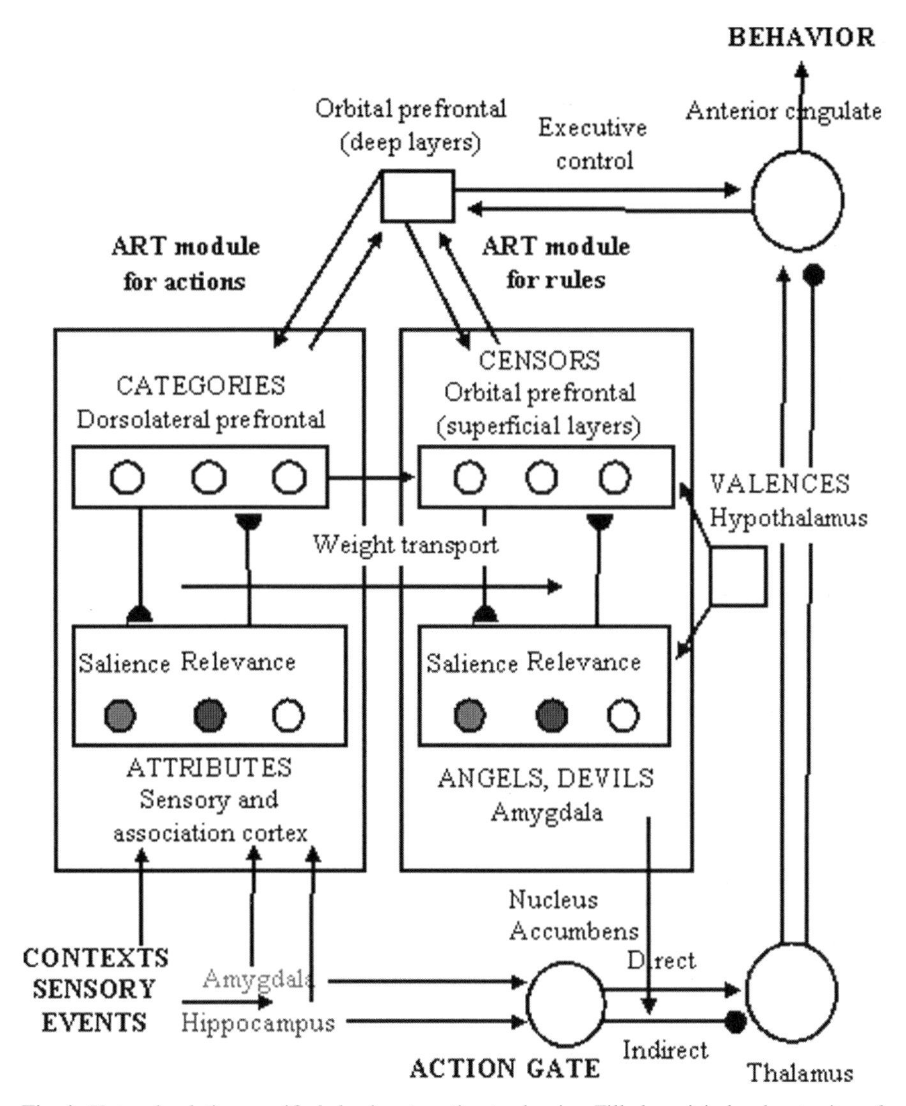

Fig. 4 Network relating specific behaviors to action tendencies. Filled semicircles denote sites of learning. "Blue-to-blue" and "red-to-red" connections are assumed be stronger than others at the action gate: that is, amygdalar activation selectively enhances the attribute of emotional salience, whereas hippocampal activation selectively enhances attributes related to task relevance. (Adapted from [39] with permission from Karger Publishers)

rule module is proposed as a possible mechanism for generalization from individual action categories to abstract censors that cover a range of actions. (Jani and Levine [29] discuss weight transport as a mechanism for analogy learning. The idea is often regarded as biologically implausible, but it might be implemented using neurons that combine inputs from two sources via multiplicative gating.)

Second, changes either in the current context or set of requirements created by the current task or environment can lead to temporary or permanent changes in the relative weightings of different lower-level attributes. This weight change in turn alters the boundaries of relevant categories. Of the wide variety of possible significant attributes, two are particularly important: task relevance and emotional salience. The influence of each of those two attributes in the network of Fig. 4 is biased by inputs to the two ART modules from other brain regions; notably, by inputs from the hippocampus and the amygdala.

The hippocampal inputs bias accumbens gating toward stimuli or actions that are appropriate for the current task. The amygdalar inputs, on the other hand, bias gating toward stimuli or actions that evoke strong emotions — whether such emotions arise from rational cognitive appraisal or from more primary, bodily-based desires or fears. Hence, prolonged or intense stress might shift the balance of the gating away from hippocampus-mediated actions based on task appropriateness toward amygdala-mediated actions based on short-term emotional salience.

3 Changes in Network Dynamics

The network of Fig. 4 is a first approximation, and may not be the entirely correct network for bringing together all the known functions of the brain areas involved in rule learning and selection. Note that some of the brain areas (such as the amygdala and the orbitofrontal cortex) play at least two separate roles, which thus far I have not been able to integrate. So it is possible that as more is known about subregions and laterality within each area, a more nuanced description will be available.

But taking this network as a first approximation to reality, we already note that the network is complex enough to provide a huge amount of flexibility in rule implementation. Which rules, or which censors, actually control behavior can change at any time, partly because of changing activities of the areas outside the ART modules. The hypothalamic valence inputs are influenced by which drives are prepotent in the need hierarchy network of Fig. 1. And there is evidence that the balance between hippocampal, relevance-related inputs and amygdalar, salience-related inputs is influenced by the organism's current stress level.

There is evidence from human and animal studies that many forms of stress (e.g., physical restraint, exposure to a predator, or social abuse) have short- or long-term effects on various brain structures [56]. Chronic or severe stress tends to reduce neural plasticity in the hippocampus, the site of consolidating memories for objective information. At the same time, stress increases neural plasticity and enhances neural activity in the amygdala, a site of more primitive emotional processing. Note from Fig. 2 that the hippocampus and amygdala both send inputs to the gates at the nucleus accumbens, which is the final common pathway for the angels and devils of [39]. It seems likely that the same individual may even be more "hippocampal-influenced" during periods of relaxation and more "amygdalar-influenced" during periods of time pressure.

Cloninger [12] describes the psychiatric process by which people ideally move toward more creativity with greater maturity, a process which Leven [37] related to the various prefrontal neural systems described herein. As described in [39], the process of personality change toward greater creativity does not mean we abandon censors, but it means we develop more creative and life-enhancing censors. For example, in a less creative state we might avoid extramarital sex because of a censor against disobeying a (human or divine) authority, whereas in a more creative state we might avoid the same behavior because of a censor against harming a healthy marital relationship.

Also as one develops greater mental maturity, the tripartite executive (or executive/legislative/judicial) system of the DLPFC, OFC, and ACC allows one to override our established censors under unusual circumstances. This can happen in contexts of strong cooperative needs (conjectured to be expressed via connections to orbitofrontal from dorsolateral prefrontal). An example concerns a man who normally refrains from stealing but would steal drugs if needed to save his wife's life [33]. It can also happen in contexts of strong self-transcendence (conjectured to be via connections to orbitofrontal from anterior cingulate) — as when a mystical experience overrides censors against outward emotional expression.

But developmental changes toward more complex angels and devils are not always total or permanent. They may be reversed under stress, or may depend on a specific mood or context for their manifestation. As Fig. 4 indicates, feedback between angels, devils, and censors means that more "amygdala-based" instead of "hippocampal-based" angels and devils will tend to lead to less creative censors. These shifts between "hippocampal" emphasis under low stress and "amygdalar" emphasis under high stress can be modeled by the combination of ART with attribute-selective attentional biases [38].

Hence, the rule network of Fig. 4 (or the much simpler needs network of Fig. 1 which influences it) could have a large number of possible attracting states representing the types of contextually influenced behavioral rules it tends to follow (*ceteris paribus*). Our discussion of personality development hints that some of these attracting states represent a higher and more creative level of development than others; hence, those states can be regarded as more optimal.

All this suggests that the problem can be studied, on an abstract level, using mathematical techniques for studying transitions from less optimal to more optimal attractors. A preliminary neural theory of transitions between rule attractors is discussed in the Appendix.

4 Conclusions

The theory outlined here for rule formation, learning, and selection is still very much a work in progress. The theory's refinement and verification await many further results in brain imaging, animal and clinical lesion analysis, and (once the interactions are more specified) neural network modeling using realistic shunting nonlinear differential equations. But I believe the theory creates a fundamentally plausible picture based on a wealth of known results about brain regions.

The marvelous plasticity of our brains has been called a double-edged sword [26]. If one of us is at any given moment operating in a manner that follows rules which are thoughtless and inconsiderate of our own or others' long-term needs, our plasticity gives us hope that at another time, with less stress or more maturity, we can act is a more constructive fashion. But our plasticity also tells us the opposite: that no matter how effectively and morally we are acting at any given moment, physical or emotional stress can later induce us to revert to more primitive rules and censors that will reduce our effectiveness.

Overall, though, the lessons from complex brain-based cognitive modeling are optimistic ones. The more we learn about our brains, and the more our neural network structures capture our brains' complexity, the more we develop a picture of human nature that is flexible, rich, and complex in ways many humanists or clinicians may not have dreamed of. To the therapist, educator, or social engineer, for example, neural network exercises such as this chapter should lead to an enormous respect for human potential and diversity. This is exactly the opposite of the reductionism that many people fear will result from biological and mathematical study of high-order cognition and behavior!

References

1. J. L. Armony, D. Servan-Schreiber, J. D. Cohen, and J. E. LeDoux. An anatomically constrained neural network model of fear conditioning. *Behavioral Neuroscience*, 109:246–257, 1995.
2. J. L. Armony, D. Servan-Schreiber, J. D. Cohen, and J. E. LeDoux. Computational modeling of emotion: Explorations through the anatomy and physiology of fear conditioning. *Trends in Cognitive Sciences*, 1:28–34, 1997.
3. J. F. Bates. Multiple information processing domains in prefrontal cortex of rhesus monkey. *Unpublished doctoral dissertation, Yale University*, 1994.
4. M. M. Botvinick, T. S. Braver, D. M. Barch, C. S. Carter, and J. D. Cohen. Conflict monitoring and cognitive control. *Psychological Review*, 108:624–652, 2001.
5. J. W. Brown and T. S. Braver. Learned predictions of error likelihood in the anterior cingulate cortex. *Science*, 307:1118–1121, 2005.
6. J. W. Brown, D. Bullock, and S. Grossberg. How the basal ganglia use parallel excitatory and inhibitory learning pathways to selectively respond to unexpected rewarding cues. *Journal of Neuroscience*, 19:10502–10511, 1999.
7. J. W. Brown, D. Bullock, and S. Grossberg. How laminar frontal cortex and basal ganglia circuits interact to control planned and reactive saccades. *Neural Networks*, 17:471–510, 2004.
8. R. M. Buijs and C. G. Van Eden. The integration of stress by the hypothalamus, amygdala, and prefrontal cortex: Balance between the autonomic nervous system and the neuroendocrine system. *Progress in Brain Research*, 127:117–132, 2000.
9. G. Bush, P. Luu, and M. I. Posner. Cognitive and emotional influences in anterior cingulate cortex. *Trends in Cognitive Science*, 4:215–222, 2000.
10. G. A. Carpenter and S. Grossberg. A massively parallel architecture for a self-organizing neural pattern recognition machine. *Computer Vision, Graphics, and Image Processing*, 37:54–115, 1987.
11. M. M. Cho, C. DeVries, J. R. Williams, and C. S. Carter. The effects of oxytocin and vasopressin on partner preferences in male and female prairie voles (microtus ochrogaster). *Behavioral Neuroscience*, 113:1071–1079, 1999.

12. R. Cloninger. A new conceptual paradigm from genetics and psychobiology for the science of mental health. *Australia and New Zealand Journal of Psychiatry*, 33:174–186, 1999.
13. M. A. Cohen and S. Grossberg. Absolute stability of global pattern formation and parallel memory storage by competitive neural networks. *IEEE Transactions on Systems, Man, and Cybernetics, SMC-13*, pp. 815–826, 1983.
14. M. Csikszentmihalyi. Flow: The psychology of optimal experience. *New York, Harper and Row*, 1990.
15. A. R. Damasio. Descartes? error: Emotion, reason, and the human brain. *New York: Grosset/Putnam*, 1994.
16. S. Dehaene and J. P. Changeux. The wisconsin card sorting test: Theoretical analysis and modeling in a neural network. *Cerebral Cortex*, 1:62–79, 1991.
17. R. Dias, T. W. Robbins, and A. C. Roberts. Dissociation in prefrontal cortex of affective and attentional shifts. *Nature*, 380:69–72, 1996.
18. R. Eisler and D. S. Levine. Nurture, nature, and caring: We are not prisoners of our genes. *Brain and Mind*, 3:9–52, 2002.
19. J. A. Fodor and Z. W. Pylyshyn. Connectionism and cognitive architecture: a critical analysis. *In S. Pinker and J. Mehler (Editors), Connections and Symbols, Cambridge, MA: MIT Press*, pp. 3–71, 1988.
20. M. J. Frank and E. D. Claus. Anatomy of a decision: Striato-orbitofrontal interactions in reinforcement learning, decision making, and reversal. *Psychological Review*, 113:300–326, 2006.
21. M. J. Frank, B. Loughry, and R. C. OReilly. Interactions between frontal cortex and basal ganglia in working memory. *Cognitive, Affective, and Behavioral Neuroscience*, 1:137–160, 2001.
22. D. Gaffan and E. A. Murray. Amygdalar interaction with the mediodorsal nucleus of the thalamus and the ventromedial prefrontal cortex in stimulus-reward associative learning in the monkey. *Journal of Neuroscience*, 10:3479–3493, 1990.
23. S. Grossberg. A neural model of attention, reinforcement, and discrimination learning. *International Review of Neurobiology*, 18:263–327, 1975.
24. S. Grossberg and D. S. Levine. Some developmental and attentional biases in the contrast enhancement and short-term memory of recurrent neural networks. *Journal of Theoretical Biology*, 53:341–380, 1975.
25. S. Grossberg and D. Seidman. Neural dynamics of autistic behaviors: Cognitive, emotional, and timing substrates. *Psychological Review*, 113:483–525, 2006.
26. J. M. Healy. Endangered minds: Why children don?t think and what we can do about it. *New York: Simon and Schuster*, 1999.
27. G. E. Hinton and T. J. Sejnowski, Learning and relearning in Boltzmann machines. In D. E. Rumelhart J. L. McClelland (Editors), *Parallel Distributed Processing*, Cambridge, MA, MIT Press, 1:282–317, 1986.
28. C. B. Holroyd and M. G. H. Coles. The neural basis of human error processing: Reinforcement learning, dopamine, and the error-related negativity. *Psychological Review*, 109: 679–709, 2002.
29. N. G. Jani and D. S. Levine. A neural network theory of proportional analogy-making. *Neural Networks*, 13:149–183, 2000.
30. W. Kilmer, W. S. McCulloch, and J. Blum. A model of the vertebrate central command system. *International Journal of Man-Machine Studies*, 1:279–309, 1969.
31. S. Kirkpatrick, Jr. C. D. Gelatt, and M. P. Vecchi. Optimization by simulated annealing. *Science*, 220:671–680, 1983.
32. A. H. Klopf. *The Hedonistic Neuron*. Washington, DC: Hemisphere, 1982.
33. L. Kohlberg. Essays on moral development: Vol. 1: The philosophy of moral development. *San Francisco: Harper and Row*, 1981.
34. G. F. Koob. Corticotropin-releasing factor, norepinephrine, and stress. *Biological Psychiatry*, 46:1167–1180, 1999.

35. J. E. LeDoux. *The Emotional Brain*. New York: Simon and Schuster, 1996.
36. J. E. LeDoux. Emotion circuits in the brain. *Annual Review of Neuroscience*, 23:155–184, 2000.
37. S. J. Leven. Creativity: Reframed as a biological process. *In K. H. Pribram (Editor), Brain and Values: Is a Biological Science of Values Possible?, Mahwah, NJ: Erlbaum*, pp. 427–470, 1998.
38. S. J. Leven and D. S. Levine. Multiattribute decision making in context: A dynamic neural network methodology. *Cognitive Science*, 20:271–299, 1996.
39. D. S. Levine. Angels, devils, and censors in the brain. *ComPlexus*, 2:35–59, 2005.
40. D. S. Levine, B. A. Mills, and S. Estrada. Modeling emotional influences on human decision making under risk. *In: Proceedings of International Joint Conference on Neural Networks*, pp. 1657–1662, Aug 2005.
41. P. D. MacLean. The triune brain, emotion, and scientific bias. *In F. Schmitt (Editor), The Neurosciences Second Study Program, New York: Rockefeller University Press*, pp. 336–349, 1970.
42. A. H. Maslow. Toward a psychology of being. *New York: Van Nostrand*, 1968.
43. S. M. McClure, M. S. Gilzenrat, and J. D. Cohen. An exploration-exploitation model based on norepinephrine and dopamine activity. *Presentation at the annual conference of the Psychonomic Society*, 2006.
44. P. R. Montague and G. S. Berns. Neural economics and the biological substrates of valuation. *Neuron*, 36:265–284, 2002.
45. W. J. H. Nauta. The problem of the frontal lobe: A reinterpretation. *Journal of Psychiatric Research*, 8:167–187, 1971.
46. J. Newman and A. A. Grace. Binding across time: The selective gating of frontal and hippocampal systems modulating working memory and attentional states. *Consciousness and Cognition,*, 8:196–212, 1999.
47. P. O'Donnell and A. A. Grace. Synaptic interactions among excitatory afferents to nucleus accumbens neurons: Hippocampal gating of prefrontal cortical input. *Journal of Neuroscience*, 15:3622–3639, 2005.
48. J. Olds. Physiological mechanisms of reward. *In M. Jones (Editor), Nebraska Symposium on Motivation, Lincoln: University of Nebraska Press*, pp. 73–142, 1955.
49. D. N. Pandya and E. H. Yeterian. Morphological correlates of human and monkey frontal lobe. *In A. R. Damasio, H. Damasio, and Y. Christen (Editors), Neurobiology of Decision Making, Berlin: Springer*, pp. 13–46, 1995.
50. J. W. Papez. A proposed mechanism of emotion. *Archives of Neurology and Psychiatry*, 38:725–743, 1937.
51. L. I. Perlovsky. Toward physics of the mind: Concepts, emotions, consciousness, and symbols. *Physics of Life Reviews*, 3:23–55, 2006.
52. L. Pessoa, S. Kastner, and L. G. Ungerleider. Attentional control of the processing of neutral and emotional stimuli. *Brain Research: Cognitive Brain Research*, 15:31–45, 2002.
53. L. Pessoa, M. McKenna, E. Gutierrez, and L. G. Ungerleider. Neural processing of emotional faces requires attention. *Proceedings of the National Academy of Sciences*, 99:11458–11465, 2002.
54. M. Posner and S. Petersen. The attention system of the human brain. *Annual Review of Neuroscience*, 13:25–42, 1990.
55. E. T. Rolls. The orbitofrontal cortex and reward. *Cerebral Cortex*, 10:284–294, 2000.
56. R. M. Sapolsky. Stress and plasticity in the limbic system. *Neurochemical Research*, 28:1735–1742.
57. J. G. Taylor and N. F. Fragopanagos. The interaction of attention and emotion. *Neural Networks*, 18:353–369, 2005.
58. H. Yamasaki, K. LaBar, and G. McCarthy. Dissociable prefrontal brain systems for attention and emotion. *Proceedings of the National Academy of Sciences*, 99:11447–11451, 2002.

Appendix

Rule Changing and Selection via Continuous Simulated Annealing

Transitions from less optimal to more optimal attractors (such as transitions between attractors that represent rules of conduct or categories of such rules) have been studied in neural networks using the well-known technique of *simulated annealing* [27, 31]. In simulated annealing, noise perturbs the system when it is close to a nonoptimal equilibrium, causing it to move with some probability away from that equilibrium and eventually toward the one that is most optimal. However, the simulated annealing algorithms so far developed have been mainly applicable to discrete time systems of difference equations. In preliminary work with two colleagues (Leon Hardy and Nilendu Jani), I am developing simulated annealing algorithms for shunting continuous differential equations, such as are utilized in the most biologically realistic neural network models.

An optimal attractor is frequently treated as a global minimum for some function that is decreasing along trajectories of the system, with the less optimal attractors being local minima of that function. For continuous dynamical systems, such a function is called a *Lyapunov function*. A system with a Lyapunov function will always approach some steady state, and never approach a limit cycle or a chaotic solution. The best known general neural system that possesses a Lyapunov function is the symmetric competitive network of Cohen and Grossberg [13].

Denote the right hand side of each of the Cohen-Grossberg equations by $F_i(x)$. Let x_0 be the optimal state and

$$\mathbf{x} = (x_1, x_2, \ldots, x_n)$$

be the current state. Let V be the Lyapunov function. Then the "annealed" Cohen-Grossberg equations (which are stochastic, not deterministic) are

$$dx_i = F_i(\mathbf{x})dt + G_i(\mathbf{x}, t)dW_i(t),$$

where W_i are the components of a *Wiener process* ("unit" random noise), and the G_i are increasing functions of $T = (V(\mathbf{x}) - V(\mathbf{x}_0)N(t)$. T is the continuous analog of what in the discrete case [31] is called *temperature*: it measures the tendency to escape from an attracting state that the network perceives as "unsatisfying." This temperature is the product of the deviation of current value of the Lyapunov function V from its global minimum and a time-varying signal gain function $N(t)$. The function $N(t)$, labeled "initiative," can vary with mood, interpersonal context, or level of the neurotransmitter norepinephrine which is associated with exploration [43].

An open mathematical question is the following. Under what conditions on the gain function $N(t)$ does the "annealed Cohen-Grossberg" stochastic dynamical system converge globally to the optimal attractor $\mathbf{x}_0$? Also unknown is what is a biological characterization of the Lyapunov function for that system. The Cohen-Grossberg system describes a network of arbitrarily many nodes connected by self-excitation and (symmetric) lateral inhibition, such as is utilized by perceptual systems but also, very likely, by the needs network of Fig. 1. But the Lyapunov function does not yet have a natural neurobiological characterization; nor is there yet a succinct biological description of what is different between its global and local minimum states.

Hence we are studying the annealing process in a simpler scalar equation where the global and local states are easy to characterize. That scalar equation is:

$$\frac{dx}{dt} = -(x-1)(x-2)(x-4) + T(x)N(t)(rand), \tag{1}$$

where T is temperature (a function of x), N is initiative (a function of t), and *rand* is a time-varying random number, normally distributed at any given time between 0 and 1.

The deterministic part of (1) is

$$\frac{dx}{dt} = -(x-1)(x-2)(x-4) \tag{2}$$

Equation 2 has three equilibrium points, 1, 2, and 4, of which 1 and 4 are stable and 2 is unstable. To determine which of 1 or 4 is "optimal," we first need to find the Lyapunov function for (2), which is the negative integral of its right hand side, namely

$$V(x) = \frac{1}{4}x^4 - \frac{7}{3}x^3 + 7x^2 - 8x.$$

Since $V(4) = -65.33$ is less than $V(1) = -3.083$, the attractor at 4 is the global minimum. Hence the temperature T of (1) is set to be the squared distance of the current value of x from 4, that is,

$$T(x) = (x-4)^2.$$

We simulated (1), in MATLAB R2006a, for initiative functions that were either constant at positive values or alternated in pulses between positive values and 0. The pattern of results is illustrated in Fig. 5. If the variable x started near the "nonoptimal" attractor 1, it would oscillate up and down around 1 for a time and then suddenly jump out of the range and toward the "optimal" attractor 4. Because of the randomness and variability of the value *rand*, the timing of getting out of the range was unpredictable. However, our early results confirmed the likelihood of global convergence to the optimal attractor if pulses in the initiative function occur, with at least some uniform positive amplitude, for arbitrarily large times.

We hope to extend these results to the Cohen-Grossberg and other multivariable neural systems that possess Lyapunov functions. Such results should provide insights about how often, and in what form, a system (or person) needs to be perturbed

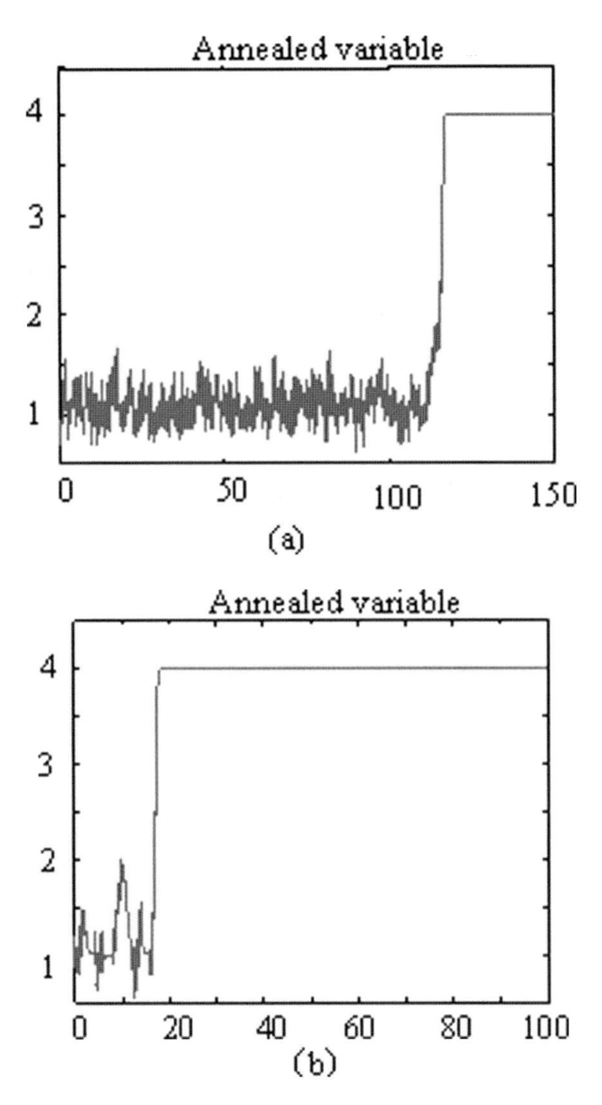

Fig. 5 Simulations of (1) with the temperature term equal to $(x-4)^2$ and the initiative term $N(t)$ equal to (a) 2 for all times; (b) alternating between 5 and 0 over time

to bring it to convergence toward an optimal set of rules. They should also illuminate possible biological characterizations of such optimal sets of rules at the level of the network of Fig. 4 (particularly the prefrontal components of that network.)

Part II
Cognitive Computing for Sensory Perception

Shape Recognition Through Dynamic Motor Representations

Navendu Misra and Yoonsuck Choe

Abstract How can agents, natural or artificial, learn about the external environment based only on its internal state (such as the activation patterns in the brain)? There are two problems involved here: first, forming the internal state based on sensory data to reflect reality, and second, forming thoughts and desires based on these internal states. (Aristotle termed these passive and active intellect, respectively [1].) How are these to be accomplished? Chapters in this book consider mechanisms of the instinct for learning (chapter PERLOVSKY) and reinforcement learning (chapter IFTEKHARUDDIN; chapter WERBOS), which modify the mind's representation for better fitting sensory data. Our approach (as those in chapters FREEMAN and KOZMA) emphasizes the importance of action in this process. Action plays a key role in recovering sensory stimulus properties that are represented by the internal state. Generating the right kind of action is essential to decoding the internal state. Action that maintains invariance in the internal state are important as it will have the same property as that of the represented sensory stimulus. However, such an approach alone does not address how it can be generalized to learn more complex object concepts. We emphasize that the limitation is due to the reactive nature of the sensorimotor interaction in the agent: lack of long-term memory prevents learning beyond the basic stimulus properties such as orientation of the input. Adding memory can help the learning of complex object concepts, but what kind of memory should be used and why? The main aim of this chapter is to assess the merit of memory of action sequence linked with a particular spatiotemporal pattern (skill memory), as compared to explicit memory of visual form (visual memory), all within an object recognition domain. Our results indicate that skill memory is (1) better than visual memory in terms of recognition performance, (2) robust to noise and variations, and (3) better suited as a flexible internal representation. These results suggest that the dynamic nature of skill memory, with its involvement in the closure of the agent-environment loop, provides a strong basis for robust and autonomous object concept learning.

L. I. Perlovsky, R. Kozma, *Neurodynamics of Cognition and Consciousness.*

1 Introduction

What does the pattern of activity in the brain mean? This is related to the problem of semantics [2] (also see chapters FREEMAN and KOZMA, this volume) or symbol grounding [3]. The question, as straight-forward as it seems, becomes quite complex as soon as we realize that there can be two different interpretations, as shown in Fig. 1. In (a), the task is to understand what is the meaning of the internal brain state of someone else's brain, while in (b), one wishes to understand, sitting within ones own brain, what the internal state means. The first task seems feasible, and it reflects how neuroscientists conduct their research. The second task seems impossible at first (it is reminiscent of Plato's allegory of the cave [4] or Searle's Chinese room [5]), but since this is how the brain operates, it should not be a problem at all. What is missing from this picture? In our previous work, we have argued that action plays an important role in recovering sensory stimulus properties conveyed only by the internal state (Fig. 2(a)) [6, 7]. Furthermore, we showed that action that maintains invariance in the internal state will have the same property as that of the represented sensory stimulus (Fig. 2(b)). Thus, generating the right kind of action (the kind that maintains internal state invariance) amounts to decoding the internal state.

One limitation of the sensorimotor agent is how the approach can generalize to learning more complex object concepts, since moment-to-moment invariance cannot be maintained while traversing the contour of a complex object. The limitation is due to the reactive nature of the sensorimotor interaction in the agent, i.e., the lack of long-term memory. Memory is needed for learning beyond the local orientation of the input. We will investigate how adding memory can help the learning of complex object concepts.

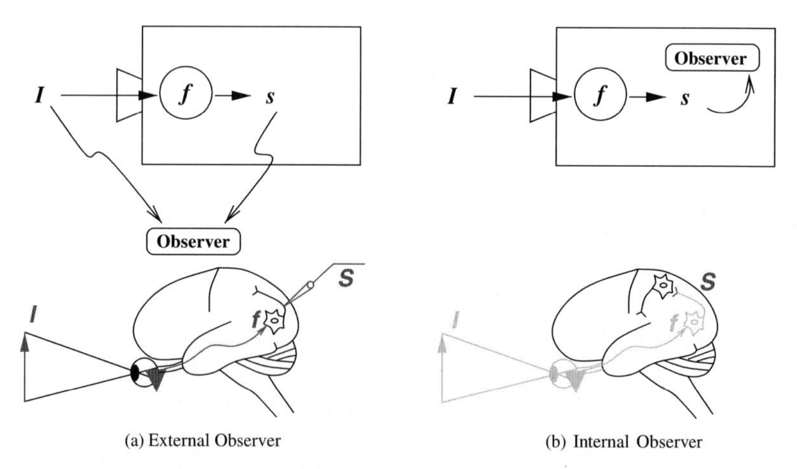

Fig. 1 External vs. internal perspective on understanding internal brain state **(a)** The diagram for external observer, on the left, demonstrates how the input I creates a spike pattern s. Since the observer has access to both of these one can infer what stimulus property is conveyed by s. **(b)** However, in the internal observer model, on the right, there is no direct access to the outside environment and as such one can only monitor ones internal sensory state or spike pattern. (Adapted from [7])

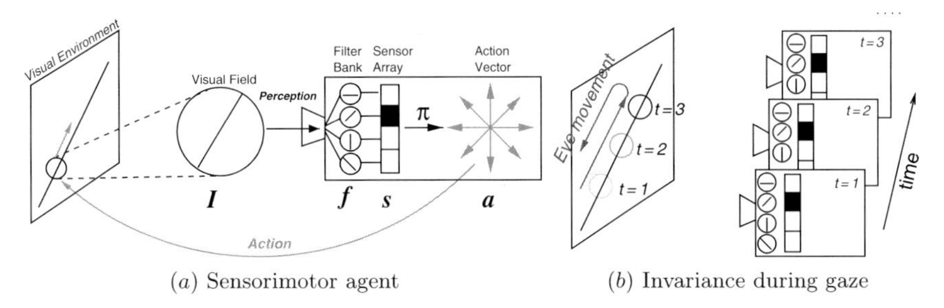

(*a*) Sensorimotor agent (*b*) Invariance during gaze

Fig. 2 Understanding internal state in a sensorimotor agent (**a**) The visual field I receives stimulus from a small region of the visual environment. This is fed to the filter bank f that activates the sensor array (of orientation filters). The agent then performs certain actions a based on the sensory state s, which may in turn affect the sensory state in the next time step. (**b**) By moving in the diagonal direction, the internal state stays invariant over time. The property of the action (traversing diagonally) exactly matches the stimulus property (diagonal orientation) signaled by the invariant internal state. (Adapted from [7])

Let us consider what the internal state activation pattern will look like over time as the sensorimotor agent traverses the contour of a form, an octagon for example (Fig. 3). As we can see from the activation pattern of the internal state in (*b*) and (*d*), moment-to-moment invariance is maintained only when a straight stretch of

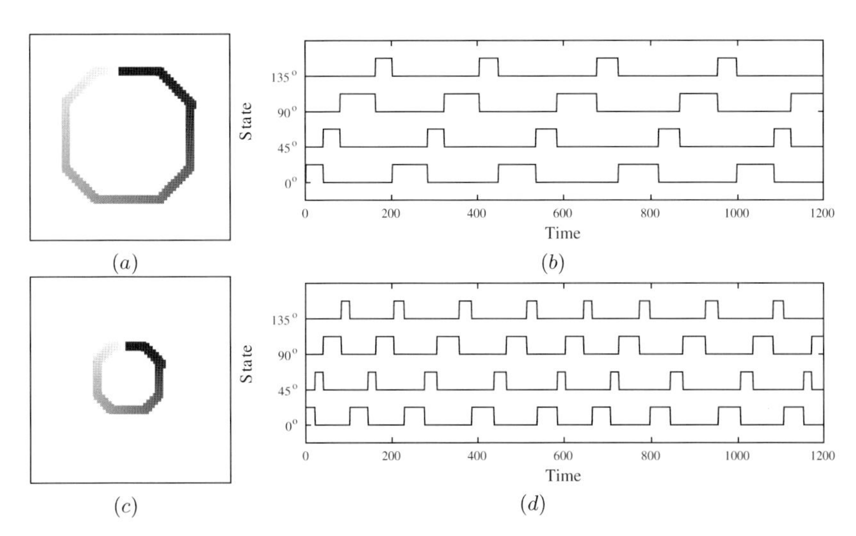

(*a*) (*b*) (*c*) (*d*)

Fig. 3 Tracing a complex object. The sensorimotor agent?s gaze over time as it traces an octagon input and its corresponding internal state is shown. (**a**) The agent?s gaze is plotted over 250 time steps, where the grayscale indicates the time step (white is time 0, and black is time 249). (**b**) The activation state of the four neurons in the agent is shown over 1200 time steps. As the agent repeatedly directs its gaze around the octagon, the internal state also shows a repeating pattern, from 0°, 45°, 90°, 135°, back to 0°, for example (the pattern in the interval [200, 449]). Note that the trace in (**a**) corresponds to the interval [0, 249] in (**b**). (**c**) and (**d**) show the same information as in (**a**) and (**b**), but for a smaller input. Note that the period of the activity is shorter in (**d**) than in (**b**), as the length to traverse is shorter in (**c**) compared to (**a**)

the octagon is being traversed. One observation here is that if the changing spatiotemporal pattern of the internal state is treated as a representational unit, then a corresponding complex action sequence maintaining invariance in this repeating spatiotemporal pattern can be found that has a property congruent with that of the stimulus object (in this case, an octagon). Memory plays an important role here, since such an invariance in the spatiotemporal pattern has to be detected: You need to know your own action sequence over time, and the corresponding spatiotemporal pattern.

An important issue is to assess the merit of memory of such a action sequence linked with a particular spatiotemporal pattern (skill memory or motor representation; or "fixed action pattern'' [8]), as compared to explicit memory of visual form (visual memory or sensory representation), all within an object recognition domain. (Note that these concepts are analogous to episodic vs. procedural memory in psychology research [9].) As we can see from Fig. 3(*b*) and (*d*), the spatiotemporal pattern and its associated action sequence, if linearly scaled in time, will exactly match each other. Properties like these can be beneficial when serving as a representational basis for invariant object recognition.

In order to assess the relative merit of skill memory, we tested object recognition performance based on skill memory vs. visual memory, and tested how easy it is to map from an arbitrary representation to either skill memory or visual memory representation. Our results indicate that skill memory is (1) better than visual memory in terms of recognition performance, (2) robust to noise, and (3) better suited as a flexible internal representation. These results indicate that the dynamic nature of skill memory, with its involvement in the closure of the agent-environment loop, provides a strong basis for robust object concept learning and generalization.

The rest of this chapter is organized as follows: The following Sect. 2 briefly reviews related work, and Sect. 3 provides details about input preparation and training procedure. Section 4 presents the main computational experiments and results, and we finally conclude (Sect. 6) after a brief discussion (Sect. 5).

2 Background

The importance of action and interaction in perception has been identified early on, as documented in the works of Lashley on motor equivalence, the concept that different effectors can be used to generate the same kinematic pattern, [10] (also see [11, 12]) and those of Gibson on ecological perception [13]. The ensuing active vision movement [14, 15, 16] and embodied robotics [17] continued investigation in this area. In this section, we will review related works putting an emphasis on the role of action in sensory perception.

As discussed in the introduction, an action-oriented approach may be able to solve the problem of internal understanding. Even when the stimulus is not directly accessible, through action and the associated change in the internal state, key properties of the stimulus can be recovered [6, 7].

Experimental results suggest that action and the motor cortex actually play an important role in perception [18, 19, 20, 21, 22], providing support for the above idea. The implication is that sensorimotor coordination may be a necessity for autonomous learning of sensory properties conveyed through sensory signals.

There are ongoing research efforts in the theories of the sensorimotor loop [23, 24, 25, 26, 27, 28], developmental robotics [29, 30, 31, 32], bootstrap learning [33, 34], natural semantics [35], dynamical systems approach to cognition [36, 37, 38], embodied cognition [39], imitation in autonomous agents [40, 41, 42, 43, 44, 45, 46, 47], etc. that touch upon the issue of sensorimotor relationship. More recent works have specifically looked at the role of action in perception and perceptual organization [48, 49], information flow [50], and the role of internal models in action selection and behavior [51]. However, these approaches have not focused on the question of how the brain can understand itself. (The works by Freeman [2, 52] and Ziemke and Sharkey [53] provide some insights on this issue.)

3 Methods

3.1 Input Preparation

The input to the neural network was prepared by randomly generating the action sequence and the 2D array representation for three different shapes (circles, triangles, and squares). Each of the representations was defined to have the same input dimension.

3.1.1 Shape Generation

The generation of the three types of shape was done by following a simple algorithm. These algorithms were constructed using a LOGO-like language [54]. In this language instructions for navigating a space is provided by a fixed range of actions, i.e., “turn left”, “move forward”, and so on that are then plotted. This language construct was adapted for the formation of the shape generation algorithms in the following manner. Initially a starting point was chosen for each of the figures. Then a series of steps were produced to allow for a full rendition of the entire shape. The algorithm was parameterized so as to produce images that were scaled and translated. As different coordinates were traversed the corresponding 2D array points were marked. The benefit of the LOGO-like algorithm was that it allowed for the easy capture of the action sequence for the particular shape. This was the case because the actions produced to traverse the object could be captured as the direct representative action sequence for a figure. Since each of the algorithms was parameterized, a sequence of random values for scaling and translating the images were provided.

3.1.2 Visual Memory – 2D Array

Visual memory representation is a direct copy of the sensory data. As a result, when the 2D array representation for the figure is visualized it appears like the figure itself. All the shapes for the 2D arrays were generated using the algorithm described above. The output was a two dimensional array with the pixels constituting the contour of the figure set to one and the background to zero. On this a Gaussian filter was applied. This caused the values in the array to have more continuous values. The primary reason for this last step was to have a more fair comparison between 2D array and action sequence. The normalized range of values was between 0 and 1 and the resultant size of the array was 30×30. Figures 4, 5 and 6 provide examples of the different shapes in the 2D array representation.

3.1.3 Skill Memory – Action Sequence

Skill memory representation, action sequence, involves the retention of actions that an agent may have performed while navigating the environment. As explained earlier the action sequence was generated by utilizing an algorithm based on the LOGO language. The produced output action sequence had four actions; motion north, motion south, motion east, and motion west. This was represented by values 0, 90, 180, and 270 degrees. The values were subsequently normalized to lie between the range 0 to 1. Before normalization the action vectors were smoothed. Smoothing was accomplished by taking an average of values representing the neighboring action vectors, as specified by the size of the smoothing window. This resulted in a less discrete change of action vectors. This difference is illustrated in Fig. 7. In (*a*) the action sequence is not smoothed and the sequence of actions constructed has stairstep-like variations. However, in (*b*) the actions get averaged to form a new action that is formed by averaging the neighboring action vectors. Figure 7 shows the action vectors at the coordinates where these actions were performed. Figures 8, 9 show the same smoothing effect but for triangles and squares.

Another issue with the action sequence was that its length (number of individual action steps in the sequence) was not fixed, unlike the visual representation where it was fixed to $30 \times 30 = 900$. This meant that resizing of the action sequence had to

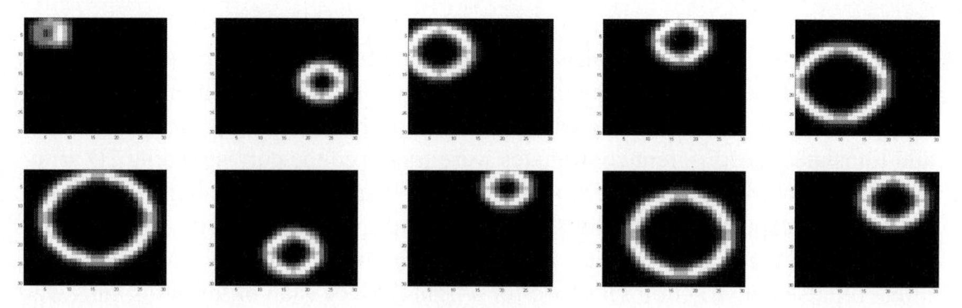

Fig. 4 The visual representation of circles. This sequence of figures illustrates the range of variations that are performed on the circle shape in the visual memory (2D array) representation

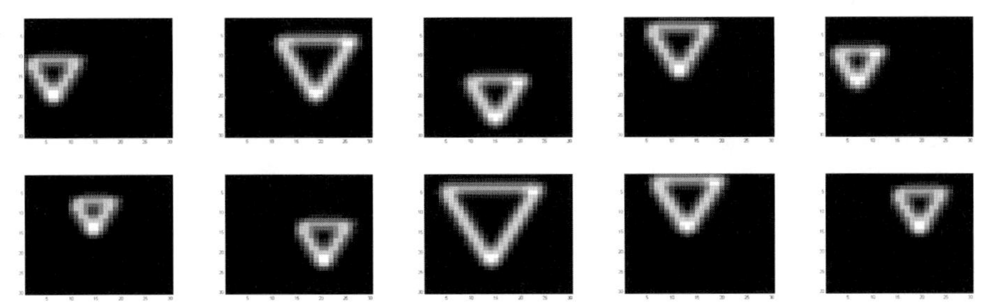

Fig. 5 The visual representation of triangles. This sequence of figures illustrates the range of variations that are performed on the triangle shape in the visual memory (2D array) representation

be done to match the input dimension ($30 \times 30 = 900$). A very simplistic algorithm was devised to resize or stretch the action sequence. The end result was that each action sequence size was of 900 dimensions.

3.2 Object Recognition and Representation Mapping

The experiments were formulated by creating one thousand randomly scaled and randomly translated figures for each shape category (triangle, circle, and square) with their corresponding action sequence and 2D array representations. The patterns generated for the action sequence and 2D array representations were stored in their respective data sets. A portion of this data set (75%) was used for training a neural network and the rest was used for testing (25%). Using the testing data set, the performance of each of the representations was recorded. A detailed explanation is given in the following sections.

There were a total of ten runs for each experiment. For each of these trials the training set and the test set were chosen at random from the data set for each class (circle, square, and triangle). This was composed of a thousand points for action sequence and 2D array for each of the three figures, resulting in action sequence and 2D array data set of three thousand each. For each run, at random, 75% of

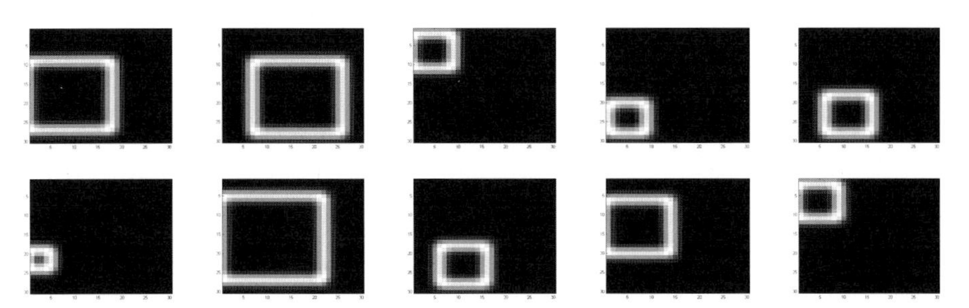

Fig. 6 The visual representation of squares. This sequence of figures illustrates the range of variations that are performed on the square shape in the visual memory (2D array) representation

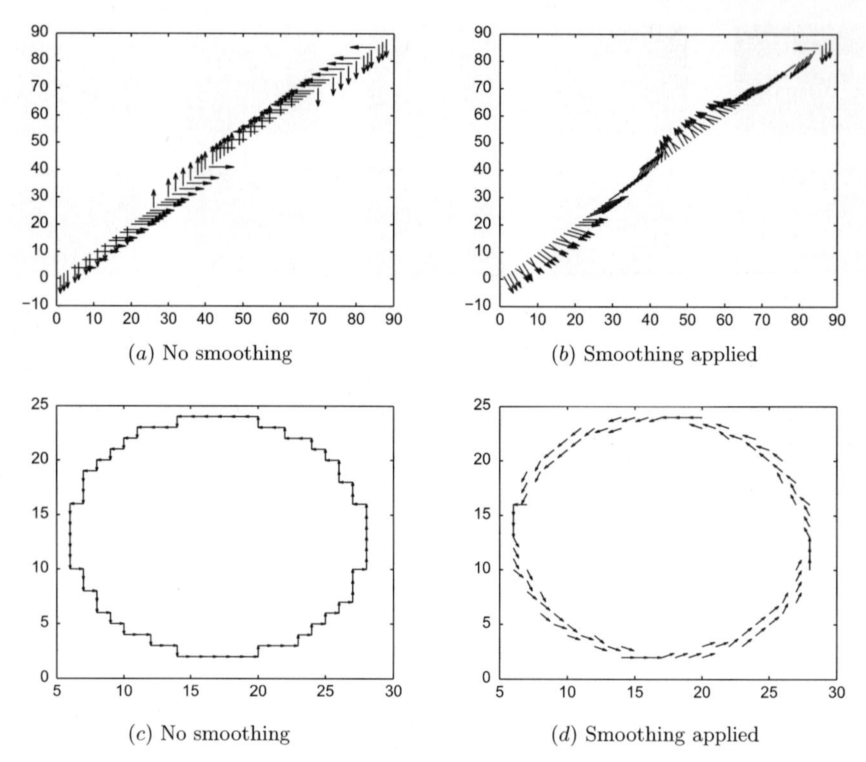

Fig. 7 The action sequence for a circle shape. (**a**) This plot demonstrates a linear ordering of the action sequence for a circle shape before any smoothing is applied. Note in this figure the discrete change in action vectors. (**b**) This plot demonstrates a linear ordering of the action sequence for a circle shape after smoothing is applied. Here the action vectors have been smoothed to display a continuous variation in actions. (**c**) This plot demonstrates a 2 dimensional view of the action sequence for a circle shape before any smoothing is applied. Note in this figure the discrete change in action vectors in the stairstep-like formation. (**d**) This plot demonstrates a 2 dimensional view of the action sequence for a circle shape after smoothing is applied. The smoothed action vectors show a more intuitive sequence of actions

the data set was chosen as training data and the rest was used as the test data. The training set was provided to the neural network, and trained using backpropagation [55, 56, 57]. The neural network had 900 input neurons, 10 hidden neurons in the single hidden layer and 3 output neurons corresponding to the 3 shape classes. For the representation mapping experiment, the target vectors were modified to be the actual visual and skill memory representations. In this case the number of output neurons was increased to 900.

Average classification rate on the test set was a measure that was used to gauge the relative performance for both visual and skill memory. The classification rate for each trial recorded the average number of times the actual output deviated from the target vector of the three output neurons. A threshold of 0.5 was set so that if the deviation of the output neuron activation was within this value then the particular

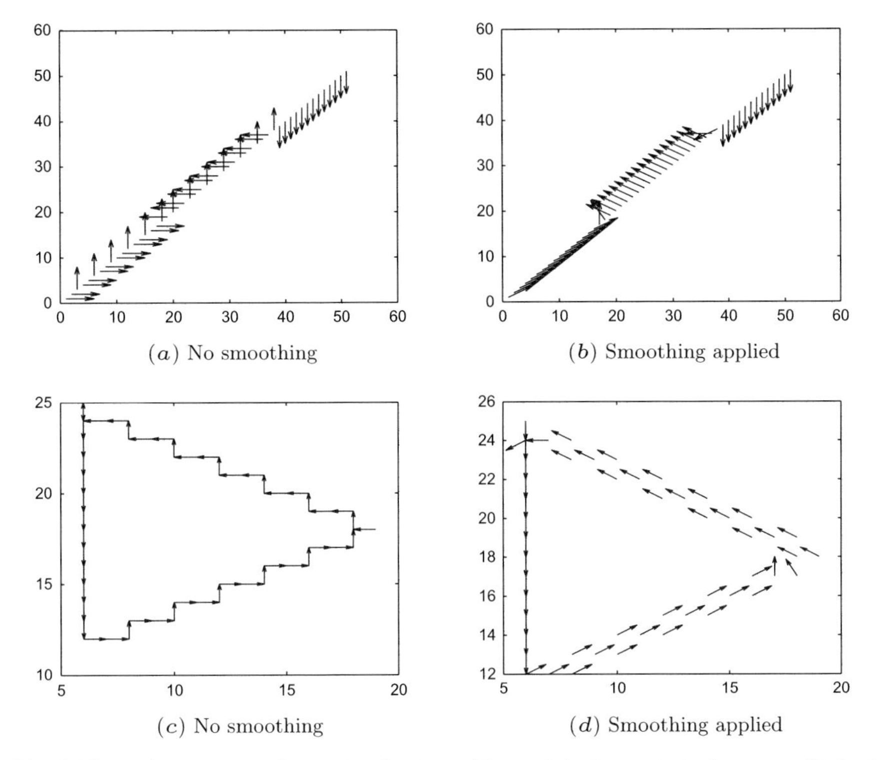

(a) No smoothing (b) Smoothing applied

(c) No smoothing (d) Smoothing applied

Fig. 8 The action sequence for a triangle shape. These plots demonstrate the same effect of smoothing as mentioned in Fig. 7, except that they show the case of a triangle

input was claimed to be properly classified. The average classification rate was then acquired by running the experiments ten times and taking the mean and standard deviation of the values. Student's t-test was used to measure the significance of the differences [58].

Another measure is the mean squared error (MSE). This value represents the average of all the squared deviations of the output values from the exact target values. MSE gives a general idea of how well the mapping was learned in case hard classification is not possible.

4 Computational Experiments and Results

In order to evaluate the effectiveness of each of the memory representations there needs to be a comprehensive evaluation of each of the memory systems with respect to the performance measures specified in the previous section. These performance measures were used to primarily demonstrate the relative difference between the two

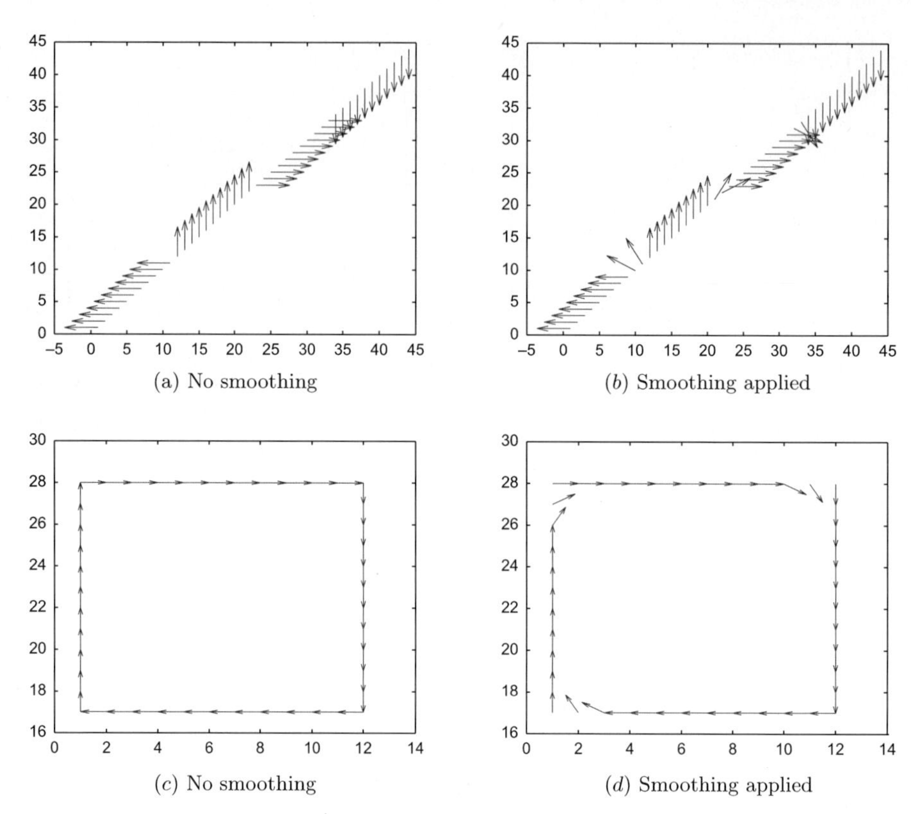

Fig. 9 The action sequence for a square shape. These plots demonstrate the same effect of smoothing as mentioned in Fig. 7, except that they show the case of a square

memory systems rather than provide a mechanism for absolute comparison with a general pattern recognition approach that may seek to maximize the performance.

4.1 Visual Memory vs. Skill Memory in Recognition Tasks

The overall speed of learning, measured using MSE, is illustrated in Fig. 10. MSE values for each of the curves were calculated by taking an average of ten trials for each of the two memory representations. The neural network was allowed to train for one thousand epochs. As can be seen from Fig. 10, the error rate for skill memory is consistently lower than that of visual memory. Also after about 200 epochs the MSE comes close to zero for skill memory while visual memory can only reach an MSE value of about 0.1 after the full period of one thousand epochs. The results clearly demonstrate that the neural network can more easily learn the various action sequences in skill memory.

The differences between skill memory and visual memory are further emphasized in Fig. 11. Here the average classification rate on the test sets is shown using

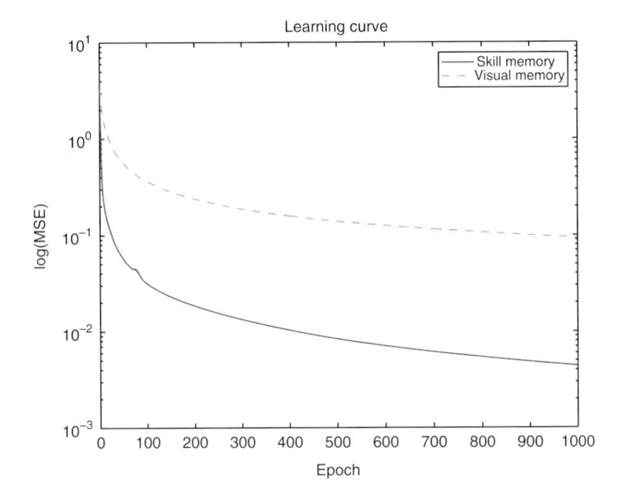

Fig. 10 Learning curve for visual and skill memory. This plot shows the average learning curve for both skill and visual memory (10 trials each). From this plot we can see that skill memory is learned faster and more accurately. On the other hand visual memory still has a higher MSE even after 1,000 epochs

the bar chart with the error bars representing the 95% confidence intervals. The average classification rate for visual memory was 0.28 while for skill memory it was almost four times higher, close to 0.97. The difference was significant under t-test ($p = 0, n = 10$). In sum, the action-based skill memory was significantly easier to learn than the bitmap-based visual memory, both in terms of speed and accuracy.

4.2 Skill Memory with Variations

The performance of skill memory was measured under variations in the formation of the action sequence. These variations included (1) changes in the smoothing window size, (2) variations to the number of starting points in object traversal for a particular action sequence, and (3) noise in action sequence.

4.2.1 Smoothing Window Size

Smoothing was applied to all the action sequence that was generated. The effect of this has already been illustrated in the previous section. The average classification rate increases slightly with an increase in the window size. However, a further increase in widow size causes the average classification rate to decrease slightly. As a result the default smoothing window size was chosen to be three. Figure 12 shows how smoothing affects the average classification rate for skill memory. These values were averages taken by performing ten trials. Difference between window

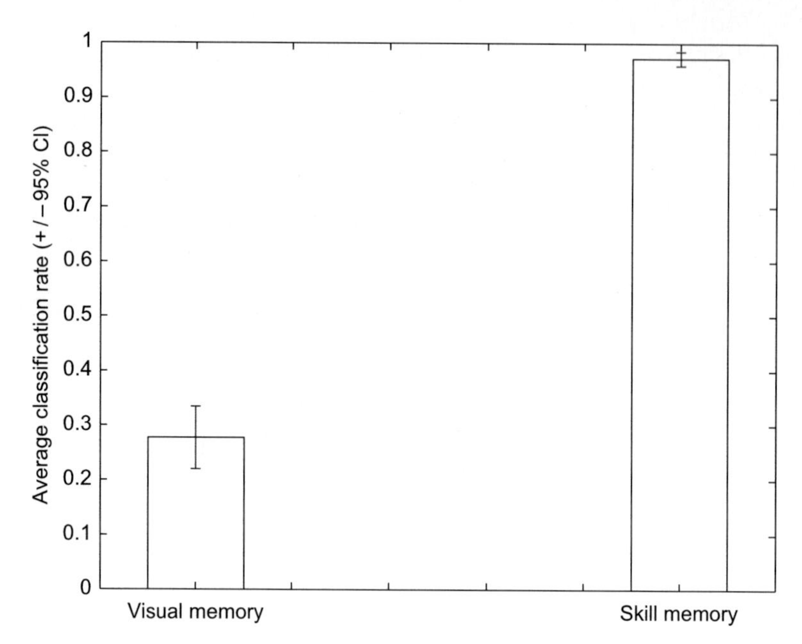

Fig. 11 The average classification rate of visual and skill memory. This bar chart shows the average classification rate of skill and visual memory on the test set (± 95% confidence interval). Skill memory has a smaller variance and higher average classification rate representing a more consistently good performance as opposed to visual memory

size 3 and size 0 was significant (t-test, $p < 0.0002, n = 10$), but the difference between window size 6 and 3 was not (t-test, $p > 0.35, n = 10$).

4.2.2 Random Starting Points

Another variation was to test how the classification rate was impacted by different starting points on the input object chosen for the action sequence generation. This was implemented by varying the number of starting locations for each action sequence. Having different starting points is an important variation because the way the memory representation system was originally setup, all the action sequences were generated by having the trajectory start at the same relative location on the shape and as a consequence the action sequence generated would not have much variation. Figure 13 shows a 2D view of the action sequence for a square and possible positions where the action sequence sequence may start. Figure 14 shows the corresponding 1D view.

The overall effect of adding different starting points was that the average classification rate decreased with increasing number of alternative starting points (shown in Fig. 15). However, even with a high variation in the possible starting points the average classification rate for skill memory was still higher than visual memory. These values were averages taken by performing ten trials with varying training and test data and a constant smoothing window size of three. Differences

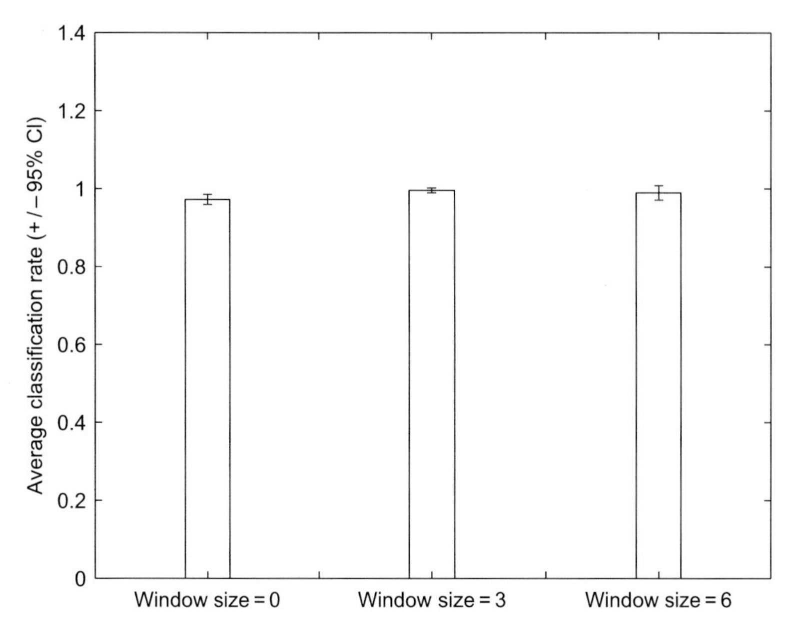

Fig. 12 The effect of smoothing on the classification rates for skill memory. This bar chart shows the effect of varying window size on the average classification rate (± 95% confidence interval). Window size 3 yielded the most optimal performance with the lowest variation

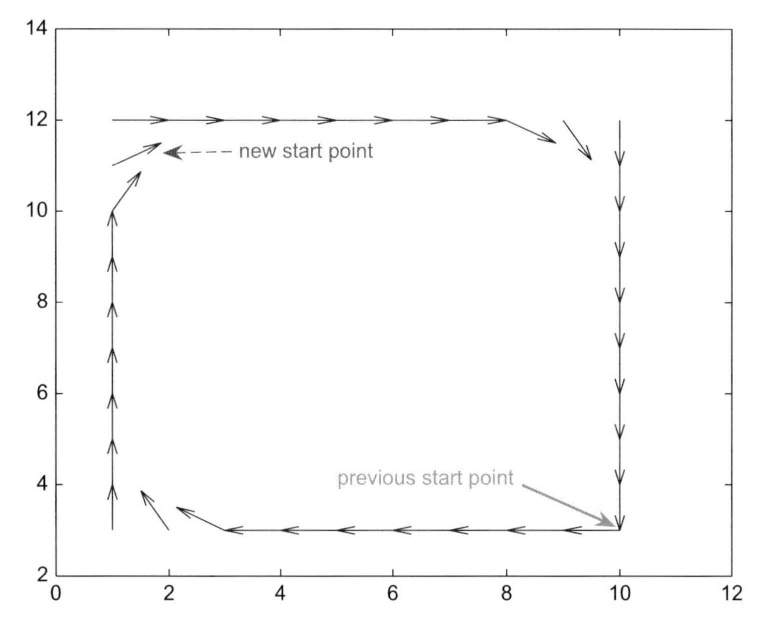

Fig. 13 The 2D view of the action sequence for a square with smoothing. The two arrows point to the two different locations where the action sequences may have started

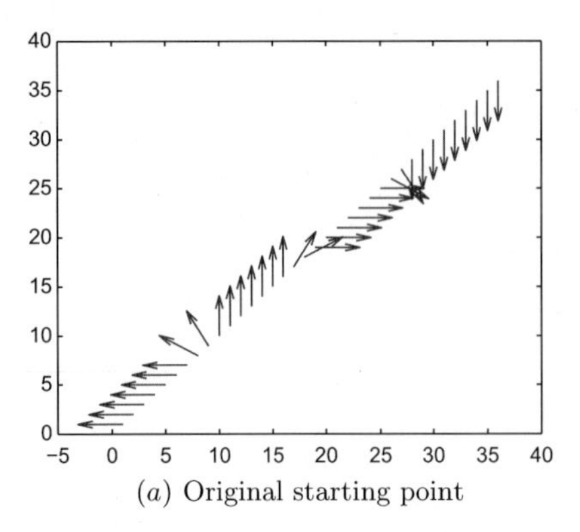

(*a*) Original starting point

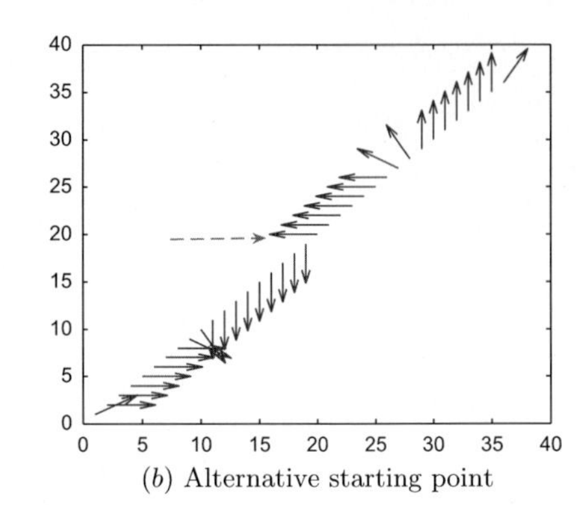

(*b*) Alternative starting point

Fig. 14 The linear ordering of the action sequence for a square with differing starting points. The plot in (**a**) shows how the action sequence will appear if the action sequence generation started from the original starting point. The plot in (**b**) shows how the action sequence will appear if the action sequence generation started from the new starting point (shown in Fig. 13). The dashed arrow in (**b**) points to the original starting location. The only difference with the original version is that the action sequence is shifted, but that is enough to affect the classification accuracy

between the visual against all the skill memory trials were significant under t-test ($p < 0.00015, n = 10$). In sum, skill memory was significantly easier to learn than visual memory, even when the task was made harder for skill memory. Note that the performance would suffer greatly if action sequence generation can start from an arbitrary point on the shape. However, humans typically start their trajectories from a stereotypical position, so the starting point may not be too arbitrary in practice. Here, the purpose was mainly to show how skill-based memory performs under typical conditions.

4.2.3 Motor Noise

In humans, tracing the contour of a form usually results in small variations in the traced trajectory. The last variation we tried was the introduction of noise in the action sequence. This means that a random error at some point in time during the creation of the action sequence occurred causing the trajectory to deviate from its normal course. The action sequences were generated as before, however, at random an angle between 0 and 360 was added based on the magnitude of the noise factor. An example of noise in the action sequence is shown in Fig. 16. In this figure a noise factor of 0.1 was used.

The noise factor is the probability that affects the magnitude by which an action vector's angle in space may be affected. Hence, a larger noise factor will mean a larger deformation of the shape that a particular action sequence may trace. Figure 17 shows how the classification rate decreases with the increase in noise. However, even at higher noise levels, skill memory is still able to outperform visual memory. Differences between visual against all skill memory trials were significant under t-test ($p < 0.002, n = 10$). This demonstrates that skill-based memory is

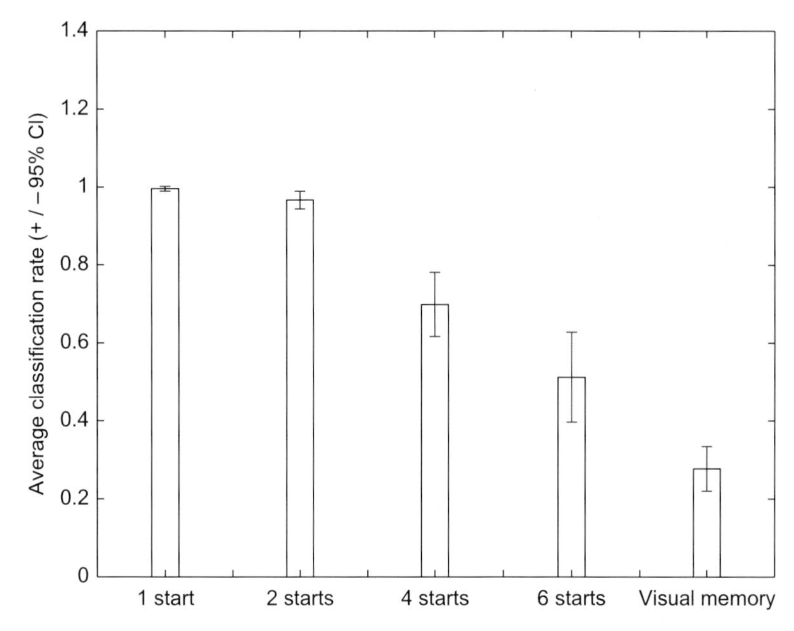

Fig. 15 Effect of increasing the number of alternative starting points in skill memory on the classification rate. This bar chart shows the change in the average classification rate as the number of trajectory starting points is increased (± 95% confidence interval). As the number of random starting points is increased, the average classification rate steadily drops, but in all cases skill memory performs better than visual memory

resilient to noise in motor sequence. The reason for the robustness is due to the fact that despite the noise, the components of skill memory (action vectors) do not change in number or position. Only their orientations change. In visual memory on the other hand, noise may increase or decrease the number and change the position of active pixels.

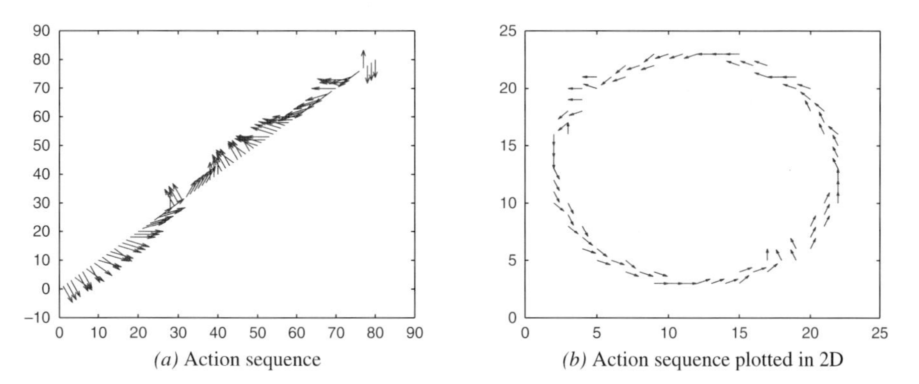

Fig. 16 The action sequence with noise for a circle shape. The plot shows a 1D and a 2D view of the action sequence for a circle after the application of random noise (noise factor 0.1) and after smoothing

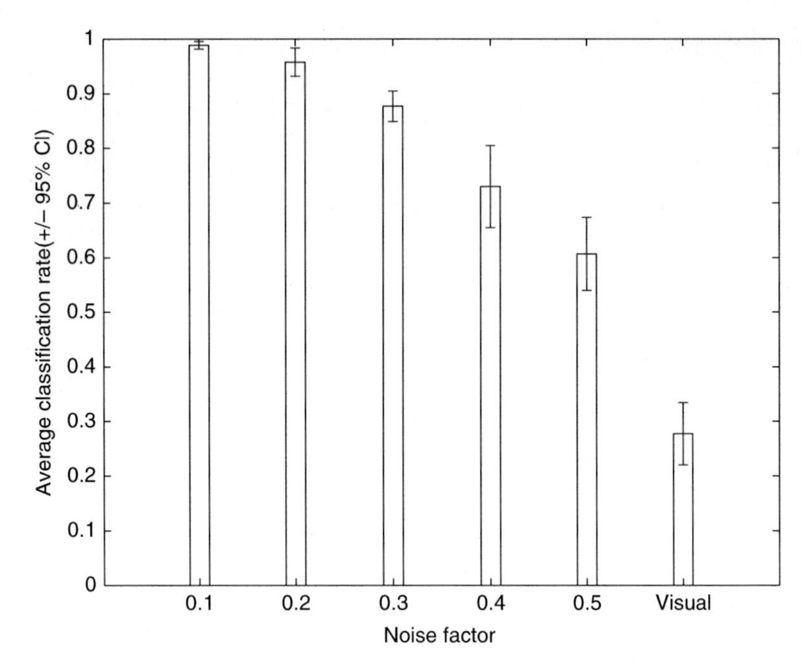

Fig. 17 The effect of motor noise on classification rate. This bar chart shows the effect of increasing noise on the average classification rate of skill memory ($\pm$ 95% confidence interval). Visual memory is shown here as a baseline. The effect of noise can be observed clearly from this bar chart. However, notice that skill memory is quite resilient to noise and outperforms visual memory at noise factor of 0.5

4.3 Representation Mapping: Action as an Intermediate Representation

In order to test the hypothesis that action sequence (skill memory) may serve as a good intermediate representation of sensory information, the following test was devised. The different mappings were; action to action, visual to action, action to visual, and visual to visual representation. If the learning for visual to action is easier with respect to sensory to sensory mapping (e.g. visual to visual), then that would indicate that in fact sensory information can be easily represented in terms of action. This idea coupled with the primary view that action-based memory may perform better at object recognition tasks further supports the idea that skill memory is a more ideal form of memory. The reason for this is that if the sensory to action mapping was very difficult then the performance advantage that skill memory holds may become less pronounced and there by limit the applicability of action-based memory.

To verify this tests involving the different mappings were conducted. This test was carried out on a neural network with 900 inputs and 900 outputs, corresponding to the 900 dimensional input for each representation. Figure 18 shows the results of the experiment. The figure shows that the learning curve for visual to action mapping

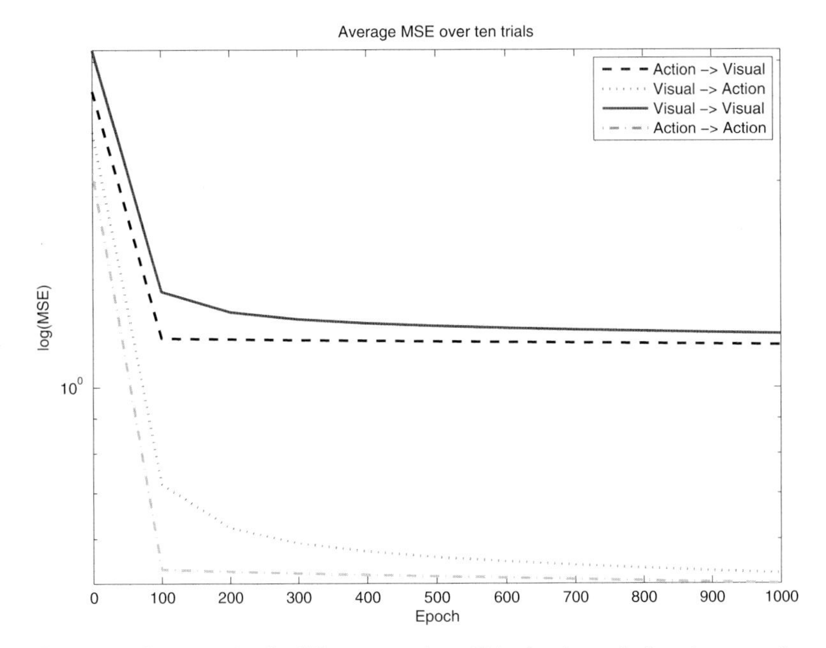

Fig. 18 The learning curve for the different mappings. This plot shows the learning curve for each representation mapping. From the learning curves we can infer how well each mapping performs with respect to other mappings. As expected mapping to action representation (skill memory) is easier to learn

is as low as action to action mapping ($p = 0.37$, n = 10). The figure also shows that the action to visual memory is slightly easier to learn than visual to visual mapping (however, t-test showed that $p = 0.82$, n = 10, indicating that the difference was not significant). All other differences were significant under t-test ($p < 0.026$, $n = 10$). This bolsters our idea that action may be a good intermediate representation for sensory data. This support makes action-based memory more appealing. As we will discuss in detail in Sect. 5.1, the demonstrated superiority of action-based memory is independent of the particular neural network learning algorithm used. So, we expect the use of any standard supervised learning algorithm to give similar results as reported above. Furthermore, psychological experiments can be conducted to test our main claim, that action-based representation help improve object recognition performance. See Sect. 5.2 for an expanded discussion on this matter.

5 Discussion

Analysis of the results in the previous section indicates that skill-based memory representation performs better than visual memory in recognition tasks. It has been additionally demonstrated that even under quite severe variations skill-based memory is able to yield results which indicates its merits. It has been further demonstrated that action may serve as a good intermediate representation for sensory information. In the following, we will discuss why we believe these results may hold independent of the particular learning algorithm used (Sect. 5.1) and how our general claims can be verified in psychological experiments (Sect. 5.2). We will also discuss the relation of our work to human memory research, and provide some future directions.

5.1 Properties of Action Sequence

The primary reason why skill memory yields such impressive results is because of its ability to capture the core discriminating property of the respective shapes. This is the case because aspects such as size and location of the figure do not cause variations in the resized action sequence. Hence the action sequence for different-sized shapes was similar in the end. This makes it easy for the neural network to learn the skill-based representation. Variations introduced to the action sequence to compensate for the apparent advantage, i.e., using noise and random start methods did not affect the results. However, even large variations did not cause visual memory to outperform skill-based memory.

The properties of the action sequence and the 2D array representations can be further analyzed as in Fig. 19. In this figure, the Principal Components Analysis (PCA) plots along two principal component axes are shown for the data points in the action sequence and the visual representations. Figure 19(b) shows that the PCA plot of skill-based representation has three distinct clusters for the three classes of input shapes. On the other hand, the PCA plot for visual memory (Fig. 19(a)) has all the data points almost uniformly scattered and overlapping, indicating that making proper class distinctions may be difficult. Such an analysis provides some insights on why skill memory performs better than visual memory in object recognition tasks.

Figure 20 shows that with the introduction of noise the class boundaries become less pronounced for skill memory. With low noise the three distinct clusters for the corresponding classes are maintained. As the noise factor is increased the clusters become less compact and it becomes slightly harder to determine the class boundaries. However, even with high noise the class boundaries can still be determined more or less. These plots help us understand why the neural network was marginally less able to properly recognize skill-based representations when the noise was high.

With the introduction of varying starting points many more clusters appear in the PCA plot, as illustrated in Fig. 21. However, it is interesting to note that the local clusters are more compact as opposed to the broader clusters that emerge with the addition of noise in Fig. 20. That is, data points from the same class are scattered around but they locally form tight non-overlapping clusters.

5.2 Related Psychological Experiments

One of the main assumptions of this research is that action can be represented along a time series that are scaled to be of the same length. However, one may question

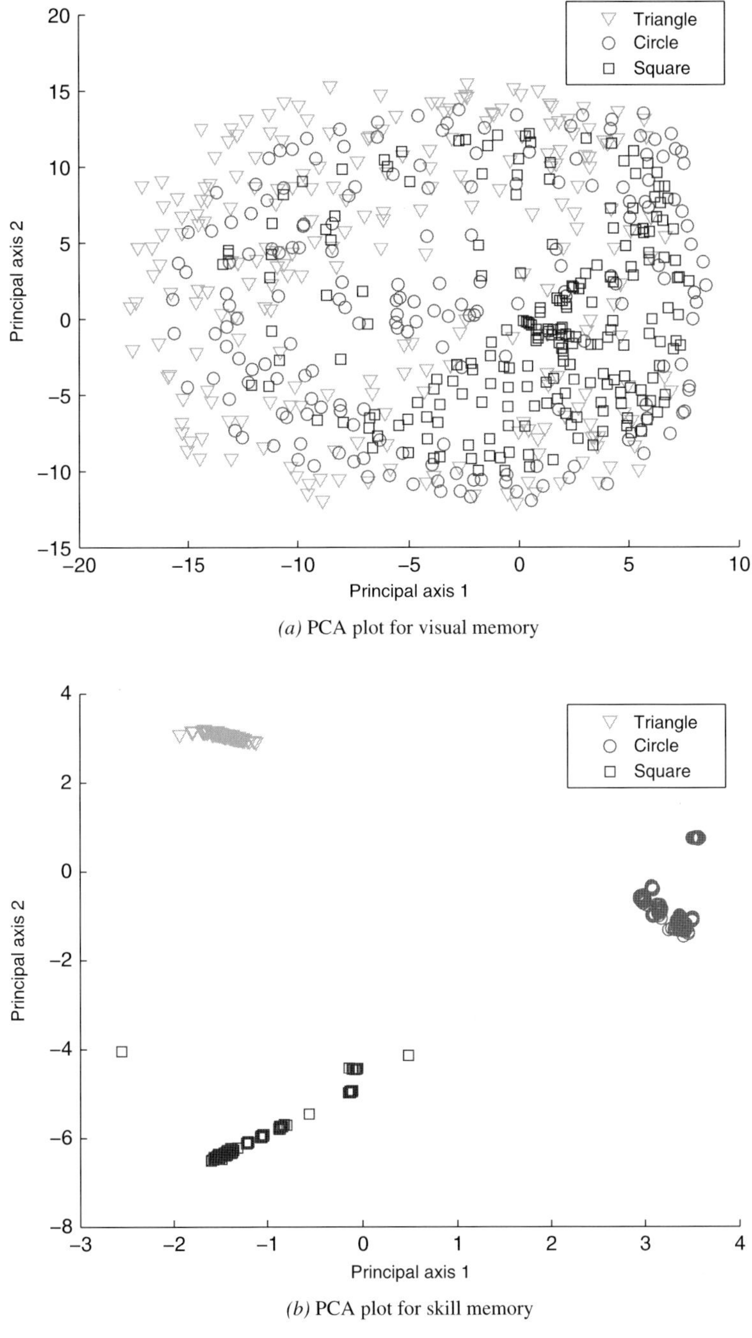

(*a*) PCA plot for visual memory

(*b*) PCA plot for skill memory

Fig. 19 The plot of PCA projection for visual and skill memory. (**a**) PCA projection for 2D array representation (visual memory) of the input data along the first two principal axes is shown. (**b**) PCA projection for action sequence representation (skill memory) of the input data along the first two principal axes is shown. Skill memory shows a much more separable clustering than visual memory

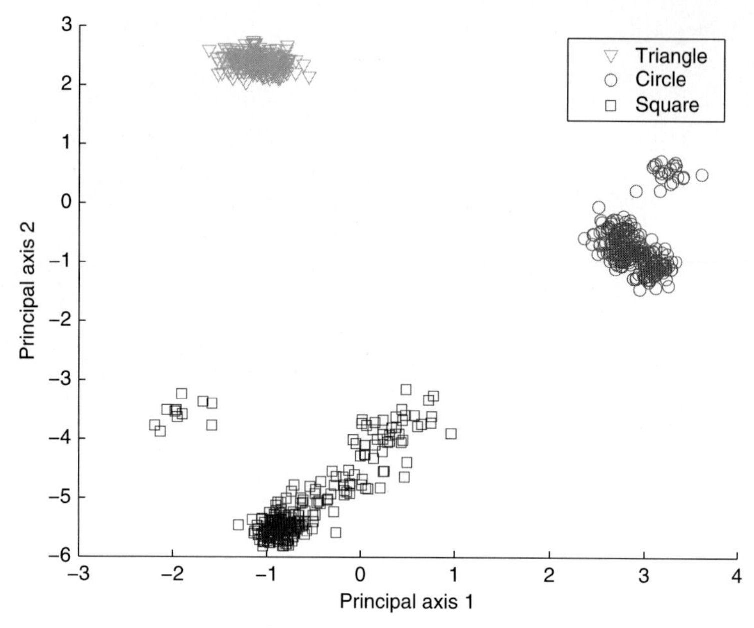

(a) PCA plot for skill memory with low noise

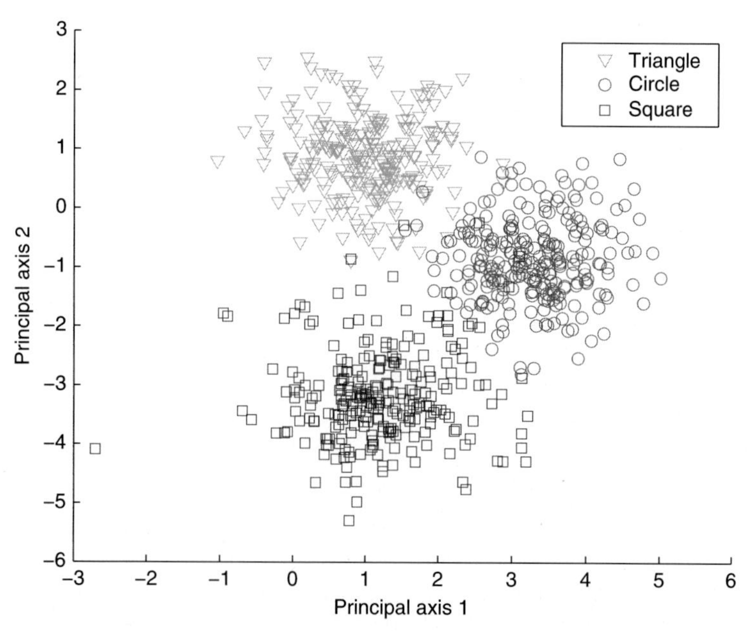

(b) PCA plot for skill memory with high noise

Fig. 20 The projection on the two principal axes for skill memory with varying noise factor. (**a**) PCA projection for skill memory with a lower noise factor of 0.1 along the first two principal axes is shown. (**b**) PCA projection for skill memory with a high noise factor of 0.5 along the first two principal axes is shown. Even at high noise, the clusters are still quite separable

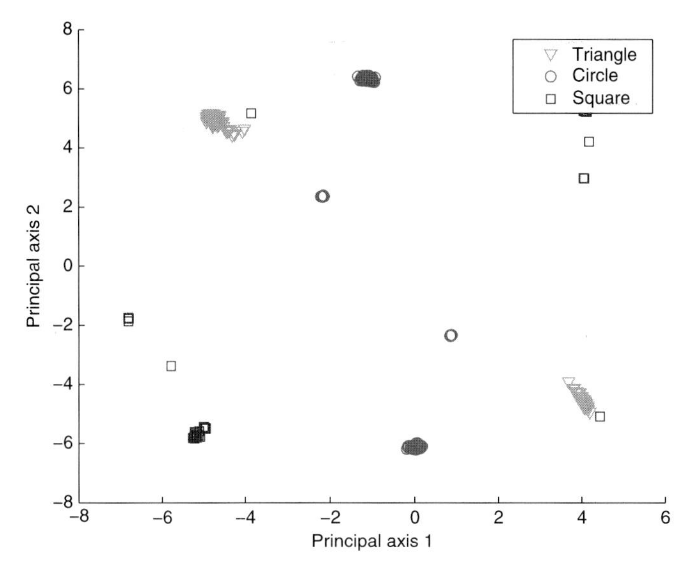

(a) PCA plot for skill memory with few starting points

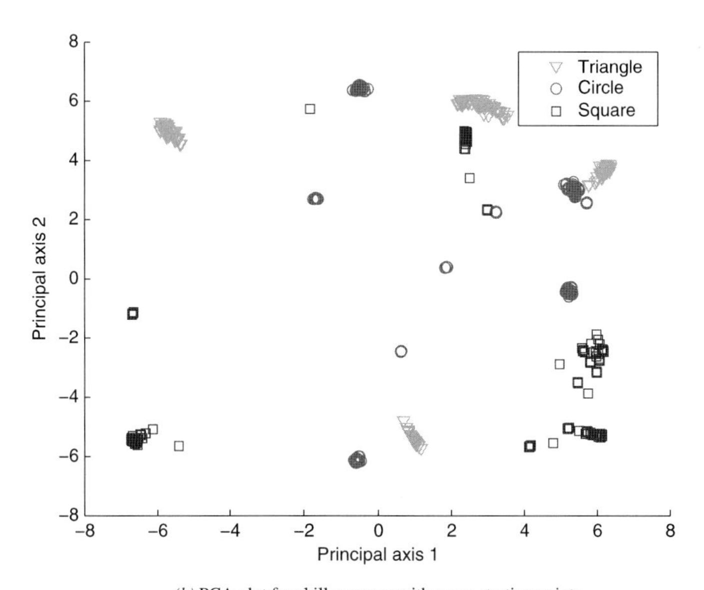

(b) PCA plot for skill memory with many starting points

Fig. 21 The projection on the two principal axes for skill memory with varying number of random starting points. (**a**) PCA projection for skill memory with two random starting points for action sequence generation along the first two principal axes is shown. (**b**) PCA projection for skill memory with four random starting points for action sequence generation along the first two principal axes is shown. With an increase in the number of starting points, the number of subclusters for each category increases, but there is minimal overlap across categories (at least locally)

the validity of creating an action sequence and scaling such an action sequence. This also leads one to question, at what intervals are the actions stored and should this interval be long or short. All of these questions can be answered by the recent experiments performed by Conditt et al. [59]. The result of their experiment suggests that when humans are asked to perform a series of actions, the actions tend to be represented as time invariant. This means that humans do not store the actions parameterized by absolute time. More precisely, humans do not have a timing mechanism that stores the exact duration between actions. This form of representation allows humans to counteract disturbances in the environment. A disturbance can result in the delay in the completion of an action for a given sequence. However, most humans are able to go along and complete the rest of action. Such experimental evidence allows us to be more confident about representing action sequence in the way we did in this chapter.

The action sequences produced in these experiments seem to be the same for each shape, thus making the job of learning trivial and therefore it appears that skill memory had an unfair advantage. Rather than this be a criticism against the validity of the research, it points out the fact that action sequence is not affected by size and translation of object. The similarity in action sequences further point out the core thesis of this chapter, that action sequence as represented in skill memory may be an inherently superior representation scheme than the raw sensory information as in visual memory, because of its ability to capture properties of the object when time can be scaled with ease.

Finally, we can devise psychological experiments to test whether action-based skill memory can improve object recognition performance. One way to test the influence of the motor system in recognition is to immobilize the motor system during recognition tasks, or to dynamically eliminate motor-induced refreshing of sensory input (e.g., using eye-tracking devices). We predict that with the motor system temporarily turned off, object recognition performance will degrade significantly. There exists indirect evidence that the motor system plays an important role in perceptual tasks. Hecht et al. showed that pure motor learning enhanced perceptual performance, through an action-perception transfer [18]. Naito et al., on the other hand, showed that the primary motor cortex is involved in somatic perception of limb movement [19]. These results suggest an intimate role for the motor system in perceptual tasks.

5.3 Relation to Memory in Humans

The two memories, visual and skill, are analogous in many ways to the types of memory employed by natural agents. Natural agents have episodic and procedural memory [9]. Episodic memory is fact-based where certain information about events is stored. It is currently believed that these events are temporarily stored in the hippocampus [60] [61]. This may have similarities to visual memory described above. On the other hand, skill memory can be thought of as being similar to procedural memory. Procedural memory deals with the ability to recall sequential steps required for a particular task that an agent may perform [62]. It may be interesting to investigate if the main results derived in this chapter applies to the understanding of human memory and recognition.

Note, however, that the results presented here are insufficient to explain how the human memory actually works. Rather, what is presented here only suggests that skill-based memory may have theoretical virtue regarding perceptual understanding, as compared to visual memory.

5.4 Future Work

The future expansion of this research topic involves the actual implementation of the skill-based memory system in an autonomous agent, as well as psychological experiments to systematically test the merits of skill-based memory. Other variations to the action sequence format can be studied such as methods that retain only the changes in the sequence of actions, i.e. only when there is a certain change in action, rather than retaining the total sequence. The resultant action sequence for shapes like square will have only four points, since in the traversal of a square the action vectors will need to be only changed four times. Another possibility is to use the relative difference between successive action directions, which will give rotation invariance as well as the other two invariances (translation and scale) already achieved by our approach.

5.5 Contributions

The primary contribution of this research is the demonstration that skill-based memory has beneficial properties that can aid in perceptual understanding. These properties are in line with other research that suggested that action is a fundamental component for learning simple properties. However, in this chapter we were able to demonstrate that action plays an important role in learning complex objects when the system was allowed to have memory. This research clearly demonstrates how action can be incorporated into a powerful autonomous learning system. Another important observation is that when things are represented dynamically, then certain invariant properties can be naturally stored, i.e., simply changing the time scale of action generation is sufficient.

6 Conclusion

The study of memory systems, arose from the desire to develop a memory system that would allow autonomous agents to learn about complex object properties. The most basic memory system that an agent can have is the direct (raw) storage of the sensory data (such as visual memory). Another system is skill-based memory, which primarily involves the retaining of action sequences performed during a task. Skill memory was anticipated to be a better representation because of the crucial role action played in simple perceptual understanding [6, 7]. To test this hypothesis, we compared the two memory representations in object recognition tasks. The two primary performance measures, average classification rate and MSE, revealed the superior properties of skill memory in recognizing objects. Additionally, a related experiment demonstrated convincingly that action can serve as a good intermediate representation for sensory data. This result provides support for the idea that various sensory modalities may be represented in terms of action (cf. [63, 26]).

Based on the above results, we conclude that the importance of action in simple perceptual understanding of objects can successfully be extended to that of more complex objects when some form of memory capability is included. In the future, the understanding we gained here is expected to help us build memory systems that are based on the dynamics of action that enable intrinsic perceptual understanding.

Acknowledgment The main results presented here is largely based on an unpublished thesis by NM [64]. The text was substantially revised for this chapter by the authors. We would like to thank the editors of this volume for their constructive suggestions.

References

1. Barnes, J.: Aristotle. In Gregory, R.L., ed.: The Oxford Companion to the Mind. Oxford University Press, Oxford, UK (2004) 45–46
2. Freeman, W.J.: How Brains Make Up Their Minds. Wiedenfeld and Nicolson Ltd., London, UK (1999) Reprinted by Columbia University Press (2001).
3. Harnad, S.: The symbol grounding problem. Physica D **42** (1990) 335–346
4. Plato: Plato's Republic. Hackett Publishing Company, Indianapolis (1974) Translated by G. M. A. Grube.
5. Searle, J.: Is the brain a digital computer? In Grim, P., Mar, G., Williams, P., eds.: The Philosopher's Annual, Atascadero, CA, Ridgeview Publishing Company (1990) Presidential address (American Philosophical Association, 1990).
6. Choe, Y., Bhamidipati, S.K.: Autonomous acquisition of the meaning of sensory states through sensory-invariance driven action. In Ijspeert, A.J., Murata, M., Wakamiya, N., eds.: Biologically Inspired Approaches to Advanced Information Technology. Lecture Notes in Computer Science 3141, Berlin, Springer (2004) 176–188
7. Choe, Y., Smith, N.H.: Motion-based autonomous grounding: Inferring external world properties from internal sensory states alone. In Gil, Y., Mooney, R., eds.: Proceedings of the 21st National Conference on Artificial Intelligence. (2006) 936–941.
8. Llinás, R.R.: I of the Vortex. MIT Press, Cambridge, MA (2001)
9. Silberman, Y., Miikkulainen, R., Bentin, S.: Semantic effect on episodic associations. Proceedings of the Twenty-Third Annual Conference of the Cognitive Science Society **23** (1996) 934–939
10. Lashley, K.S.: The problem of serial order in behavior. In Jeffress, L.A., ed.: Cerebral Mechanisms in Behavior. Wiley, New York (1951) 112–146
11. Moore, F.C.T.: Bergson: Thinking Backwards. Cambridge University Press, Cambridge, UK (1996)

12. Bergson, H.: Matter and Memory. Zone Books, New York, NY (1988) Translated by Nancy Margaret Paul and W. Scott Palmer.
13. Gibson, J.J.: The Perception of the Visual World. Houghton Mifflin, Boston (1950)
14. Aloimonos, J.Y., Weiss, I., Bandopadhay, A.: Active vision. International Journal on Computer Vision **1** (1988) 333–356
15. Bajcsy, R.: Active perception. Proceedings of the IEEE **76** (1988) 996–1006
16. Ballard, D.H.: Animate vision. Artificial Intelligence **48** (1991) 57–86
17. Brooks, R.A.: Intelligence without representation. Artificial Intelligence **47** (1991) 139–159
18. Hecht, H., Vogt, S., Prinz, W.: Motor learning enhances perceptual judgment: A case for action-perception transfer. Psychological Research **65** (2001) 3–14
19. Naito, E., Roland, P.E., Ehrsson, H.H.: I felt my hand moving: A new role of the primary motor cortex in somatic perception of limb movement. Neuron **36** (2002) 979–988
20. Held, R., Hein, A.: Movement-produced stimulation in the development of visually guided behavior. Journal of Comparative and Physiological Psychology **56** (1963) 872–876
21. Bach y Rita, P.: Brain Mechanisms in Sensory Substitution. Academic Press, New York (1972)
22. Bach y Rita, P.: Tactile vision substitution: Past and future. International Journal of Neuroscience **19** (1983) 29–36
23. O'Regan, J.K., Noë, A.: A sensorimotor account of vision and visual consciousness. Behavioral and Brain Sciences **24(5)** (2001) 883–917
24. Philipona, D., O'Regan, J.K., Nadal, J.P.: Is there something out there? Inferring space from sensorimotor dependencies. Neural Computation **15** (2003) 2029–2050
25. Philipona, D., O'Regan, J.K., Nadal, J.P., Coenen, O.J.M.D.: Perception of the structure of the physical world using unknown multimodal sensors and effectors. In Thrun, S., Saul, L., Schölkopf, B., eds.: Advances in Neural Information Processing Systems 16, Cambridge, MA, MIT Press (2004) 945–952
26. Humphrey, N.: A History of the Mind. HarperCollins, New York (1992)
27. Hurley, S.: Perception and action: Alternative views. Synthese **129** (2001) 3–40
28. Granlund, G.H.: Does vision inveitably have to be active? In: Proceedings of the 11th Scandinavian Conference on Image Analysis. (1999) 11–19
29. Weng, J., McClelland, J.L., Pentland, A., Sporns, O., Stockman, I., Sur, M., Thelen, E.: Autonomous mental development by robots and animals. Science **291** (2001) 599–600
30. Lugarella, M., Metta, G., Pfeifer, R., Sandini, G.: Developmental robotics: A survey. Connection Science **15** (2003) 151–190
31. Pfeifer, R., Scheier, C.: Understanding Intelligence. MIT Press, Cambridge, MA (1999)
32. Almássy, N., Sporns, O.: Perceptual invariance and categorization in an embodied model of the visual system. In Webb, B., Consi, T.R., eds.: Biorobotics: Methods and Applications. AAAI Press/MIT Press, Menlo Park, CA (2001) 123–143
33. Pierce, D.M., Kuipers, B.J.: Map learning with uninterpreted sensors and effectors. Artificial Intelligence **92** (1997) 162–227
34. Kuipers, B., Beeson, P., Modayil, J., Provost, J.: Bootstrap learning of foundational representations. Connection Science **18** (2006) 145–158
35. Cohen, P.R., Beal, C.R.: Natural semantics for a mobile robot. In: Proceedings of the European Conferenec on Cognitive Science. (1999)
36. Beer, R.D.: Dynamical approaches to cognitive science. Trends in Cognitive Sciences **4** (2000) 91–99
37. Cariani, P.: Symbols and dynamics in the brain. Biosystems **60** (2001) 59–83
38. Kozma, R., Freeman, W.J.: Basic principles of the KIV model and its application to the navigation problem. Journal of Integrative Neuroscience **2** (2003) 125–145
39. Varela, F.J., Thompson, E., Rosch, E.: The Embodied Mind: Cognitive Science and Human Experience. MIT Press, Cambridge, MA (1993)
40. Schaal, S., Ijspeert, A.J., Billard, A.J.: Computational approaches to motor learning by imitation. Philosophical Transactions of the Royal Society: Biological Sciences **358** (2003) 537–547
41. Billard, A.: Imitation. In Arbib, M.A., ed.: The Handbook of Brain Theory and Neural Networks. 2nd edn. MIT Press, Cambridge, MA (2003) 566–569

42. Breazeal, C., Scassellati, B.: Robots that imitate humans. Trends in Cognitive Sciences **6** (2002) 481–487
43. Rao, R.P.N., Shon, A.P., Meltzoff, A.N.: A bayesian model of imitation in infants and robots. Cambridge University Press, Cambridge, UK (2004) In press.
44. Matarić, M.J.: Sensory-motor primitives as a basis for imitation: Linking perception to action and biology to robotics. In Dautenhahn, K., Nehaniv, C., eds.: Imitation in Animals and Artifacts. MIT Press, Cambridge, MA (2001) 391–422
45. Billard, A., Matari¢, M.J.: A biologically inspired robotic model for learning by imitation. In: International Conference on Autonomous Agents: Proceedings of the Fourth International Conference on Autonomous Agents, New York, ACM Press (2000) 373–380
46. Ikegami, T., Taiji, M.: Imitation and cooperation in coupled dynamical recognizers. In: Proceedings of the 5th European Conference on Advances in Artificial Life: Lecture Notes in Computer Science 1674. Springer, London (1999) 545–554
47. Ito, M., Tani, J.: On-line imitative interaction with a humanoid robot using a dynamic neural network model of a mirror system. Adaptive Behavior **12** (2004) 93–115
48. Wyss, R., König, P., Verschure, P.F.M.J.: A model of the ventral visual system based on temporal stability and local memory. PLoS Biology **4** (2006) e120
49. Floreano, D., Suzuki, M., , Mattiussi, C.: Active vision and receptive field development in evolutionary robots. Evolutionary Computation **13** (2005) 527–544
50. Lungarella, M., Sporns, O.: Mapping information flow in sensorimotor networks. PLoS Computaitonal Biology **2** (2006) 1301–1312
51. Bongard, J., Zykov, V., Lipson, H.: Reslient machines through continuous self-modeling. Science **314** (2006) 1118–1121
52. Freeman, W.J.: A neurobiological theory of meaning in perception. In: Proceedings of the International Joint Conference on Neural N etworks, IEEE (2003) 1373–1378
53. Ziemke, T., Sharkey, N.E.: A stroll through the worlds of robots and animals: Applying Jakob von Uexküll's theory of meaning to adaptive robots and artificial life. Semiotica **134** (2001) 701–746
54. Abelson, H.: LOGO for the Apple II. 1st edn. McGraw-Hill, Peterborough, N.H. (1982)
55. Werbos, P.J.: Beyond Regression: New Tools for Prediction and Analysis in the Behavioral Sciences. PhD thesis, Department of Applied Mathematics, Harvard University, Cambridge, MA (1974)
56. Rumelhart, D.E., McClelland, J.L., eds.: Parallel Distributed Processing: Explorations in the Microstructure of Cognition, Vol. 1: Foundations. MIT Press, Cambridge, MA (1986)
57. Werbos, P.J.: Backpropagation through time: What it does and how to do it. Proceedings of the IEEE **78** (1990) 1550–1560
58. Press, W.H., Teukolsky, S.A., Vetterling, W.T., Flannery, B.P.: Numerical Recipes in C. Second edn. Cambridge University Press (1992)
59. Conditt, M.A., Mussa-Ivaldi, F.A.: Central representation of time during motor learning. Journal of Neurobiology **20** (1999) 11625–11630
60. Tulving, E., Markowitsch, H.J.: Episodic and declarative memory: role of the hippocampus. Hippocampus **8** (1996) 198–204
61. Buckner, R.L.: Neural origins of 'i remember'. Nature Neuroscience **3** (2000) 1068–1069
62. Wise, S.P.: The role of the basal ganglia in procedural memory. Seminars in Neuroscience **8** (1996) 39–46
63. Humphrey, N.: Seeing Red. Harvard University Press, Cambridge, MA (2006)
64. Misra, N.: Comparison of motor-based versus visual representations in object recognition tasks. Master's thesis, Department of Computer Science, Texas A&M University, College Station, Texas (2005)

A Biologically Inspired Dynamic Model for Object Recognition

Khan M. Iftekharuddin, Yaqin Li, and Faraz Siddiqui

Abstract Biological vision provides an excellent promise in the design of automated object recognition (AOR) system. A particularly important unresolved issue is that of learning. This topic is also explored in Chapters (Choe) and (Perlovsky), where learning is related to actions (Choe) and to the knowledge instinct (Perlovsky). Reinforcement learning (RL) is part of procedural learning that is routinely employed in biological vision. The RL in biology appears to be crucial for attentive decision making process in a stochastic dynamic environment. RL is a learning mechanism that does not need explicit teacher or training samples, but learns from an external reinforcement. The idea of RL is related to the knowledge instinct explored in Chapter (Perlovsky), which provides internal motivations for matching models to sensor signals. The model in this chapter implements RL through neural networks in an adaptive critic design (ACD) framework in automatic recognition of objects. An ACD approximates the neuro-dynamic programming employing an action and a critic network, respectively. Two ACDs such as Heuristic Dynamic Programming (HDP) and Dual Heuristic dynamic Programming (DHP) are both exploited in implementing the RL model. We explore the plausibility of RL for distortion-related object recognition inspired by principles of biological vision. We test and evaluate these two designs using simulated transformations as well as face authentication problems. Our simulations show promising results for both designs for transformation-invariant AOR.

Key words: Biologically inspired model · Reinforcement learning (RL) · Automated object recognition (AOR) · Heuristic dynamic programming (HDP) · Dual Heuristic Dynamic programming (DHP) · Image transformation · Face authentication

1 Introduction

In this chapter, we are going to discuss a biologically inspired vision model for automated object recognition (AOR) of distorted images. Modern AOR systems have witnessed an unprecedented growth in complexity over the last few decades. In general, an AOR is defined as a system that minimizes and/or removes the

L. I. Perlovsky, R. Kozma, *Neurodynamics of Cognition and Consciousness.*

human-teacher role such that the acquisition, processing and classification of object(s) can be performed automatically [1, 2]. The on-going advancements in increased computer processing capabilities, new understanding of low-level sensor phenomenology, improved signal processing and information fusion algorithms, and an expanded communication infrastructure have contributed to complexity of current AOR systems. Substantial technical challenges have arisen in terms of the design, operation, and management of these AOR systems. The robust operation of a complex AOR system containing numerous, disparate image sensing and operating conditions necessitates a requirement for efficient processing, optimization and classification of images. Notably, biological organisms routinely accomplish difficult sensing, processing and decision-making tasks in a holistic manner, enabling them to scale up to larger problems and implying some efficient underlying mechanisms to process information.

The biological vision, in particular, consists of highly coordinated, distributed hierarchical modular functional units that perform vast and complex sensory and motor activities. Even though the biological vision system occasionally miscues, for the most part it is accurate and fast enough to accomplish its required functions. Among many sensory recognition functionalities that are performed by the visual system such as identification of shape, color, depth, orientation, scale, rotation, spatial and temporal dynamics and communication among sensory organs, we are primarily interested in distortions invariant object recognition. One of the nontrivial problems in object recognition that continue to persist is the development of algorithms and heuristics to support recognition that is transformation invariant with respect to object translation, scale, aspect, interposition, range, rotation, perspective and illumination level [3]. Transformation invariant AOR has also been an active research area due to its widespread applications in a variety of fields such as military operations, robotics, medical practices, geographic scene, analysis and many others [4, 5, 6]. The unsurpassed recognition ability of a biological vision system in transient and nonlinear environments may become an important inspiration for design of recognition and learning systems. Biological vision systems are known for its robustness in extended operating conditions [4]. Therefore, biologically inspired model may provide a great incentive for design of robust object recognition systems. The primary goal for this research is automatic object detection and recognition in the presence of image transformations such as resolution, rotation, translation, scale and occlusion. The parallel computing capacity of neural networks (NN) makes it an effective tool for implementation of transformation invariant object recognition. The NN is preferable because it automates the training based on different transformations of images. In a classical training-testing setup, the performance is largely dependent on the range of transformation or orientation involved in training [7]. However, a serious dilemma in the setup is that there may not be enough training data available at the beginning of the learning process, or even no training data at all. To alleviate this problem, a biologically inspired reinforcement learning (RL) approach, exploiting minimal training data, may be preferable for transformation invariant object recognition [8, 9, 10, 11, 12]. In the RL approach, the model of the environment is not known a priori [13], and the learning is performed online while interacting with the environment [14]. Compared to static AOR systems, the dynamic AOR with the RL ability can adapt better in a changing and unpredictable environment. The RL has made major advancement by implementation of the temporal difference (TD) learning methods [15, 16, 17, 18, 19], and has various successful applications in neuro-computing, control problems, and in multi-resolution object recognition [8, 9, 10, 11, 12, 20, 21, 22, 23]. The Adaptive critic design (ACD) is a neural network centric approach that is designed based on the dynamic programming principles [24]. Dynamic programming attempts to optimize the Bellman equation for adaptive control of a system. Though the Bellman equation provides the theoretical basis, it suffers from the well known 'curse of dimensionality'. The ACDs prove to be a powerful tool for approximating the Bellman equation. The ACD also successfully incorporates the concepts of reinforcement learning (RL). An ACD uses the neuro-dynamic programming employing an action and a critic networks, respectively [24]. The critic network attempts to optimize cost function, commonly known as the Bellman equation, due to error generated by the combination of plant and action network. ACD is categorized into three major families such as Heuristic Dynamic Programming (HDP), Dual Heuristic dynamic Programming (DHP), and Globalized DHP (GDHP) [24]. In HDP, the critic estimates the cost-to-go function using Bellman equation [11, 25] as discussed in previous section. In DHP, the critic network estimates the derivative of the cost function with respect to the states of the system [12]. A GDHP combines both HDP and DHP for optimization of the cost function. Each family has its action dependent (AD) form if the action network is directly connected to the critic network. Action dependent form of the three original designs are denoted by an abbreviation with a prefix AD before each specific structure, for example, ADHDP and ADDHP denote "action dependent HDP" and "action dependent DHP", respectively. A comprehensive discussion of ACD is provided in [24].

In this chapter, two ACD algorithms such as a Heuristic Dynamic Programming (HDP) and a Dual Heuristic dynamic Programming (DHP) are implemented respectively. In HDP, the critic network estimates the cost-to-go function represented by Bellman equation of dynamic programming; while in DHP, the critic network estimates the derivative of the cost function with respect to the states of the system. A more detailed description on HDP and DHP is provided in the next section.

This chapter is organized as follows. After a brief description of the theoretical background, the two ACD implementations of the reinforcement model for the object recognition system are provided in Sect. 2, followed by the simulation results and numerical analysis in Sect. 3. Finally, the conclusion and ideas for future research are presented in Sect. 4.

1.1 Biology of Vision

The brain senses through many different modalities by extracting relevant patterns (shapes, sounds, odors, and so on) from a noisy, non-stationary and often, unpredictable environment [4, 5, 8]. The primate visual organs make use of retina, a complex tangle of interconnected neurons, which transform visual stimuli into nerve impulses representing tonic (static) and phasic (changing) temporal imagery. Multiple neuroscience, biological and physiological studies suggest that motion perception and anticipation is achieved in retina. The key for the mechanism of attentional selection may lie in the functional subdivisions of the afferent visual pathways. The three primary visual pathways, such as P, M and K-koniocellular that terminate at the striate visual cortex (V1), process information in parallel [26]. Each of these areas in primary visual cortex (V's) maintains one processed, but topographically correct image map that falls on the retina. The M-dominated stream that extends from V1 and certain compartments of V2 to middle temporal area (MT/V5) and further to the parietal cortex (PP) processes visual attributes such as movement, depth and position (space). Similarly, the P-dominated ventral pathway that extends from V1 to areas such as V2, V4 and the inferotemporal cortex (IT) involves object discriminations based on features such as color, form and texture. There are some cross-talks between the channels at various cortical levels as shown in Fig. 1. According to 'columnar organization' concept, the neighboring neurons in the visual cortex have similar orientation tunings and comprise an orientation column [26]. Neuropsychological findings suggest that invariance emerges in the hierarchy of visual areas in the primate visual system in a series of successive stages under the influence of the attentional window. One important aspect of invariant processing is 'feature binding' such that the local spatial arrangements of features should be preserved. Further, extensive anatomical, physiological and theoretical evidences show that the cerebellum is a specialized organism for supervised learning, the basal ganglia are for reinforcement learning and the cerebral cortex is for unsupervised learning for a rapid parallel forward recognition. Evidence also exists that the outcome of the learning process may also drive a feedback controlled refinement loop all the way back to the retina [13, 26].

In classical neuroscience, predictive learning occurs whenever a stimulus is paired with a reward or punishment. However, more recent analysis of associative learning argue that a 'prediction error' between the prediction stimulus and the actual reinforcer is also required [26]. Thus, the reinforcer pairings play an important role in human conditioning, casual learning, animal conditioning and artificial neural learning. Neurobiological investigations of associative learning have shown that dopamine neurons respond physically to rewards in a manner compatible with the coding of prediction errors [27]. These biological studies suggest that dopamine-like adaptive-critic reward responses may serve as effective learning and higher-level perception in primate cognitive behavior. There are various non-mathematical and mathematical theories to explain the role of prediction in reinforcement learning.

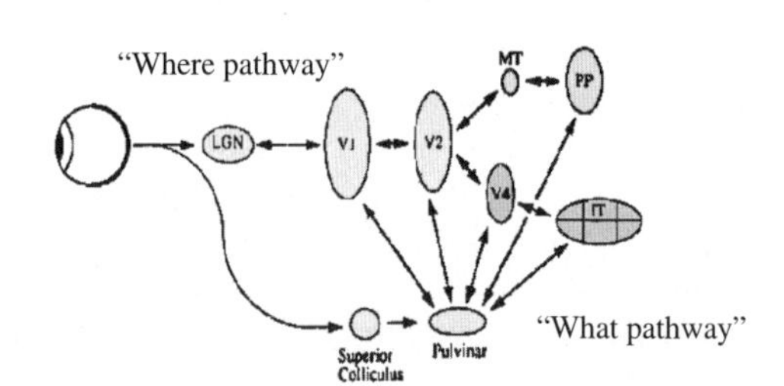

Fig. 1 Simplified model of visual pathways

A few of these mathematical theories attempt to explain attention as consequence of reinforcement learning in an ACD formulation [13, 14, 14]. The computational implementation of attention mechanism may help in gradual focusing of sensor suits for more effective AOR.

1.2 Biologically Inspired Model for Automatic Object Recognition

It is plausible that mimicking biological vision may revolutionize AOR. Specifically, the RL-based AOR can be considered as a candidate for a robust and promising object recognition system. The RL can be viewed as an optimal decision making process in a stochastic environment. The ACD, on the other hand, has provided a good approximation of the dynamic programming. A coarse-to-fine analysis that is usually performed as part of attention mechanism in a visual system can be simulated in an ACD framework.

2 The Biologically Inspired Dynamic Model for RL

2.1 Theoretical Background

The RL in the AOR implementation can be viewed from a control perspective as an adaptive optimal control problem of dynamic stochastic systems [14]. In AOR modeling, one needs to achieve optimized recognition in presence of dynamic extended operating conditions. Consequently, dynamic programming is identified as an effective technique for optimization in a stochastic and noisy environment. Dynamic programming can be dated back to Bellman [25] and Bertsekas [18], and since then, it has found extensive applications in the area of engineering, economics, operational research, and control. However, the computational complexity of the dynamic programming can be enormous for many problems, this is related to the so-called "curse of dimensionality". In addition, the choice of cost-to-go function in dynamic programming is critical for achieving optimization [19]. Also, dynamic programming requires a stochastic model of the system. To address these rather non-trivial requirements, the ACD has been explored as an approximation of the dynamic programming.

In this chapter, we implement two ACD-based AOR models for recognition of images in the presence of different transformations such as multi-resolution, scale, translation, occlusion, and rotation [8, 10, 11, 12, 21, 26]. A representative simulation model that exploits RL/ACD HDP setup [11, 21] is shown in Fig. 2. In the ACD, the critic network "critiques" the action value in order to optimize a future cost function and the action value is determined on a step-by-step basis. Artificial

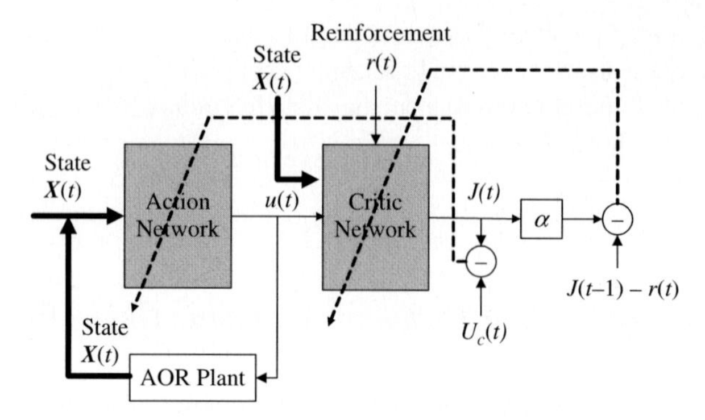

Fig. 2 Schematic diagram for the AOR system using HDP design

neural networks are applied as the learning and recognition components for both the action and critic networks in Fig. 2. In addition to the HDP-base AOR, we also exploit a DHP-based AOR model in this chapter.

2.2 Network Structure

Figure 2 provides a schematic diagram for the implementation of the HDP algorithm in the AOR system. The three major components in the HDP-based AOR are the action network, critic network and the AOR plant. The plant is the AOR that processes different distorted (transformed) images for recognition or authentication. The objective of the critic network is to approximate the discounted total cost-to-go function $R(t)$, given by

$$R(t) = r(t+1) + \alpha r(t+2) + \cdots \tag{1}$$

where $R(t)$ is the accumulative discounted future cost evaluated at time t, and $r(t)$ is the external reinforcement value at time t. The discount factor α is between 0 and 1. The weights of the critic network is adapted to approximate the cost-to-go function for performance optimization of the AOR system. This is achieved by using a modified Bellman equation given as [11, 21],

$$J^*(X(t)) = \min_{u(t)} \left\{ J^*(X(t+1)) + g(X(t), X(t+1)) - U_0 \right\} \tag{2}$$

where $g(X(t), X(t+1))$ is the immediate cost incurred by action $u(t)$ at time t.

The weight updating rule for the critic network is obtained by a gradient descending method to minimize the critic error [11, 21]

$$E_c(t) = \frac{1}{2} e_c^2(t) = \frac{1}{2} \{\alpha J(t) - [J(t-1) - r(t)]\}^2 . \tag{3}$$

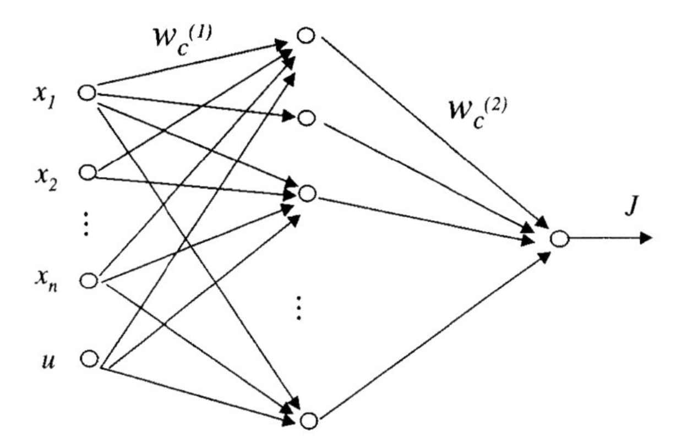

Fig. 3 Neural network structure of the critic network

Similarly, for the action network, the weight updating rule is to minimize the action error, which is defined as [11, 12, 21],

$$E_a(t) = \frac{1}{2}\, e_a^2(t) = \frac{1}{2}\, [J(t) - U_c(t)]^2 . \tag{4}$$

ANN is chosen for implementation of the action and critic networks for its fast processing speed and parallel computing capability. The specific structure of the NN for both the action and the critic networks are implemented as a nonlinear multi-layer feedforward NN with one hidden layer. The schematic diagram for the critic network is shown in Fig. 3 [11, 21]. A detailed mathematical treatment of the HDP dynamics can be found in Appendix A.

Figure 4 shows the schematic diagram for DHP-based AOR system. Compared to HDP, DHP is an incremental RL algorithm. The ACD network of DHP is similar to that of the HDP. In implementation of DHP, the critic network is to approximate the derivative of the cost function with respect to the plant state. Gradient descending

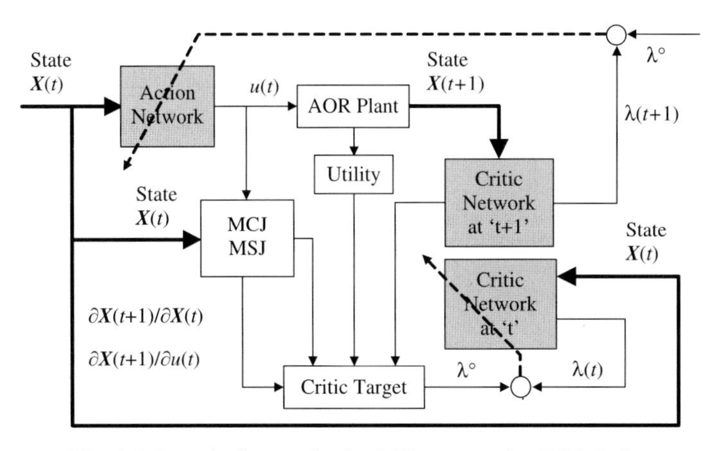

Fig. 4 Schematic diagram for the AOR system using DHP design

method is also use for the optimization. For DHP algorithm, the mathematical details can be found in Appendix B.

3 Simulation Results

In this section, we discuss the performance of HDP and DHP-based object recognition systems to explore the role of various transformations in object authentication and recognition using 2-D images. For both the HDP and DHP-based AOR computational model, as shown in Figs. 2 and 4 respectively, the following selections of parameters apply. The object transformations (distortions) investigated in this chapter are, specifically, multi-resolution, rotation, scaling, translation and occlusion. The AOR plant processes these distorted images for subsequent object recognition or authentication. The plant updates the state according to the action decision made by RL in the action network. In our simulation, the action network can offer output of $u(t)$ to the plant depending on the dynamics of the AOR system. If the action value, $u(t)$, is less than or equal to 0, the plant proceeds one step forward to the next state $i+1$; otherwise, it will go one step backwards to the previous state $i-1$. The reinforcement signal $r(t)$ takes a simple binary form of '0' and '-1' corresponding to 'reward' and 'punishment'. The value of the desired objective utility function $U_c(t)$ for the action network is set to 0. The initial learning rate $l_c(0)$ and $l_a(0)$ is set to 0.25, and it decreases as $l_c(t) = \frac{1}{t} l_c(0)$ and $l_a(t) = \frac{1}{t} l_a(0)$ for the critic network and action network respectively. The numbers of nodes in the hidden layer of both active and critic network are set to 6. The discount factor α is chosen as 0.95. Some of the sample runs and statistical evaluation of the two algorithms are demonstrated in this section.

3.1 Multi-resolution

The image transformation with resolution variation is shown in Fig. 5. For an object recognition system these multiresolution images can be considered as the incoming unknown test images with only one varying parameter, that is, the resolution. Multi-resolution evolution of image is a good approximation of the attention driven cognition in biological systems. Similar RL-based AOR design that exploits DHP is discussed in [12]. Both HDP- and DHP-based AOR models are investigated for

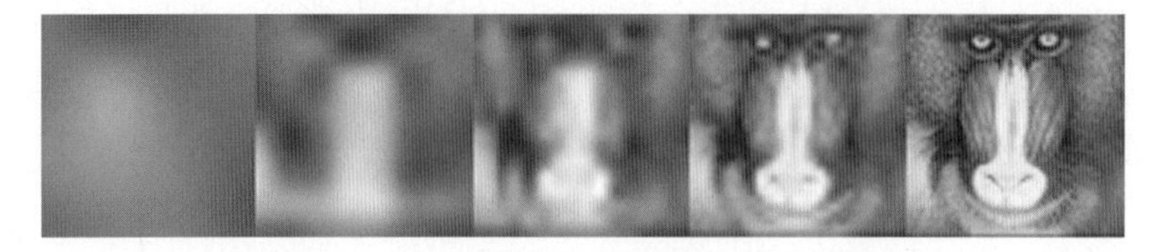

Fig. 5 Multi-resolution image as input to the system

multiresolution distortions in an image of an object [11, 12]. In our experiment, the image bank is obtained by application of Daubechie's level 4 wavelet analysis [11]. The resulting image bank consists of five images of gradually increasing resolution level, as shown in Fig. 5. These images are used as input into the plant as states. Thus, for a typical run, one of the reduced resolution images is used as the incoming input to the AOR system's plant module. The AOR network is then allowed to implement its dynamic simulation. Once the learning process is completed, the system finally converges to the highest resolution reference image. At this time, the AOR system has either recognized or authenticated the incoming reduced resolution image. The similar system dynamics also applies to all other transformation (distortion) problems discussed in the subsequent sections. For our implementation of the object recognition system, the resolutions are numbered from 0 for the lowest resolution to 4 for the highest one. The image at the highest resolution serves as the reference signal to the system. The initial state of the system is at resolution level 0, and the state of the system evolves according to the action value, $u(t)$, throughout the learning processes of the AOR system, Figs. 6 and 7 show a typical run of HDP and DHP for recognition of multi-resolution images, respectively. The change of resolution level and the critic error are shown as functions of iteration number. In both learning algorithms, as the iteration progresses, the network increases resolution level and maintains the states after it reaches the reference level. In both HDP and DHP, the ACD networks have successfully recognized the object and locked at the reference level once it accomplishes successful learning. Figure 6(a) shows that in this run of HDP design, the system takes 6 iterations to recognize the highest resolution level, while in the DHP run illustrated in Fig. 7(a), it takes 7 iterations to reach the reference level. Note that even though we use the resolution levels (or other transformations such as rotations, scales etc.) as states for simplicity, the recognition/authentication problem is not merely recognizing/authenticating counters that identify different states. Rather, the AOR system dynamics considers the whole image as the input to system to attain desired recognition/authentication. Further, if the AOR system starts somewhere at the middle of the states, for an ex-

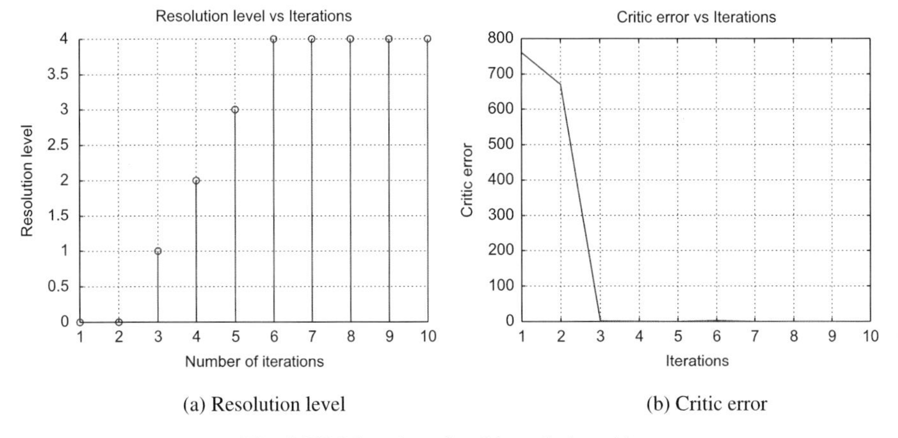

(a) Resolution level (b) Critic error

Fig. 6 HDP learning of multi-resolution object

Fig. 7 DHP learning of multi-resolution object

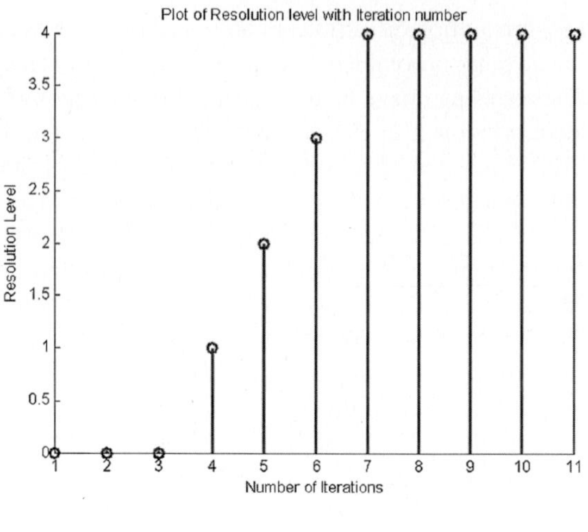

(a) Resolution level

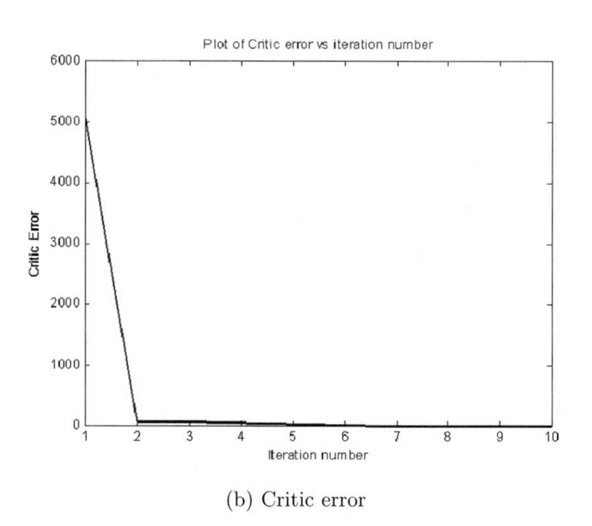

(b) Critic error

ample, at level 4 in multiresolution in Fig. 5; the system still goes through the same dynamic simulation to arrive at the target reference image. For this case, however, the iteration to reach the desired target would be less when compared to an input at resolution level less than 4 for the plant.

The critic errors, as they change with each iterations, for both HDP and DHP are shown in Figs. 6(b) and 7(b), respectively. In both plots, the initial error is very high, and the critic error decrease rapidly once the system starts the learning process. In both designs, once the system recognizes the object, the critic error goes to zero and remains at zero thereafter when the system locks to the object. A statistical comparison of the HDP and DHP based learning process is provided in Sect. 3.4.

3.2 Rotation

For recognition of images in different rotation angle, the image bank used in the simulation consists of a set of images with different rotation angles. In this section, both 90° and 180° in-plane rotation are considered.

The image bank for 180° degree rotation is shown in Fig. 8. The rotation levels are also number from 0 for the leftmost image to 9 for the rightmost image, and the rightmost image is the reference signal to the system.

The learning processes in 90° and 180° cases show similar behavior. Figures 9 and 10 provide an example run for 180° rotation. In both learning algorithms, as the iteration progresses, the ACD networks successfully recognize the image and lock the object once they accomplish successful learning.

The critic errors as they change with each iterations for both HDP and DHP, are shown in Fig. 9(b) and Fig. 10(b), respectively. In both plots, the critic errors decrease rapidly once the system starts the learning process. In both designs, once the system recognizes the image, the critic error goes to zero and remains at zero thereafter when the system locks the object. While in the HDP case, the critic error shows some fluctuations during the learning process, the critic error in DHP decreases monotonously.

3.3 Scale, Occlusion and Translation

Simulation procedures are also applied to scaled images, images with different degree of occlusion in the center, and translated images. The image banks used in the three different cases are demonstrated in Figs. 11, 12 and 13.

The learning processes for scaled, occluded and translated images demonstrate similar behaviors as shown in Sect. 3.1 and 3.2. Both HDP and DHP learning algorithms performs the recognition or authentication task successfully, and they can also lock the object once successful learning is accomplished.

3.4 Statistical Comparison of HDP- and DHP-based AOR Systems

To compare the performance of HDP- and DHP-based AOR systems statistically, a batch learning experiment is performed. For each individual transformation, the batch learning program consists of 100 runs. In one run, a maximum of three consecutive trials are permitted. A run is considered successful if it successfully learns

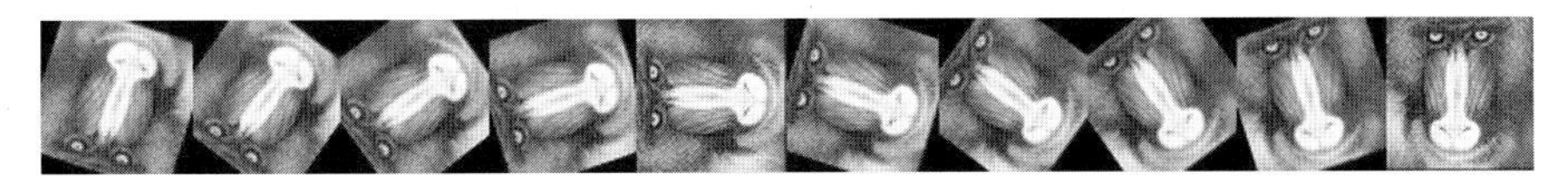

Fig. 8 Image bank for rotation up to 180°

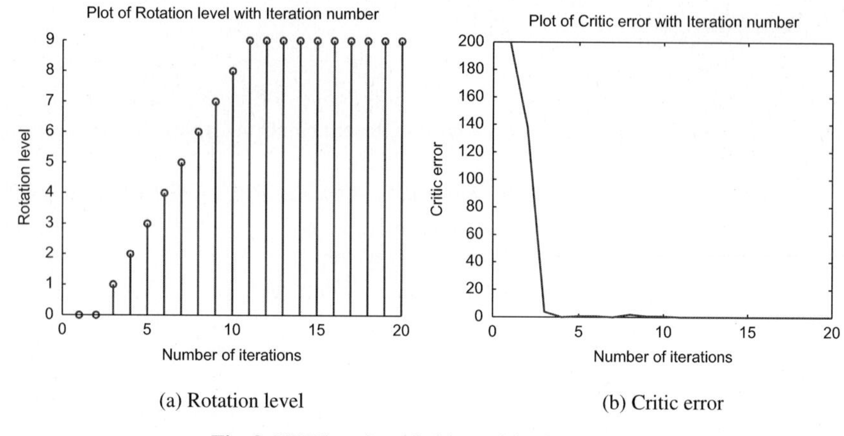

(a) Rotation level (b) Critic error

Fig. 9 HDP learning of object with 180° rotation

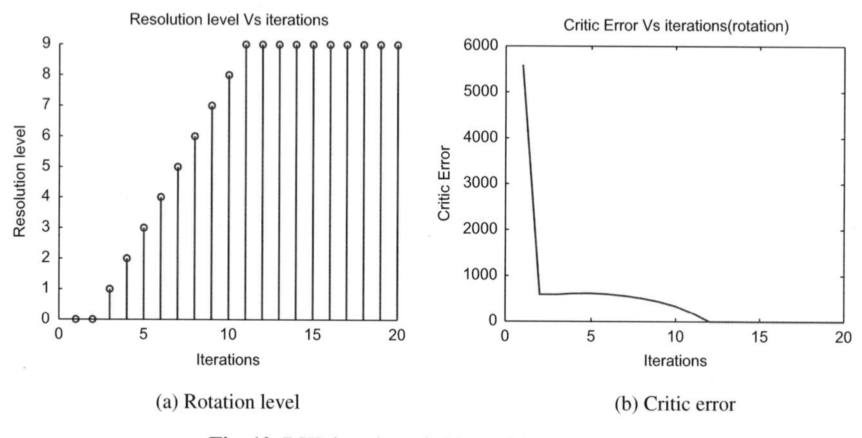

(a) Rotation level (b) Critic error

Fig. 10 DHP learning of object with 180° rotation

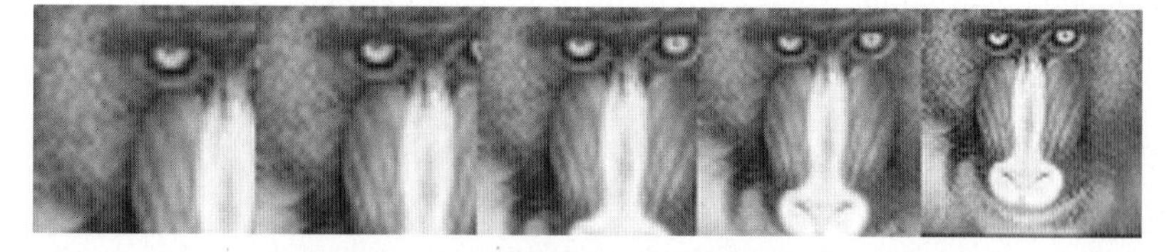

Fig. 11 Image bank for scaling image

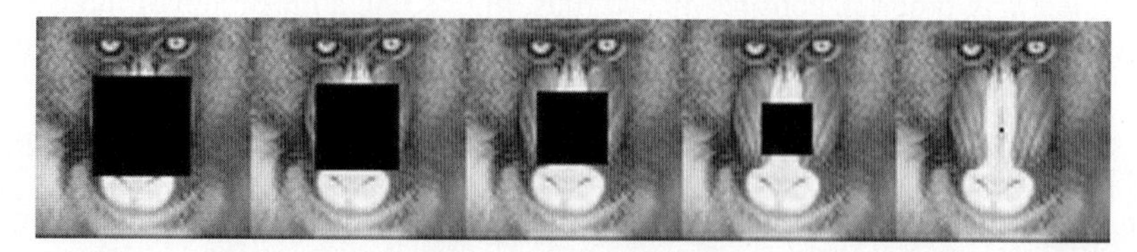

Fig. 12 Image bank for occlusion

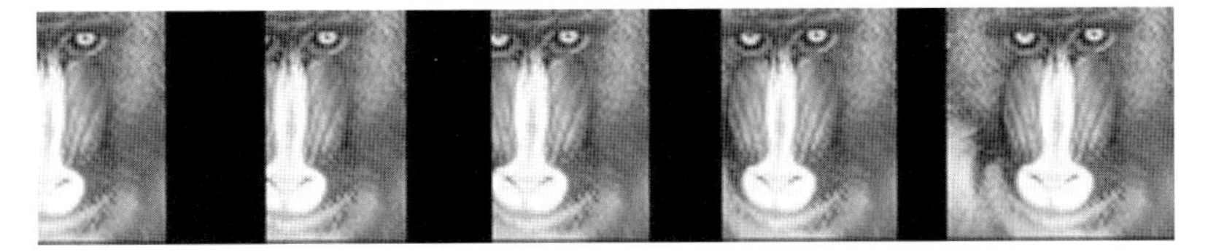

Fig. 13 Image bank for translated image

and locks at the reference level within three trials, otherwise it is considered unsuccessful. Within one trial, the AOR system is considered to recognize the subject if it can maintain the state in the reference level for 5 steps. In the 100 runs, the network weights are set at random initial conditions at the beginning of each learning process. If a run is successful, the number of trials it takes to perform such a task is recorded. The number of trials in the table is obtained by averaging over all the successful runs. If a trial is successful, the number of steps it takes to perform a successful learning is also recorded, and the average number of steps in the table corresponds to the one averaged over all of the successful trials. The success rate is also shown as the percentage of successful runs out of 100. A good learning system is the one with a high percentage of success rates as well as a low average number of trials and steps needed to perform the recognition task. The statistical results for both designs are summarized in Table 1. From Table 1, we can observe that, overall the success rate of HDP is higher; only in the multi-resolution case, the DHP performs slightly better. In terms of the average trials it takes to perform a successful learning, HDP is better in that on the average it takes less trials that DHP. Comparing the average number of steps to perform a successful run, throughout all the cases, the average steps for DHP is less than that of HDP. It is consistent with the characteristics of the DHP in that an incremental learning algorithm, and can learn

Table 1 Comparison of success rate for different transformations in image

Algorithm	Transformation	Success rate	Average Trials	Average Steps
HDP	Multi-Resolution	98%	1.04	5.56
	Scale	97%	1.15	6.47
	Occlusion	100%	1.00	5.50
	Translation	100%	1.06	5.74
	90° rotation	100%	1.04	10.20
	180° rotation	100%	1.10	11.50
DHP	Multi-resolution	100%	1.80	5.04
	Scale	74%	2.18	5.12
	Occlusion	72%	2.27	5.15
	Translation	76%	2.09	5.96
	90° rotation	78%	2.04	9.69
	180° rotation	68%	2.54	10.30

faster than HDP. The comparison can be viewed as a trade-off in design of learning algorithms between speed and accuracy.

3.5 Evaluation of HDP and DHP for Face Authentication Using UMIST Database

In this section, we evaluate our HDP- and DHP-based AOR designs for face authentication using the UMIST face database [28]. Precise and fast face authentication is critical in applications involving personal identification and verification. Data acquisitions for personal authentication purposes include 2-D and 3-D face data, and various biometrics. Incorporating data from different sources in the process can significantly improve the performance of face authentication [29]. Face authentication using 2-D data has been explored by a broad range of researchers [29]. In this section, the 2-D face data in the UMIST database are used.

The UMIST database consists of 2-D face data from 20 people, each covering a range of poses from profile to frontal views. The subjects are labeled as '1a', '1b' through '1t'. The original face images are in PGM format, approximately 220×220 pixels in 256 gray levels [28]. The face images are cropped into a uniform size of 112×92 pixels. The cropped version of the face database is exploited to evaluate the performance of both HDP and DHP designs for face authentication with out-of-plane rotation. In order to manage the face dataset better (i.e., to make the data arrangement comparable to our setup as described above), we group the sequence in a pattern of 90° rotation as follows. The sequences with a rotation of more than 90° are truncated; while the sequences with 180° rotation are split into two subsequences of 90° rotation. The resulting image arrays are then sorted in a 90° rotation from profile view to frontal view before it is supplied to the AOR plant as system states. The image sequence used for subject '1a' is shown in Fig. 14. The original images for this subject can be found in http://images.ee.umist.ac.uk/danny/database.html. The images used in this chapter is a truncated sequence consisting of the first 34 images of subject '1a', which capture the 90° rotation from right profile to frontal view of the subject.

The statistical results of the simulation experiments for different subjects in the UMIST database are summarized in Table 2. Comparing the performance of both our designs for face authentication using UMIST database, we observe that, on average, the overall success rates of HDP across all subjects are higher. However, DHP-based AOR achieves a 100% success rate for more often (higher frequency).

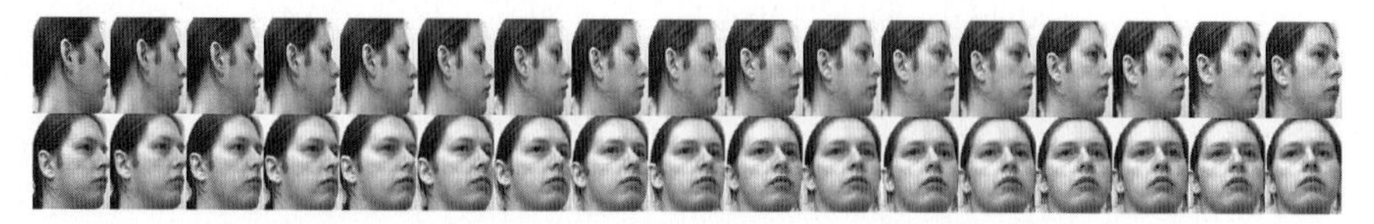

Fig. 14 Sample image sequence for subject '1a' in UMIST database [28]

Table 2 DHP success rate for UMIST database with no noise

Algorithm	Subject	Success rate	Average trials	Average Steps
HDP	1a	89%	1.22	34.03
	1b	98%	1.06	10.39
	1c	95%	1.11	26.42
	1d	93%	1.23	27.02
	1e	99%	1.02	26.10
	1f	94%	1.12	23.04
	1ga (subsequence 1)	99%	1.05	8.51
	1gb (subsequence 2)	100%	1.00	8.08
	1h	98%	1.05	22.30
	1ia (subsequence 1)	100%	1.00	12.06
	1ib (subsequence 2)	98%	1.05	8.19
	1j	98%	1.06	24.73
	1k	94%	1.14	34.93
	1l	87%	1.70	54.08
	1m	86%	1.28	26.01
	1na (subsequence 1)	92%	1.16	8.04
	1nb (subsequence 2)	91%	1.18	11.02
	1o	97%	1.06	14.02
	1p	94%	1.16	22.87
	1q	94%	1.28	31.69
	1r	92%	1.16	33.05
	1s	91%	1.19	48.16
	1t	97%	1.06	24.11
DHP	1a	100%	1.81	60.96
	1b	74%	2.04	26.79
	1c	100%	2.11	53.13
	1d	100%	2.07	49.95
	1e	100%	1.03	26.23
	1f	98%	2.18	51.21
	1ga (subsequence 1)	36%	2.28	9.31
	1gb (subsequence 2)	39%	2.22	9.67
	1h	100%	2.24	51.87
	1ia (subsequence 1)	48%	2.14	17.60
	1ib (subsequence 2)	43%	2.24	9.21
	1j	97%	2.28	54.37
	1k	100%	1.85	63.45
	1l	100%	1.85	62.31
	1m	98%	2.32	57.57
	1na (subsequence 1)	45%	2.10	9.69
	1nb (subsequence 2)	47%	2.08	13.21
	1o	76%	2.10	33.96
	1p	98%	2.24	51.65
	1q	97%	2.15	53.54
	1r	97%	1.98	64.28
	1s	100%	1.61	76.96
	1t	98%	2.18	52.40

Throughout the batch simulations for all the different subjects, the average number of trials to perform a successful run in DHP is more than that of HDP, and the average number of steps needed for DHP to finish a successful run is also larger than that of HDP. Thus, overall HDP is faster and more robust than the DHP throughout the variation of subject data. However, DHP offers more frequent 100% accurate face authentication for more of subjects in the database. This also demonstrates a trade-off in learning between robustness in general and accuracy in one particular case.

4 Conclusion and Future Research

In this chapter, we discussed a novel biologically-inspired neural network model for dynamic AOR. The proposed AOR exploits ACD framework using RL approach to the transformation invariant object recognition and face authentication. Two RL algorithms such as HDP and DHP are implemented and analyzed. The simulation results demonstrate that RL is an effective tool for adaptive transformation invariant object recognition and authentication. Comparing the two algorithms, DHP-based AOR outperforms that of HDP in specific cases, but is not as robust as HDP as far as success rate is concerned. Directions for future research include improving the current simplified plant dynamic to a more realistic model, such that the AOR plant may process multiple transformations in object simultaneously. Further, a feature-based AOR rather than the current image-based system may also be desirable for reduced computational complexity.

Acknowledgment The authors wish to acknowledge partial support through Herff endowment funds, and informal discussions with colleagues at Wright Patterson Air Force Research Laboratory, Dayton, OH.

References

1. Casasent, D., Psaltis, D.: Position, rotation and scale invariant optical correlator. Appl. Opt. **15** (1976) 1795–1799
2. Iftekharuddin, K. M., Ahmed, F., Karim, M.A.: Amplitude-coupled mace for automatic target recognition applications. Opt. Engr. **35** (1996) 1009–1014
3. Fuji, H., Almeida, S.P., Dowling, J.E.: Rotational matched spatial filter for biological pattern recognition. Appl. Opt. **19** (1980) 1190–1195
4. Diab, S.L., Karim, M.A., Iftekharuddin, K.M.: Multi-object detection of targets with fine details, scale and translation variations. Opt. Engr. **37** (1998) 876–883
5. Wei, G.Q., Qian, J., Schramm, H.: A dual dynamic programming approach to the detection of spine boundaries. Lecture Notes In Computer Science, Proceedings of the 4th International Conference on Medical Image Computing and Computer-Assisted Intervention **2208** (2001) 524–531
6. Schmid, C., Mohr, R.: Matching by local invariants. Reseach report, robotics, image and vision N2644, INRIA-Rhône-Alps, 46 Ave Felix Viallet, 38031 Grenoble Cedex 1 (France) (1995)
7. Bryant, M.L., Worrell, S.W., Dixon, A.C.: MSE template size analysis for MSTAR data. In

Zelnio, E.G., ed.: Algorithms for synthetic aperture radar imagery V, Proc. of SPIE. Volume 3370. (1998) 396–405
8. Iftekharuddin, K.M., Power, G.: A biological model for distortion invariant target recognition. In: Proc. of IJCNN, Washington DC, U.S.A., IEEE Press (2001) 559–564
9. Iftekharuddin, K.M., Rentala, C., Dani, A.: Determination of exact rotation angle and discrimination for rotated images. Optics & Laser Technology **34** (2002) 313–327
10. Iftekharuddin, K.M., Malhotra, R.P.: Role of multiresolution attention in automated object recognition. In: Proc. of IJCNN, Hawaii, U.S.A., IEEE Press (2002) 2225–2230
11. Iftekharuddin, K.M., Widjanarko, T.: Reinforcement learning in multiresolution object recognition. In: Proc. of IJCNN. Volume 2., Hawaii, U.S.A., IEEE Press (2004) 1085–1090
12. Siddiqui, F., Iftekharuddin, K.M.: Multiresolution object recognition using dual heuristic programming. In: Proc. of IJCNN. (2006)
13. Ge, Z.: Automated target recognition by reinforcement learning without a prior model. Ph.D. thesis, Texas Technological University, Lubbock, TX, USA (1997)
14. Sutton, R.S., Barto, A.G., Williams, R.J.: Reinforcement learning is direct adaptive optimal control. In: Proceedings of 1991 IEEE American Control Conference. (1991)
15. Tsitsiklis, J.N., Van Roy, B.: Feature-based methods for large scale dynamic programming. In: Machine Learning. Volume 22. Springer Netherlands (1994) 59–94
16. Marbach, P., Tsitsiklis, J.N.: Simulation-based optimization of markov reward processes. IEEE Trans. Auto. Contr. **46** (2) (2001) 191–209
17. Van Roy, B.: Neuro-dynamic programming: overview and recent trends. Handbook of Markov Decision Processes: Methods andApplications (2001)
18. Bertsekas, D.P.: Dynamic programming: deterministic and stochastic models. Printice Hall, Englewood Cliffs, NJ, U.S.A. (1987)
19. Bertsekas, D.P., Tsitsiklis, J.N.: Neuro-dynamic programming. Athena Scientific, Belmont, MA (1996)
20. Ens, R., Si, J.: Apache helicopter stabilization using neuro dynamic programming. Report G6316, AIAA (2000)
21. Si, J., Wang, Y.T.: On-line learning control by association and reinforcement. IEEE Trans. Neural Networks **12**(2) (2001) 264–276
22. Venayagamoorthy, G.K., Harley, R.G., Wunsch, D.C.: Implementation of adaptive critic based neurocontrollers for turbogenerators in a multimachine power system. IEEE Trans. Neural Networks **15**(5) (2003) 1047–1064
23. Liu, W., Venayagamoorthy, G.K., Wunsch, D.C.: A heuristic dynamic programming based power system stabilizer for a turbogenerator in a single machine power system. In: IEEE-IAS 38th Annual Meeting. (2003)
24. Prokhorov, D., Wunsch, D.C.: Adaptive critic designs. IEEE Trans. Neural Networks **8** (5) (1997) 997–1007
25. Bellman, R.E.: Dynamic programming. Princeton University Press, Princeton, NJ, U.S.A. (1957)
26. Iftekharuddin, K.M.: Attention-driven object recognition. Final report AFSOR Grant# F33615-01-C-1985, Wright Patterson AF Lab (2001)
27. Schultz, W.: Predictive reward signal of dopamine neuron. J. Neurophysiology **80** (1998) 1–27
28. Graham, D.B., Allinson, N.M.: Characterizing virtual eigensignatures for general purpose face recognition. In H. Wechsler, P. J. Phillips, V.B.F.F.S., Huang, T.S., eds.: Face Recognition: From Theory to Applications. Volume 163 of F, Computer and Systems Sciences. NATO ASI (1998) 446–456
29. K. Chang, K.W.B., Flynn, P.J.: Face recognition using 2d and 3d facial data. In: ACM Workshop on Multimodal User Authentication. (2003) 25–32

Appendix A: Mathematics of the HDP Dynamics

The objective of the critic network is to approximate the discounted total cost-to-go function $R(t)$, given by

$$R(t) = r(t+1) + \alpha r(t+2) + \cdots \quad (5)$$

where $R(t)$ is the accumulative discounted future cost evaluated at time t, and $r(t)$ is the external reinforcement value at time t. The discount factor α is between 0 and 1.

The weights of the critic network is adapted to approximate the cost-to-go function. This is achieved by using a modified Bellman equation given by [11, 21]

$$J^*(X(t)) = \min_{u(t)} \left\{ J^*(X(t+1)) + g(X(t), X(t+1)) - U_0 \right\} \quad (6)$$

where $g(X(t), X(t+1))$ is the immediate cost incurred by action $u(t)$ at time t. For the critic network, its objective is to be minimized the critic error [11, 21],

$$E_c(t) = \frac{1}{2} e_c^2(t) = \frac{1}{2} \{\alpha J(t) - [J(t-1) - r(t)]\}^2 . \quad (7)$$

The weight updating rule for the critic network is [11, 21],

$$\mathrm{w}_c(t+1) = \mathrm{w}_c(t) + \Delta \mathrm{w}_c(t) \quad (8)$$

where,

$$\Delta \mathrm{w}_c(t) = l_c(t) \left[-\frac{\partial E_c(t)}{\partial \mathrm{w}_c(t)} \right] \quad (9)$$

and $l_c(t)$ is the learning rate of the critic network at time t.

The weight updating for the action network follows similarly [21, 11, 12],

$$\mathrm{w}_a(t+1) = \mathrm{w}_a(t) + \Delta \mathrm{w}_a(t) \quad (10)$$

where,

$$\Delta \mathrm{w}_a(t) = l_a(t) \left[-\frac{\partial E_a(t)}{\partial \mathrm{w}_a(t)} \right] \quad (11)$$

and $l_a(t)$ is the learning rate of the action network at time t, and the instant error to be minimized is,

$$E_a(t) = \frac{1}{2} e_a^2(t) = \frac{1}{2} [J(t) - U_c(t)]^2 . \quad (12)$$

For the critic network shown in Fig. 3 the output $J(t)$, is [11, 21],

$$J(t) = \sum_{i=1}^{N_h} w_{c_i}^{(2)}(t) p_i(t) \tag{13}$$

where,

$$p_i(t) = \frac{1 - \exp^{-q_i(t)}}{1 + \exp^{-q_i(t)}}, \quad i = 1, 2, \cdots, N_h \quad \text{and} \tag{14}$$

$$q_i(t) = \sum_{j=1}^{n+1} w_{c_{ij}}^{(1)}(t) x_j(t), \quad i = 1, 2, \cdots, N_h; \tag{15}$$

where,

q_i is the ith hidden node input to the critic network
p_i is the ith corresponding output of the hidden node
N_h is the number of hidden nodes in the critic network
$n + 1$ is the total number of inputs into the critic network

By applying the chain rule, the weight updating rule for the critic network is as follows:

$$\Delta \mathrm{w}_c(t) = l_c(t) \left[-\frac{\partial E_c(t)}{\partial \mathrm{w}_c(t)} \right] \tag{16}$$

where,

$$\frac{\partial E_c(t)}{\partial \mathrm{w}_c(t)} = \frac{\partial E_c(t)}{\partial J(t)} \frac{\partial J(t)}{\partial \mathrm{w}_c(t)} = \alpha e_c(t) \frac{\partial J(t)}{\partial \mathrm{w}_c(t)}; \tag{17}$$

and the weight updating rule is,

$\Delta w_{c_i}^{(2)}$: from hidden to output layer

$$\Delta w_{c_i}^{(2)}(t) = -l_c(t) \alpha e_c(t) \frac{\partial J(t)}{\partial w_{c_i}^{(2)}(t)} \tag{18}$$

$$\frac{\partial J(t)}{\partial w_{c_i}^{(2)}(t)} = p_i(t) \tag{19}$$

$\Delta w_{c_{ij}}^{(1)}$: from input to hidden layer

$$\Delta w_{c_{ij}}^{(1)}(t) = -l_c(t)\alpha\, e_c(t)\frac{\partial J(t)}{\partial w_{c_i}^{(1)}(t)} \tag{20}$$

$$\frac{\partial J(t)}{\partial w_{c_i}^{(1)}(t)} = \frac{\partial J(t)}{\partial p_i(t)}\frac{\partial p_i(t)}{\partial q_i(t)}\frac{\partial q_i(t)}{\partial w_{c_{ij}}^{(1)}(t)} = w_{c_i}^{(2)}(t)\left[\frac{1}{2}(1-p_i^2(t))\right]x_j(t) \tag{21}$$

The action network is implemented similarly by a feedforward neural network, its input is the n-dimensional state, and the output is the action value $u(t)$ at time t[11, 12, 21].

$$u(t) = \frac{1-\exp^{-v(t)}}{1+\exp^{-v(t)}} \tag{22}$$

where,

$$v(t) = \sum_{i=1}^{N_h} w_{a_i}^{(2)}(t)g_i(t), \tag{23}$$

$$g_i(t) = \frac{1-\exp^{-h_i(t)}}{1+\exp^{-h_i(t)}}, \quad i = 1, 2, \cdots, N_h \quad \text{and} \tag{24}$$

$$h_i(t) = \sum_{j=1}^{n+1} w_{a_{ij}}^{(1)}(t)x_j(t), \quad i = 1, 2, \cdots, N_h \tag{25}$$

where v, g and h are, respectively

v	input to the action node
g_i	output of the hidden nodes
h_i	input to the hidden nodes

The weight updating rule for the action network follows similarly.

$$\Delta \mathrm{w}_a(t) = l_a(t)\left[-\frac{\partial E_a(t)}{\partial \mathrm{w}_a(t)}\right] \tag{26}$$

where,

$$\frac{\partial E_a(t)}{\partial \mathrm{w}_a(t)} = \frac{\partial E_a(t)}{\partial J(t)}\frac{\partial J(t)}{\partial \mathrm{w}_a(t)} = e_a(t)\frac{\partial J(t)}{\partial \mathrm{w}_a(t)} \tag{27}$$

and,

$\Delta w_{a_i}^{(2)}$: from hidden to output layer

$$\Delta w_{a_i}^{(2)}(t) = -l_a(t)e_a(t)\frac{\partial J(t)}{\partial w_{a_i}^{(2)}(t)} \tag{28}$$

$$\frac{\partial J(t)}{\partial w_{a_i}^{(2)}(t)} = \frac{\partial J(t)}{\partial u(t)}\frac{\partial u(t)}{\partial v(t)}\frac{\partial v(t)}{\partial w_{a_i}^{(2)}(t)} \tag{29}$$

$$= \left[\frac{1}{2}(1-u^2(t))\right] g_i(t) \sum_{i=1}^{N_h}\left[w_{c_i}^{(2)}(t)\,\frac{1}{2}\left(1-p_i^2(t)\right) w_{c_{i,n+1}}^{(1)}(t)\right] \tag{30}$$

$\Delta w_{a_{ij}}^{(1)}$: from input to hidden layer

$$\Delta w_{a_{ij}}^{(1)}(t) = -l_a(t)e_a(t)\frac{\partial J(t)}{\partial w_{a_{ij}}^{(1)}(t)} \tag{31}$$

$$\frac{\partial J(t)}{\partial w_{a_{ij}}^{(1)}(t)} = \frac{\partial J(t)}{\partial u(t)}\frac{\partial u(t)}{\partial v(t)}\frac{\partial v(t)}{\partial g_i(t)}\frac{\partial g_i(t)}{\partial h_i(t)}\frac{\partial h_i(t)}{\partial w_{a_{ij}}^{(1)}(t)} \tag{32}$$

$$= \left[\frac{1}{2}(1-u_i^2(t))\right] w_{a_i}^{(2)}(t)\left[\frac{1}{2}(1-g_i^2(t))\right] \tag{33}$$

$$x_j(t)\cdot\sum_{i=1}^{N_h}\left[w_{c_i}^{(2)}(t)\,\frac{1}{2}\left(1-p_i^2(t)\right) w_{c_{i,n+1}}^{(1)}(t)\right]. \tag{34}$$

Appendix B: Mathematics of the DHP Dynamics

For DHP, the prediction error of the critic network is defined as in [12],

$$e_c(t) = \frac{\partial J(t)}{\partial X(t)} - \alpha\frac{\partial J(t+1)}{\partial X(t)} - \frac{\partial r(t)}{\partial X(t)}, \tag{B-1}$$

and the objective function to be minimized in the critic network is,

$$E_c(t) = \sum e_c(t)e_c^T(t). \tag{B-2}$$

In training the network, a target value is generated for the output of the critic network as[12],

$$\lambda_s^o(t) = \frac{\partial J(t)}{\partial X_s(t)} = \left[\frac{\partial r(t) + \alpha\partial J(t+1)}{\partial X_s(t)}\right]. \tag{B-3}$$

Application of the chain rule offers,

$$\lambda_s^o(t) = \frac{\partial r(t)}{\partial X_s(t)} + \alpha\frac{\partial J(t+1)}{\partial X_s(t+1)}\left[\frac{\partial X_s(t+1)}{\partial X_s(t)} + \frac{\partial X_s(t+1)}{\partial u(t)}\frac{\partial u(t)}{\partial X_s(t)}\right]. \tag{B-4}$$

The weight updating rule for the critic network is[12]

$$\mathrm{w}_c(t+1) = \mathrm{w}_c(t) + \Delta \mathrm{w}_c(t) \tag{B-5}$$

where,

$$\Delta \mathrm{w}_c(t) = l_c(t)\left[-\frac{\partial E_c(t)}{\partial \mathrm{w}_c(t)}\right] \quad \text{and} \tag{B-6}$$

$$\frac{\partial E_c(t)}{\partial \mathrm{w}_c(t)} = \frac{\partial E_c(t)}{\partial J(t)}\frac{\partial J(t)}{\partial \mathrm{w}_c(t)}. \tag{B-7}$$

Here $l_c(t)$ is the learning rate of the critic network at time t. In the process, the derivatives of the state vector is approximated by its differences as the following,

$$\frac{\partial X(t+1)}{\partial X(t)} = \frac{X(t+1) - X(t)}{X(t) - X(t-1)}. \tag{B-8}$$

The structure and weight updating rule for the action network follows[12],

$$\mathrm{w}_a(t+1) = \mathrm{w}_a(t) + \Delta \mathrm{w}_a(t), \tag{B-9}$$

where,

$$\Delta \mathrm{w}_a(t) = l_a(t)\left[-\frac{\partial J(t)}{\partial \mathrm{w}_a(t)}\right], \quad \text{and,} \tag{B-10}$$

$$\frac{\partial J(t)}{\partial \mathrm{w}_a(t)} = \frac{\partial J(t)}{\partial u(t)}\frac{\partial u(t)}{\partial \mathrm{w}_a(t)}, \tag{B-11}$$

where $l_a(t)$ is the learning rate of the action network at time t,

$$\frac{\partial J(t)}{\partial u(t)} = \frac{\partial r(t)}{\partial u(t)} + \alpha \frac{\partial J(t+1)}{\partial u(t)}, \tag{B-12}$$

and

$$\frac{\partial J(t+1)}{\partial u(t)} = \frac{\partial J(t+1)}{\partial X(t+1)}\frac{\partial X(t+1)}{\partial u(t)}. \tag{B-13}$$

A Brain-Inspired Model for Recognizing Human Emotional States from Facial Expression

Jia-Jun Wong and Siu-Yeung Cho

1 Introduction

Emotions are essential in human society. The ability to understand, interpret and react to emotions is vital in maintaining relationships with friends, colleagues and family [1, 2, 3]. There is growing evidence that emotions play a part in decision making, perception and empathic understanding in a way that affects intelligent functions [4, 5, 6]. Majority of people is able to interpret emotions expressed by others most of the times, but there are people who lack this ability, such as people diagnosed along the autism spectrum [7]. For example, Capgras' syndrome patients [8] have suffered from head trauma and had the link between their visual cortex and limbic system severed, think that family and friends are replaced by impostors because they are unable to feel any emotions for these persons. This is due to the unconscious covert recognition process (involving emotions) is disconnected and the conscious recognition process (involving brain modules for identifying people) are still connected [8]. A cognitive-emotion interaction is one of the most debated research areas in psychology. Perlovsky discusses infinite variations of aesthetic emotions involved in cognitive processes [9]. Appraisal researchers have shown that emotions produced by any event depend on what is appraised [10]. Ekman has identified six basic categories of emotions [10] (i.e. fear, anger, sadness, surprise, disgust and joy). Here we concentrate on recognition of facial expression of these emotions.

Figure 1 shows a subject expressing represent the six basic emotions. Such emotions are revealed earlier through facial expression than people verbalize or even realize their emotional states [11]. Ekman has shown how to use a facial expression to identify a lie [12].

Schachter [13] showed how the brain concocts to explain the body reactions into emotions. Emotional stimuli activate sensory pathways that trigger the hypothalamus part of the brain to modulate the heartbeat, blood pressure, and respiration. Davis [14] had identified the amygdala (which is part of the limbic system) and hippocampus formation as the center of emotional cognitive level of awareness. The prefrontal cortex produces a conscious decision by information from associative and non-associative semantic memories in the neocortex and hippocampus. In a magneto-encephalography (MEG) study [15], the judgment of emotion from

J.-J. Wong, S.-Y. Cho, *Neurodynamics of Cognition and Consciousness.*

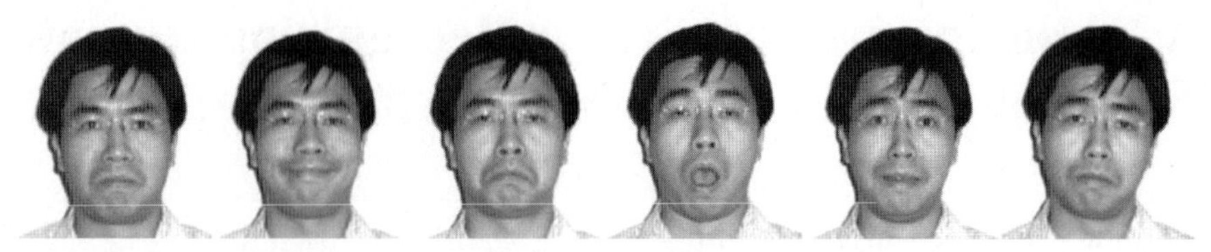

Fig. 1 Six basic emotions (from left to right): anger, joy, sadness, surprise, fear and disgust. (Images taken from NTU asian emotion database)

expression first elicited a stronger response, as compared to simple face detection, in posterior superior temporal cortex.

In this chapter, we used a framework as shown in Fig. 2. based on Wong and Cho's Emotion Recognition System in [16] which represents the neuro-cognitive structure of the brain for emotion recognition in facial expression [17]. Within this brain-inspired framework, the face detection and feature extraction block is built to model the functions of the visual cortex. Pre-processing steps attempt to detect and track the face. We used the Intel Open CV library's face and feature detector to locate the face and the 4 fiducial points (i.e. Eyes, Nose and Mouth). The Open CV face and feature detector implements the Viola and Jones algorithm [18], which uses an AdaBoost [19] classifier with Haar-like features for detecting the face and fiducial points. Gabor wavelet transform is used as a feature extractor with an approximate experimental filter response profile encountered in cortical neurons [20]. Gabor wavelet has capability to capture the properties of spatial localization, orientation selectivity, spatial frequency selectivity, and quadrature phase relationship.

The representation from the Gabor feature extractor does not take into account any possible relationships among the facial components. Therefore, we proposed to transform the Gabor features to FacE Emotion Tree Structure (FEETS) [21], which models functions of the Sensory Association Cortex [22]. Ekman's Facial Action Coding System [23] provided foundation, defining Regions of Interest (ROI) for feature extraction. The ROI features are captured in the Localized Gabor Feature (LGF) vector and transformed into a tree structure representation. By using this FEETS representation, we are able to encode feature relationship information among the face features. This relationship information is lost in a static data structure such as traditional flat-vectors or arrays features.

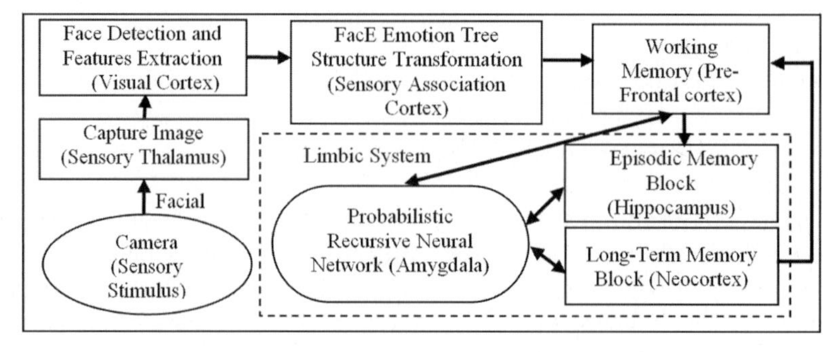

Fig. 2 Diagram of emotion recognition system in [16]

Traditional processing by machine learning methods [24, 25] was normally carried out by sequential representation. However, two major draw-backs were highlighted in [26]. The first is the poor generalization capability, which might be caused by sequential mapping of data structures. It is necessary to identify regularity inherently associated with the data structure. The second is the long-term dependency, when the number of nodes grew exponentially with the depth of the tree, which results in low learning efficiency with the large number of parameters. In this study, a Probabilistic Recursive Neural Network (PRNN) is used to model functions of human amygdala for recognizing emotions from facial expressions. The PRNN extended from Frasconi et al. [27], is used for adaptive processing of the FacE Emotion Tree Structures (EETS), PRNN parameters are learned from features and are stored in the long-term memory during the training phase; this might model information stored in hippocampus. In the recognition phase, the learned parameters are loaded into the working memory, and the emotional state is recognized from the learned parameters by PRNN; this might model processes in the amygdala.

2 FEETS Representation

This section describes the FacE Emotion Tree Structure (FEETS) transformation process.

2.1 Feature Representation

Ekman's facial action coding system (FACS) [23] provided the foundation of dominant facial components of interest for recognizing human facial expression. In this work, we used similar concept of facial components of interest to extract the facial features to represent human face. These areas of interest are essential to provide a clear indication of the expression of human face in an image. Most computer based system accepts features as a flat vector or aray input, which is very different for a human cognitive system. A human brain processes information not in a sequential manner or analyze the entire data for the cognitive response. It processes information in parallel and have networks connected at each slice of information for a global response. This is why human are more robust when it comes to image recognition, as more or less not all features are available due to rotation of the subjects.

We intend to model the human brains cognitive function by using tree structure to present the relationship information between the features from coarse to fine detail, which are represented by the Localized Gabor Feature vector. In this approach, the facial emotion can be represented by a 5 level deep tree structure model as shown in Fig. 3, the entire face region acting as a root node and localized features including upper and lower face and left, right and center of the faces became the second level branch nodes. At the third level nodes, the forehead, eyes, nose, mouth and cheek area became the corresponding branch nodes. At the fourth level, the forehead, eyes, eyebrows, nose, cheeks and mouth act as the branching nodes to the third

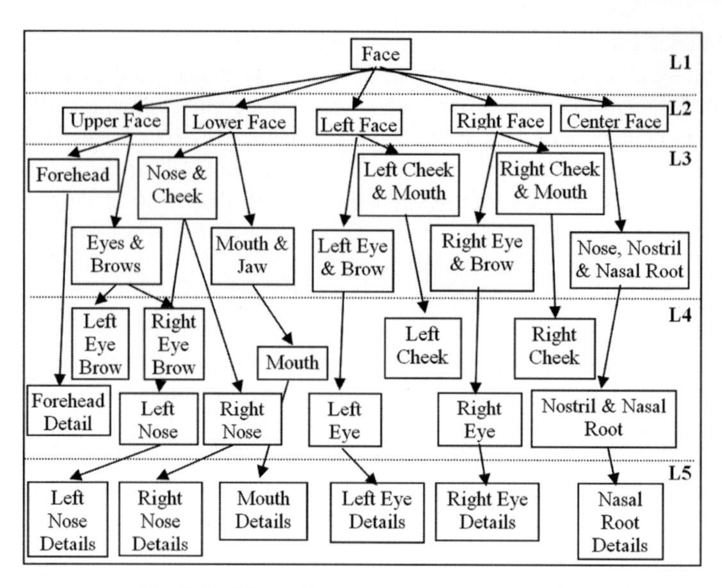

Fig. 3 Facial emotion tree structure representation

level nodes. Sub-detail features from the 4 key fiducial points form the leaves of the tree structure. The extracted features are grouped and attached to the corresponding nodes as shown in Fig. 3. The actual tree structure would have more connecting branches and arcs. The arcs between each of the nodes corresponding to the features relationship are used to represent the semantic of human facial components.

If a subject's face is rotated in half profile position, 45 degrees in any direction, this would cause any feature location engine for finding one of the eyes to fail. i.e. features related to that eye would not be able to be extracted or is extracted at the wrong location. In a FacE Emotion Tree Structure (FEETS) representation, only the branches related to this feature location would not be available. Typically affecting nodes from level 4 and 5, since these features require the correct location of the fiducial points. Level 1 to 3 are dependent on the location of the face region, which are typically easier and less affected by pose variations.

Also if a subject's face is covered by sunglasses or wears a veil, the eyes, nose and mouth regions would be undetectable by a feature location engine. But for a human to be able to recognize an emotion from such a face would not be too difficult, as the human brain does not require all the areas of the face to be reveal in order to predict/guess the emotion expressed by the person. Similarly, such approach of using a FEETS representation aims to solve the dependency of the completeness of information before an accurate recognition can be made.

2.2 Feature Extraction

We need a feature extraction method that models or approximates the human vision system. Gabor wavelets, which capture the properties of spatial localization, orientation selectivity, spatial frequency selectivity, and quadrature phase relationship, seems to be a good approximation to filter response profiles encountered experimentally in cortical neurons [20]. We used a pre-defined global filter based on the two-dimensional Gabor wavelets $g(x, y)$, which can be defined as follows [28]:

$$g(x,y) = \left(\frac{1}{2\pi\sigma_x\sigma_y}\right)\exp\left[-\frac{1}{2}\left(\frac{x^2}{\sigma_x^2}+\frac{y^2}{\sigma_y^2}\right)+2\pi jWx\right] \quad (1)$$

where parameters $W = U_h$, $\sigma_x = 2\sigma_u/\pi$ and $\sigma_y = 2\sigma_v/\pi$. $\sigma_u = \frac{(a-1)U_h}{(a+1)\sqrt{2\ln 2}}$ $\sigma_v = \tan(\frac{\pi}{2K})\left[U_h - 2\ln(\frac{\sigma_u^2}{U_h})\right]\left[2\ln 2 - \frac{(2\ln 2)^2\sigma_u^2}{U_h^2}\right]^{-0.5}$ and K is the total number of orientations, $a = \left(U_h/U_l^{-\frac{1}{s-1}}\right)$ and s is the number of scales in the multi-resolution decomposition. U_h and U_l is the lower and upper center frequencies respectively.

In order to extract features that from level 4 and 5 of the FEETS representation, we would require the system to detect four primary feature locations. The system employs Viola and Jones' Haar-like features detection algorithm [18], which will provide the coordinate location for the center of the left eye, center of the right eye, tip of the nose and the center of the lips as shown in Fig. 4. The location point of the left and right eye features is being derived from the location of the center of

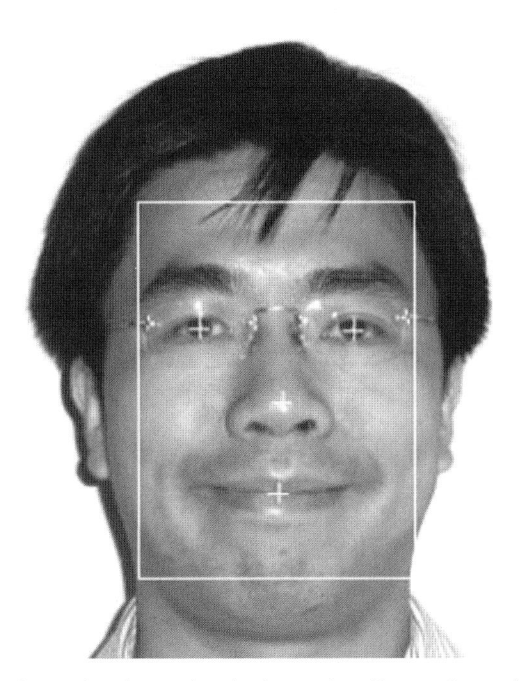

Fig. 4 Four primary feature locations and entire face region. Crosses denote the center of fiducial points. Rectangle box denotes of region of interest

the left eye and right eye denoted by the coordinates (x_{LE}, y_{LE}) and (x_{RE}, y_{RE}) respectively. The location of the nose bridge is the middle point of the left and right eye on the X-axis. The nose feature locations are derived from the location of tip of the nose denoted by the coordinates (x_{NS}, y_{NS}) . The locations of lips features are derived from the center of lips coordinates (x_{LS}, y_{LS}) .

In a vision recognition system, features need to be extracted from a filter response of an image. We have selected Gabor Wavelet as the chose filter for its similarity to the response that human visual cortical neurons produce. For a given image $I(x, y)$, the Gabor wavelet response can be defined as follows:

$$W_{mn}(x, y) = \int I(x_1, y_1) g_{mn} * (x - x_1, y - y_1) dx_1 dy_1. \tag{2}$$

The subscript m denotes the size of the filter bank in terms of number of orientation. The subscript n denotes the size of the filter bank in terms of number of scales. The localized Gabor Feature $F(x_F, y_F)$ can be expressed as a sub-matrix of the holistic Gabor wavelet output from (2),

$$F_{mn}(x_F, y_F) = W_{mn} \begin{bmatrix} (x_F, y_F) & \dots & (x_{F+S}, y_F) \\ \vdots & \ddots & \vdots \\ (x_F, y_{F+S}) & \dots & (x_{F+S}, y_{F+S}) \end{bmatrix}, \tag{3}$$

where s defines the size of the feature area. The x_F and y_F can be defined respectively as:

$$x_F = x_{RF} + c, \tag{4}$$

$$y_F = y_{RF} + c, with - 20 \leq c \prec 10, \tag{5}$$

where the subscript "RF" refers to the relative center location coordinates, i.e., "LE" for left eye, "RE" for right eye, "NS" for noise tip or "LS" for lips. The mean and standard deviation of the convolution output is used as the representation for classification purpose:

$$\mu_{mn} = \int\int |F_{mn}(x_F, y_F)|\, dxdy, and \tag{6}$$

$$\sigma_{mn} = \sqrt{\int\int (|F_{mn}(x_F, y_F)| - \mu_{mn})^2 dxdy}. \tag{7}$$

Localized Gabor Feature (LGF) is used to represent each area of interest as highlighted in the FEETS representation. The LGF vector of each of the image can be formed as:

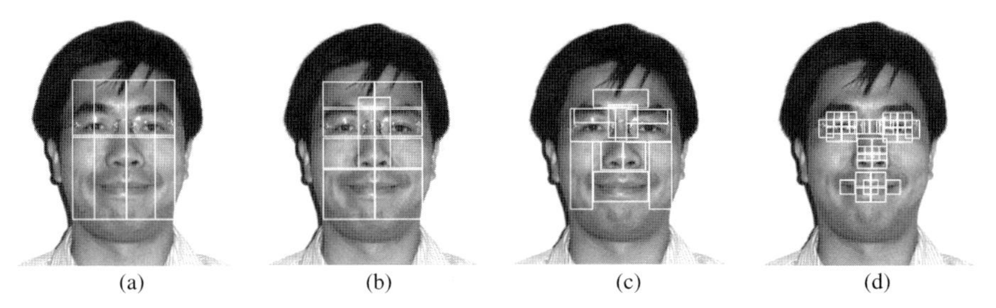

Fig. 5 60 Extended feature regions denoted by rectangle boxes at various detail levels (from left to right) **a)** Upper, lower, left, right and center region of face. **b)** Forehead, left and right eye, eyes, nose, mouth, left and right cheek and nostril. **c)** Forehead, eyebrows, details of left and right eye, left and right cheek, left and right side of nose, nostril, mouth. **d)** Details of left and right eye, details of nose, details of mouth and details of nose bridge

$$\vec{X} = \left[\vec{F}^{0}, \vec{F}^{1}, \vec{F}^{2}, \ldots, \vec{F}^{60}\right]. \tag{8}$$

Each of feature $\vec{F}^{n}$ is a vector of features extracted using (6) and (7) from the sub-matrix of the convolution output for the image with the Gabor filter bank. The superscript n denotes the set of features derive from each of the 60 feature location region.

$$\vec{F}^{n} = [\mu_{00}, \sigma_{00}, \mu_{01}, \sigma_{01}, \ldots, \mu_{mn}, \sigma_{mn}]. \tag{9}$$

Features extracted from regions shown in Fig.5(a) and Fig. 5(b) are for the Level 2 and 3 of the FEETS representation. These features are not dependent on the location of the 4 fiducial points, instead, they take their reference from the area of where the face is detected. In this approach, level 2 and 3 features (can be considered as course features) are not affected by the accuracy of the feature location engines. The accuracy or the placement of these region of interest does not need to be very accurate as the region they cover can be considered as large compared to level 4 and 5 features. Level 4 and 5 features are extracted based on their respective region of interest, which are extrapolated from the locations of the 4 fiducial points.

3 Probabilistic Based Recursive Neural Network (PRNN)

This section describes how Probabilistic Neural Networks (PNN) incorporating with Recursive Neuron can be used for processing tree structure representation of a human face. PNN embeds discriminative information in the classification model and were successfully used for clustering from the input features. Perlovsky proposed neural networks for modeling any probability distribution density as Gaussian Mixture Model [6, 9]. It was later expanded for many engineering applications and to model higher brain functions [3]. Streit and Luginbuhl [29] proposed PNN that as a four layer feed-forward PNN using a Gaussian kernel, or Parzen window, which implements homoscedastic Gaussian mixtures. Roberts and Tarassenko [30] proposed a robust method for Gaussian Mixture Models (GMMs). This method uses a GMM together with a decision threshold to reject unknown data during performing classification task. Elliptical Basis Function Networks (EBFNs) proposed in [31] determine the cluster centers of input features in their hidden layer. It was shown that EBFNs outperform radial basis function networks and vector quantization. Probabilistic Decision-Based Neural Networks (PDBNNs) can be considered as a special form of GMMs with trainable decision thresholds. In [32], a face recognition system based on this PDBNN was proposed, where a decision threshold for each class is used to reject patterns not belonging to any known classes.

In this chapter, a hybrid architecture, namely Probabilistic based Recursive Neural Network (PRNN), is given, using GMMs at the hidden layer and recursive neurons at the output layer. The learning task of this architecture is a hybrid of locally unsupervised and globally supervised learning algorithms. The detail and analysis of this model are given in [33]. The following briefly describes its architecture and learning algorithm.

3.1 Architecture

Figure 6 depicts the architecture of the proposed recursive model in which each neuron at the hidden layer is represented by a Gaussian Mixture Model (GMM) and each neuron at the output layer is represented by a sigmoid activation function model. Each parameter in this GMM has a specific interpretation and function. All weights and node thresholds are given explicitly by mathematical expressions involving estimates of the parameters of the Gaussian mixture and the *a priori* class probabilities and misclassification costs. As shown in Fig. 6(a), the structured recursive neural model is modified so that the input features as the children's output are acting as input vector of GMMs at the hidden layer. Figure 6(b) presents the architecture of each GMM in which the output of a GMM is the weighted sum of G component densities.

The basic architecture of the recursive neural model for the structural processing is adapted from the works of [34]. The recursive here implying that copies of the same neural network are used to encode every node of the structured patterns. Each recursive neuron receives two kinds of inputs that it then uses to generate its output. The first is the output of its node's children, while the second is the input attributes of related vertices that are provided by the structure of the underlying pattern. Such architecture has been shown to be useful in performing classification tasks involving structured patterns [33]. It is also flexible enough to allow the model to deal with the representations of different internal structures and of varying node numbers.

A maximum number of children is required to be assumed for each node (i.e. a maximum branch factor c). For instance, a binary tree is an example of a tree structure that has a maximum branch factor c of two (each node has only two children). Because there are no inputs arising from children at the terminal nodes (i.e. the nodes at L5 in Fig. 3), we set all the y components in the input layer to zero.

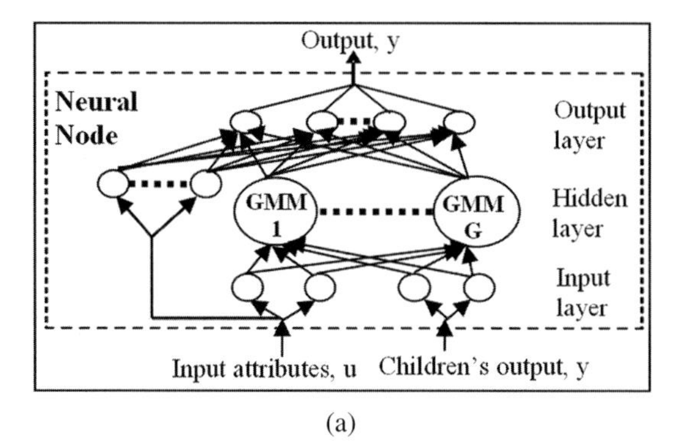

(a)

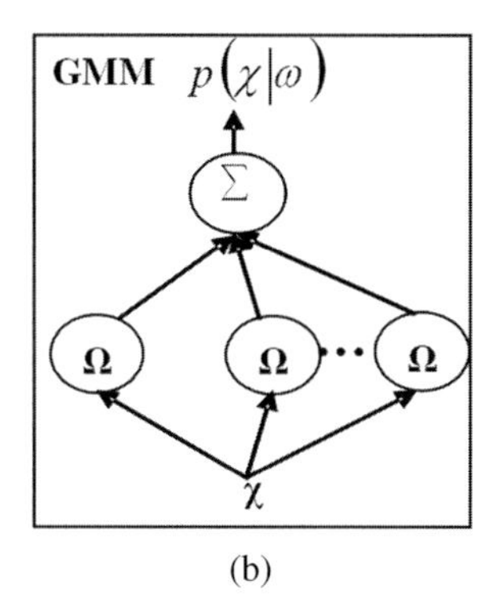

(b)

Fig. 6 **a)** Architecture of Probabilistic based recursive neural network using GMM for neural Node Representation; **b)** structure of a Gaussian Mixture Model [1] for a GMM neural node, where $\sum$ and Ω respectively denotes summation and Gaussian basis functions operations

The terminal nodes in this case are known as frontier nodes. The feed-forward processing of tree-structures goes from the frontier nodes to the root node in a bottom up fashion. Such bottom-up processing from a child node to its parent node can be denoted by an operator q^{-1}.

Now, let m denotes the dimension of the input features of each node in the tree, and p denotes the dimension of the outputs of each node. The input pattern of each GMM can be denotes as

$$\begin{aligned} \chi &= \left(\mathbf{u} \quad q^{-1}\mathbf{y}^T\right) \\ &= \{x_i; i = 1, 2, \ldots (m + p \times c)\}\text{'} \end{aligned} \tag{10}$$

where $\mathbf{u}$ and $\mathbf{y}$ are the m-dimensional input vector and the p-dimensional output vector respectively. q^{-1} is a notation which the input to the node is taken from its child so that $q^{-1}\mathbf{y}$ is equivalent to:

$$q^{-1}\mathbf{y} = \left(q_1^{-1}\mathbf{y}q_2^{-1}\mathbf{y}\cdots q_c^{-1}\mathbf{y}\right)^T. \tag{11}$$

Assuming the input pattern acts as structured pattern, χ associated with class ω that the class likelihood function $p(\chi|\omega)$ for class ω is a mixture of G components in Gaussian distribution, i.e.

$$p(\chi|\omega) = \sum_{g=1}^{G} P(\Theta_g|\omega)p(\chi|\omega,\Theta_g), \tag{12}$$

where Θ_g represents the parameters of the gth mixture component, G is the total number of mixture components. $P(\Theta_g|\omega)$ denotes the prior probability of cluster g. It is also called mixture coefficients of the gth component, by definition,

$$\sum_{g=1}^{G} P(\Theta_g|\omega) = 1, \tag{13}$$

In this approach, $p(\chi|\omega,\Theta_g)$ is the probability density function of the gth component which typically is a Gaussian distribution with mean $\boldsymbol{\mu}_g$ and $\boldsymbol{\Sigma}_g$ covariance, given by

$$p(\chi|\omega,\Theta_g) = \frac{1}{(2\pi)^{(m+p\times c)/2}|\boldsymbol{\Sigma}_g|^{1/2}} \cdot \exp\left\{-\frac{1}{2}(\chi-\boldsymbol{\mu}_g)\boldsymbol{\Sigma}_g^{-1}(\chi-\boldsymbol{\mu}_g)^T\right\}. \tag{14}$$

As shown in Fig. 6(a), the outputs of the proposed recursive network can be simply defined as:

$$\mathbf{y} = F_k(\mathbf{Wp}+\mathbf{Vu}), \tag{15}$$

where $\mathbf{p} = \begin{pmatrix} p_1(\chi|\omega) \\ \vdots \\ p_n(\chi|\omega)) \end{pmatrix}$. $F_k(.)$ is k-dimensional vector, where their elements are the nonlinear sigmoid activation function. **W** and **V** are the weighting parameters in $(p \times n)$ and $(p \times m)$ -dimensional matrices respectively at the output layer.

According to [35], a single hidden layer of this probabilistic based recursive network is sufficient to approximate any structural mapping problems. Learning task in this approach does not require *a priori* knowledge of any structured patterns or any *a priori* information concerning the internal structures of input features. However, learning this proposed model by means of the gradient-based BPTS algorithm [25] may encounter previously mentioned problems of local minima and long-term dependency [36]. It is due to the fact that as gradients vanishes; the learning information may disappear at a certain level of the tree structures before it reaches the frontier nodes [26]. The convergence may stall and poor generalization may yield. A special kind of structural learning algorithm is proposed in this chapter to overcome these problems. We present the details in the next subsection.

3.2 *Learning Algorithm*

The learning scheme is divided into two phases. In the first phase the locally unsupervised algorithm for the parameters of GMMs can adopt the Expectation-Maximization (EM) algorithms. In the second phase, globally structured supervised learning occurs for the recursive neural networks at the output layer. The penalized optimization algorithm is adopted in the structural processing during this second phase. Both learning phases require several epochs to converge and the globally structured supervised learning starts after the locally unsupervised learning is converged.

3.2.1 Locally Unsupervised Learning Phase for GMMs

The parameters of the GMMs as shown in (14) are initialized and estimated by this learning phase. This learning phase is based on unsupervised clustering to determine the parameters. The Expectation-Maximization (EM) method [29] is used for this locally unsupervised learning scheme. The EM method consists of two steps: The first step is called the expectation (E) step and the second is called the maximization (M) step. The E step computes the expectation of a likelihood function to obtain an auxiliary function and the M step maximizes the auxiliary function with respect to the parameters to be estimated. The EM algorithm in the locally unsupervised learning phase is as follows:

Using the GMM in (14), the goal of the EM learning [29] is to maximize the log likelihood of input feature set in structured pattern $\chi^* = \begin{pmatrix} \chi_1 \\ \vdots \\ \chi_{N_T} \end{pmatrix}$. Since we refer to the observable attributes χ^* as the "incomplete" data, so we define an indicator α_{jk} to specify the data belonging to which cluster and include it into the likelihood function as

$$l(\chi^*, \Theta) = \sum_{j=1}^{N_T} \sum_{g=1}^{G} \alpha_{jk} \left[log P(\Theta_g|\omega) + log p(\chi_j|\omega, \Theta_g) \right] \tag{16}$$

where $\alpha_{jk} = \begin{cases} 1, & \text{if structured pattern } \chi_j \text{ belongs to cluster } k \\ 0, & \text{otherwise} \end{cases}$.

In E step, we take the expectation of the observable data likelihood in (16), a conditional posterior probabilities can be obtained by Bayes' rule:

$$P(\Theta_g|\chi_j, \hat{\Theta}(t_L)) = \frac{P(\Theta_g|\omega) p(\chi_j|\omega, \Theta_g)}{\sum_{r=1}^{R} P(\Theta_r|\omega) p(\chi_j|\omega, \Theta_r)}, \text{ at } t_L\text{-th iteration} \tag{17}$$

In M step, the parameters of a GMM are estimated iteratively by maximizing the expectation of the observable data likelihood function with respect to Θ, we may calculate the optimized cluster mean and covariance of the GMMs models.

3.2.2 Globally Structured Supervised Learning Phase for Recursive Nodes

The structural learning through the whole trees starts after the EM learning converged in the previous phase. The goal of this learning phase is to optimize the parameters for the entire model in the structural manner. Optimization minimizes the cost function given by errors between the known target classes of emotion and the output values of the root node in the trees. In this learning phase, fine-tuning in the decision boundaries of the GMMs is carried out by utilizing the known target values. The reinforced and/or anti-reinforced learning techniques [37] are applied to update the cluster mean and covariance of each class.

In the mean time, the W and V parameters at the output layer shown in (15) are optimized to generalize the trees by minimizing the total sum-squared-error function in this learning phase. In this study, a penalized optimization is selected for this globally structured learning phase. The details of this optimization method can be referred to [38]. This algorithm is able to provide much faster convergence and avoiding the gradient vanishing in the deep trees. In case of using the penalized based method, the cost function J must be rewritten according to [38] and a Gaussian-like penalty function is introduced, which is superimposed under the weight space domain.

3.2.3 Learning Steps

Step 1. Initialization

a. Set $t_L = 0$ and $t_G = 0$, where t_L and t_G are the iteration index of learning phase 1 and phase 2 respectively.
b. Initialize, by random, the parameters A=$(W\ V)$ and Θ_g in the proposed probabilistic based recursive model.
c. Calculate the *priori* probability

$$P(\Theta_g|\omega) = \frac{\sum_{j=1}^{N_T} P\left(\Theta_g|\chi_j, \hat{\Theta}(t_L)\right)}{N_T}, \tag{18}$$

of each cluster by means of the input features and initial parameters.
d. Propagate the structured learning patterns to obtain the root output.

Step 2. Locally Unsupervised Learning for GMMs

a. Given the input attributes $\{\chi_1, \ldots, \chi_{N_T}\}$ of each node in all structured patterns.
b. For $t_L = 1 \ldots T_L$:

Calculate the mean and covariance parameters iteratively:

$$\mu_g(t_L+1) = \frac{\sum_{j=1}^{N_T} P\left(\Theta_g|\chi_j, \hat{\Theta}(t_L)\right)\chi_j}{\sum_{j=1}^{N_T} P\left(\Theta_g|\chi_j, \hat{\Theta}(t_L)\right)}, \tag{19}$$

$$\sum_g (t_L+1) = \frac{\sum_{j=1}^{N_T} P\left(\Theta_g|\chi_j, \hat{\Theta}(t_L)\right)\left(\chi_j - \mu_g(t_L+1)\right)\left(\chi_j - \mu_g(t_L+1)\right)^T}{\sum_{j=1}^{N_T} P\left(\Theta_g|\chi_j, \hat{\Theta}(t_L)\right)}. \tag{20}$$

Step 3. Globally Structural Supervised Leaning
For $t_G = 1 \ldots T_G$

a. Calculate the outputs $p(\chi|\omega)$ of each GMM in each structured pattern and class.
b. Fine-tuning the decision boundaries of each GMM using reinforced and/or anti-reinforced learning techniques.
c. Estimate the weighting parameters, $\mathbf{A} = \mathbf{a}_{k\,p\times(m+n)}$ at the output layer iteratively:

$$\mathbf{a}_k(t_G+1) = \mathbf{a}_k(t_G) + \eta\left[-\frac{\partial J}{\partial \mathbf{a}_k}\right], \text{ where } \eta \text{ is the learning rate.} \tag{21}$$

4 Asian Emotion Database

To the best of our knowledge, little investigation has been conducted on analyzing face emotion behavior among the different races in the Asian population. Most of the publicly available emotion database contains images that are captured from video recording or stored in low-resolution quality. The closest Asian emotion database is the Japanese Female Emotion Database [39], and it contains 213 images of 7 facial expressions (including neutral) posed by 10 Japanese Actresses. A 3D Facial expression database [40] by Yin et al. contains 100 subjects in various emotion from various races found in America. The development of our database was designed to capture high-resolution 2D facial images for various races, age group and gender found in the Asian population in seven emotional states and in 3 different poses. As Singapore is in the heart of Asia and has a high mixture of different races in Asia, we have collected our data from our University.

4.1 Creation of Database

We set up a photograph station in a public venue, and invited volunteers (both female and male) of various races and age group from the public to participate in our data collection exercise. The procedures are found in the Appendix.

Each subject is instructed to sit in front of the camera. They are requested to perform 7 expressions, i.e. Neutral, Anger, Joy, Sadness, Surprise, Fear, and Disgust. As the digital camera is unable to capture dynamic facial expressions, we require the subject to perform the expression for a short period. Ideally, a video clip could be used for eliciting a genuine emotion state of subjects. However, it is difficult to provide such a setup, especially for emotions such as sadness and fear [41]. Cowie et al. [42] quotes, displays of intense emotion or "pure" primary emotions rarely happened. Due to the time constraints and the limitation of setup, we have asked the subjects to perform pseudo-emotions. The subjects were also asked to position themselves, 45 degrees to the left and right, so we can capture these expressions for half profile poses as shown in Fig. 7.

4.2 Statistics of Participants

The Asian Emotion Database[1] contains around 4947 images for 7 different facial expressions and 3 different poses from 153 subjects, who participated in the data collection exercise over a period of 4 days. Of the 153 subjects, 64 gave consent to using their images in publications. The Asian Emotion Database has facial images from various races, including, Chinese, Malay, Indian, Thai, Vietnamese, Indonesian, Iranian and Caucasian. About 72% of the subject belongs to the Chinese race, 7% Indian, 7% Vietnamese, 5% Caucasian, 4% Malay and 4% others, Table 1 shows

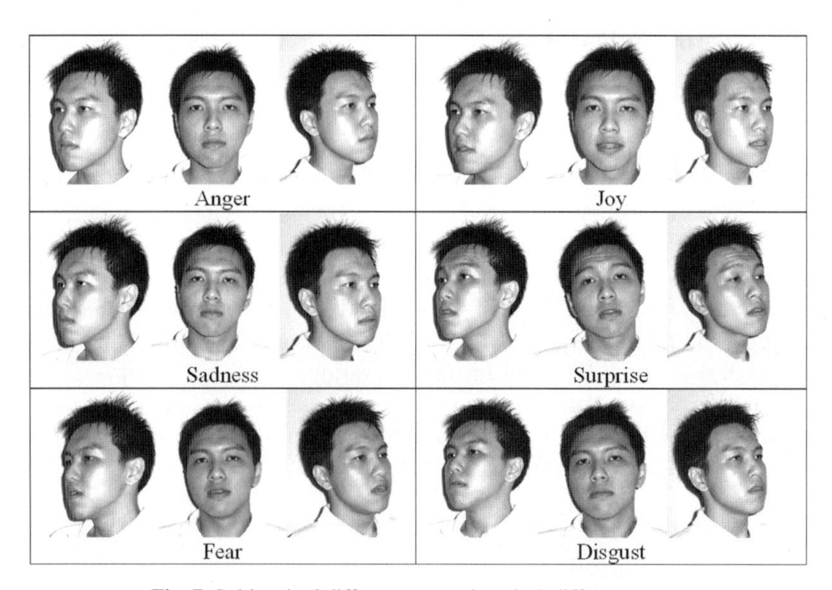

Fig. 7 Subject in 6 different expressions in 3 different poses

[1] http://forse.ntu.edu.sg/emotiondatabase.htm

Table 1 Distribution of Subjects according to Race and Gender

	Chinese	Malay	Indian	Vietnamese	Caucasian	Others	Total
Female	33	5	3	1	1	3	46
Male	77	2	8	10	7	3	107
Total	110	7	11	11	8	6	153

Table 2 Distribution of subjects according to race and age group

Age (years)	Chinese	Malay	Indian	Vietnamese	Caucasian	Others	Total
< 20	4	0	1	0	0	3	8
20-29	80	6	5	11	4	3	109
30-39	20	0	2	0	1	0	23
40-49	6	1	2	0	3	0	12
≥ 50	0	0	1	0	0	0	1
Total	110	7	11	11	8	6	153

the detail distribution of our database according to Race and Gender. Table 2 shows the majority of the subjects about 72% were of the 20–29 year old age group, as well as the detail breakdown of the database according to Race and Age group. Due to time constraints of our participants, we were only able to get 27 of them to pose for the 2 half profile poses for all expressions.

5 Experiments and Results

We evaluated our system using the Asian Emotion Database we built. We have used all the 4947 images from our database containing 153 persons in 6 basic emotions and 1 neutral emotion. For this experiment, we created 2 datasets, i.e., subject-dependent dataset (Dataset A) and subject-independent dataset (Dataset B). Subject-dependent dataset will be able to show how well the system performed when the system knows how each of the subject's pseudo emotions looked. The training images contain 3901 images of all 153 subjects in various emotions. The testing set used the remaining 1046 images. Subject-independent dataset will be able to evaluate the performance of this system when it is used in a situation where, prior knowledge of subject is not available, as the system does not know how the evaluated subject's pseudo emotions will look. We performed 5-folds cross validation in this evaluation.

We benchmarked our proposed method against other well-known classifiers such as *Support Vector Machine (SVM)* [43], *C4.5 Tree (C45)* [44], *Gaussian Radial Basis Function network (RBF network)*, *Naïve Bayesian Multimodal (NBM) classifier* [45]. We have used these classifiers from the Weka package [46]. The following parameters were used in these experiments: SVM using polynomial kernel, with complexity parameter at 1.0, gamma parameter at 0.01; C4.5 using 3-folds tree pruning with confidence factor of 0.25. All of these tested classifiers were used with flat-vectors input. Therefore, some regularity inherently associated with the tree structures of our algorithms was broken. This led to poorer generalization. Some

Table 3 Performance of FEETS/PRNN model against other classifiers

	FEETS	SVM	C45	RBF	NBM
Dataset A	62.5%	52.9%	45.8%	30.5%	16.1%
Dataset B	56.8%	50.0%	41.4%	31.7%	15.7%

of these algorithms might have suffered from poor convergence and resulted in a relatively low classification rate. Table 3 summarized the performance of our model against other classifiers for perfect feature location detection, for dataset A and B. It shows that our algorithm yields a better performance.

We evaluated how well our system performed when there was noise in the accuracy of the feature detectors, i.e., error in locating the center of features. Figure 8 shows an example of error in locating the center of features at various degree of error. Under normal situations, we should consider error levels, less than 15 pixels off the center of feature locations, as a common error in feature detection. We benchmarked the system performance for this type of noise by using the Dataset A, against other classifiers. Results in Table 4, shows that FEETS is the most robust and less subjected to this type of noise. Results show that traditional classifiers suffered from noise in the feature detection accuracy.

We further evaluated the system against extreme conditions where detection of some features would fail completely. Feature lost can be considered a high occurrence error for any facial or image recognition problem, as subjects might not always be in perfect view of the camera, for example, objects occlusion or self-occlusion. Figure 9 shows some extreme situations where feature detector will fail. Figure 9(b) shows a subject wearing a veil, which will cause nose and mouth detector to fail, similarly, sunglasses in Fig 9(c) will cause eyes detection to fail. This will evaluate the system's robustness when features are lost due to failure to detect feature locations. In this experiment, undetected features are padded with zeros in order to retain

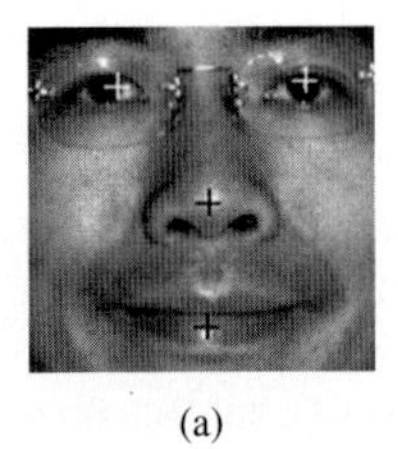
(a)

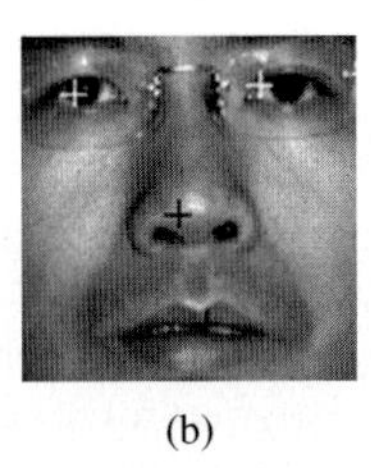
(b)

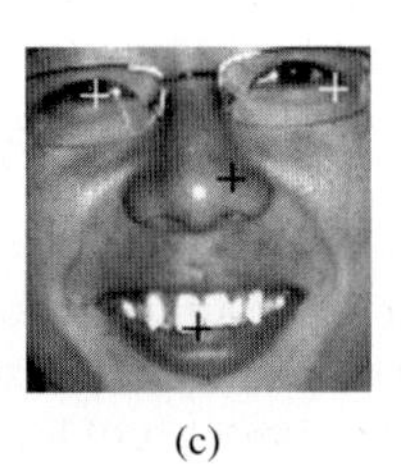
(c)

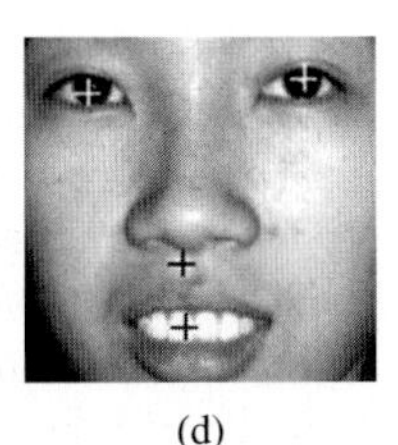
(d)

Fig. 8 **a)** Feature location is 5 pixels or less off from the ideal center of features. **b)** Features are 6 to 10 pixels off from the ideal center of features. **c)** 11 to 15 pixels off from the ideal center of features. **d)** More than 16 pixels off the ideal center of features

Table 4 Performance of FEETS/PRNN model against other classifiers

Error (pixels)	FEETS	SVM	C45	RBF	NBM
≤ 5	58.8%	42.3%	32.3%	29.7%	15.4%
≤ 10	57.9%	36.7%	27.8%	28.2%	15.4%
≤ 15	50.9%	39.6%	22.6%	30.2%	22.6%
> 15	47.1%	17.7%	29.4%	29.4%	5.9%

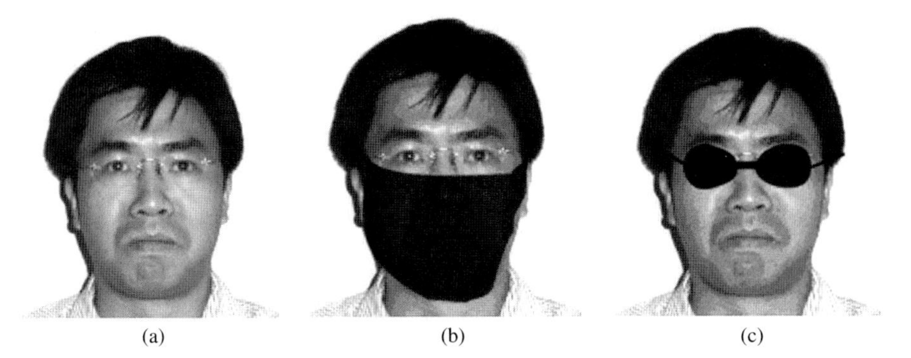

(a) (b) (c)

Fig. 9 **a**: Subject without any artifacts. **b**: Subject wearing a veil **c**: Subject wearing sunglasses

the similar length for the feature vector, as well as the shape of the FEETS. The system was evaluated against various degree of loss, i.e. single eye missing, nose missing, mouth missing. As well as multiple feature loss, i.e. both eyes are missing and nose and mouth missing. Using the perfect features location and full features as training set, we evaluated the system when tested with missing features. Figure 10(a) and 10(b) shows the performance of FEETS against other system using Dataset A and Dataset B respectively. Experiments in Dataset B is run under 5-folds cross validation, to ensure that all the images are train and tested in turns. The experiment results show that our system is more robust than the others when features are lost or while nodes are missing.

6 Conclusion and Future Works

This chapter described a framework for robust facial emotion recognition system based on replicating the functions of the human emotion recognition systems. The FacE Emotion Tree Structure (FEETS) models the function of the sensory transform associated in the visual cortex and hippocampus. The Probabilistic Recursive Neural Network (PRNN) models for adaptive processing emotion recognition function of the amygdala. Unlike conventional connectionist and statistical approaches that rely on static representations of data resulting in vectors of features, the approach in this chapter allows patterns to be properly represented by directed graphs or trees. In this chapter, we emphasize the brain-inspired computational framework offers highly adaptive and dynamical ability of emotion discrimination without *a priori* class characterization. We evaluated the system using an Asian Emotion Database that we build to represent the Asian population. For empirical studies, the performance for subject dependent and subject independent emotion recognition is benchmarked against various well-known classifiers using the conventional connectionist (C4.5, RBF) and statistical approaches (NBM, SVM). The robustness of the system is tested with errors in fiducial points as well as extreme conditions where fiducial points are missing.

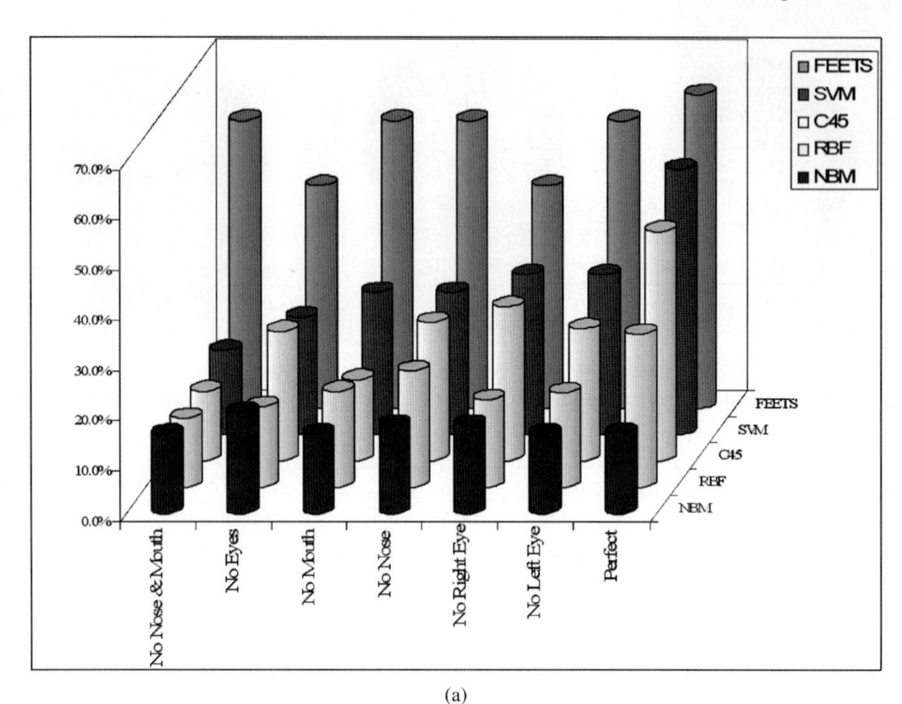

(a)

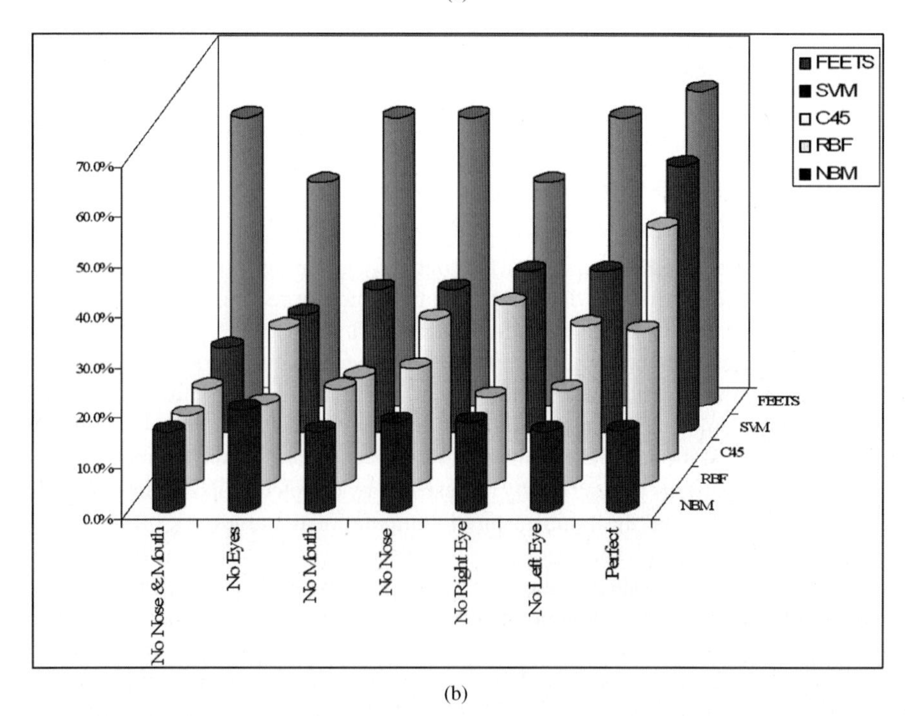

(b)

Fig. 10 3D bar chart of performance results of FEETS against other classifiers when features are missing for **a)** Dataset A. **b)** Dataset B

Further studies can be extended to develop a neuro-cognitive computational framework that is capable of perceiving and reporting on socio-emotional information in real-time human interaction. The first stage of processing emotional faces is a feed-forward sweep through primary visual cortices ending up in associative cortices. In the event of an emotional face being detected, projections at various levels of the visual (primary) and the associative cortices to the amygdala can alert the latter of the onset of an emotional face. Once the amygdala is alerted to the onset of an emotional face it can in turn draw attention back to that face so as to delineate the categorization of the face in detail. It can be done by either employing its projections back to the associative cortex to enhance processing of this face or via activation of the prefrontal cortex that can initiate a re-prioritization of the salience of this face within the prefrontal cortex area and drive attention back to the face.

Acknowledgment We would like to thank all the volunteers who have participated and contributed to the creation of the Asian Emotion Database. We would like to thank Muhammad Raihan Jumat Bin Md T. J. for processing and encoding the Asian Emotion Database.

References

1. Ekman, P., Emotions Revealed. First Owl Books. 2004, New York: Henry Holt and Company LLC.
2. Daniel Levine Chapter in this book.
3. Leonid Perlovsky Chapter in this book.
4. Bechara, A., H. Damasio, and A.R. Damasio, Emotion, Decision making and the Orbitofrontal Cortex. Cerebral Cortex, 2000.
5. Isen, A.M., Positive Affect and Decision Making, in Handbook of Emotions. 2000, Guilford Press: New York. p. 417–435.
6. Perlovsky, L.I. (2001). Neural Networks and Intellect: using model-based concepts. Oxford University Press, New York, NY (3rd printing).
7. Baron-Cohen, S., Mindblindness: An Essay on Autism and Theory of Mind. MIT PRess. 1995.
8. Ellis, H.D. and M.B. Lewis, Capgras delusion: a window on face recognition. Trends in Cognitive Science, 2001. 5: p. 149–156.
9. Perlovsky, L.I. and McManus, M.M. (1991). Maximum Likelihood Neural Networks for Sensor Fusion and Adaptive Classification. Neural Networks 4 (1), pp. 89–102.
10. Frijda, N.H., The Emotions. 1986: Cambridge: Cambridge University Press.
11. Tian, Y.-L., T. Kanade, and J.F. Cohn, Recognizing Action Units for Facial Expression Analysis. IEEE Trans. on Pattern Analysis and Machine Intelli-gence, 2001. 23(2): p. 1–18.
12. Ekman, P., Telling Lies. 1991, New York: W.W. Norton.
13. S.Schachter, The interaction of Cognitive and physiological determinants of emotional state, in Advances in Experimental Social Psychology, L. Ber-towitz, Editor. 1964, Academic Press. p. 49–80.
14. Davis, M., The role of the amygdale in fear and anxiety. Annual Rev. Neuro-science, 1932. 15: p. 353–375.
15. Streit, M., A.A. Ioannides, L. Liu, W. Wolwer, J. Dammers, J. Gross, W. Gaebel, and H.W. Muller-Gartner, Neurophysiological correlates of the rec-ognition of facial expressions of emotion as revealed by magnetoencephalo-graphy. Cognitive Brain Res., 1999. 7: p. 481–491.
16. Wong, J.-J. and S.-Y. Cho, A Brain-Inspired Framework for Emotion Recog-nition. Neural Information Processing, 2006. 10(7): p. 169–179.

17. Taylor, John G., Fragopanogos, N., Cowie, R., Douglas-Cowie, E., Fotinea, S-E., Kollias, S., An Emotion Recognition Architecture Based on Human Brain Structure, Lecture Notes in Computer Science, (2714), pp. 1133–1142, 2003.
18. Viola, P. and M. Jones. Robust Real-time Object Detection. in Second Inter-national Workshop on Statistical and Computational Theories of Vision - Modeling, Learning, Computing, and Sampling. 2001. Vancouver, Canada.
19. Freund, Y. and R.E. Schapire, A decision-theoretic generalization of online learning and an application to boosting, in Computational Learning Theory: Eurocolt '95. 1995, Springer-Verlag. p. 23–37.
20. Liu, C. and H. Wechsler, Independent Component Analysis of Gabor Features for Face Recognition. IEEE Transactions on neural networks, 2003. 14(4): p. 919–928.
21. Wong, J.-J. and S.-Y. Cho. Recognizing Human Emotion From Partial Facial Features. in IEEE World Congress on Computational Intelligence (IJCNN). 2006. Vancouver, Canada. p. 166–173.
22. Nakamura, K., Nitta, J. Takano, H., and Yamazaki, M., Consecutive Face Recognition by Association Cortex - Entorhinal Cortex - Hippocampal Formation Model, in International Joint Conference on Neural Networks, (3), pp. 1649–1654, 2003.
23. Ekman, P. and W. Friesen, Facial Action Coding System: A Technique for the Measurement of Facial Movement. 1978, Palo Alto, CA: Consulting Psy-chologist Press.
24. Fracesoni, E., P. Frasconi, M. Gori, S. Marinai, J.Q. Sheng, G. Soda, and A. Sperduti, Logo recognition by recursive neural networks, in Lecture Notes in Computer Science, R. Kasturi and K. Tombre, Editors. 1997, Springer-Verlag: New York. p. 104–117.
25. Sperduti, A. and A. Starita, Supervised neural networks for classification of structures. IEEE Trans. Neural Networks, 1997. 8: p. 714–735.
26. Cho, S.-Y., Z. Chi, W.-C. Siu, and A.C. Tsoi, An Improved Algorithm for learning long-term dependency problems in adaptive processing of data struc-tures. IEEE Trans. on Neural Networks, 2003. 14(4): p. 781–793.
27. Frasconi, P., M. Gori, and A. Sperduti, A General Framework for Adaptive Processing of Data Structures. IEEE Trans. Neural Networks, 1998. 9: p. 768–785.
28. Manjunath, B.S. and W.Y. Ma, Texture Features for Browsing and Retrieval of Image Data. IEEE Transactions on Pattern Analysis and Machine Intelli-gence, 1996. 18(8): p. 837–842.
29. Streit, D.F. and T.E. Luginhuhl, Maximum likelihood training of probabilistic neural networks. IEEE Trans. on Neural Networks, 1994. 5(5): p. 764–783.
30. Roberts, S. and L. Tarassenko, A probabilistic resource allocating network for novelty detection. Neural Computation, 1994. 6: p. 270–284.
31. Mak, M.W. and S.Y. Kung, Estimation of elliptical basis function parameters by the EM algorithms with application to speaker verification. IEEE Trans. Neural Networks, 2000. 11(4): p. 961–969.
32. Lin, S.H., S.Y. Kung, and L.J. Lin, Face recognition/detection by probabilistic decision-based neural network. IEEE Trans. on Neural Networks, Special Is-sue on Biometric Identification, 1997. 8(1): p. 114–132.
33. Cho S.Y., Probabilistic Based Recursive Model for Adaptive Processing of Data Structures, Expert Systems With Applications, in press, 2007. (http://dx.doi.org/10.1016/j.eswa.2007.01.021)
34. Tsoi, A.C., Adaptive Processing of Data Structure : An Expository Overview and Comments. 1998, Faculty Informatics, Univ. Wollongong, Wollongong: Australia.
35. Hammer, B., M. A., A. Sperduti, and S. M., A general framework for unsu-pervised processing of structured data. Neurocomputing, 2004. 57: p. 3–35.
36. Bengio, Y., P. Simard, and P. Frasconi, Learning Long Term Dependencies with Gradient Descent is difficult. IEEE Trans. on Neural Networks, 1994. 5(2): p. 157–166.
37. Kung, S.Y. and J.S. Taur, Decision-Based Neural Networks with Sig-nal/Image classification applications. IEEE Trans. Neural Networks, 1995. 6: p. 170–181.
38. Cho, S.-Y. and C.T. W.S., Training Multilayer Neural Networks Using Fast Global Learning Algorithm - Least Squares and Penalized Optimization Methods. Neurocomputing, 1999. 25(1–3): p. 115–131.

39. Lyons, M.J., J. Budynek, and S. Akamatsu, Automatic Classification of Single Facial Images. IEEE Transactions on Pattern Analysis and Machine Intelligence, 1999. 21(12): p. 1357–1362.
40. Yin, L., X. Wei, Y. Shun, J. Wang, and M.J. Rosato. A 3D Facial Expression Database for Facial Behavior Research. in 7th International Conference on Automatic Face and Gesture Recognition. 2006.
41. Sebe, N., M. Lew, I. Cohen, Y. Sun, T. Gevers, and T. Huang. Authentic facial expression analysis. in International Conference on Automatic Face and Gesture Recognition. 2004. Seoul, Korea. p. 517–522.
42. Cowie, R., E. Douglas-Cowie, N. Tsapatsoulis, G. Votsis, S. Kollias, W. Fellenz, and J. Taylor, Emotion Recognition in Human-computer Interaction. IEEE Signal Processing Magazine, 2001. 18(1): p. 32–80.
43. Platt, J., Fast Training of Support Vector Machines using Sequential Minimal Optimization, in Advances in Kernel Methods - Suppoort Vector Learning, B. Scholkopf, C. Burges, and A. Smola, Editors. 1998, MIT Press. p. 185–208.
44. Quinlan, R., C4.5: Programs for Machine Learning. 1993, San Mateo, CA: Morgan Kaufmann Publishers.
45. Mccallum, A. and K. Nigam. A Comparison of Event Models for Naive Bayes Text Classification. in International Conference on Machine Learning. 1998. p. 41–48.
46. Witten, I.H. and E. Frank, Data Mining: Practical machine learning tools and techniques. 2nd Edition. 2005: Morgan Kaufmann, San Francisco.

Appendix

Figure 11 shows the set up of the photograph station. The image is cap-tured using a 5 mega-pixels Ricoh R1V digital camera using ISO 64 set-tings with flash from the camera. Subject is sit 143 cm away against a white background from the camera. We

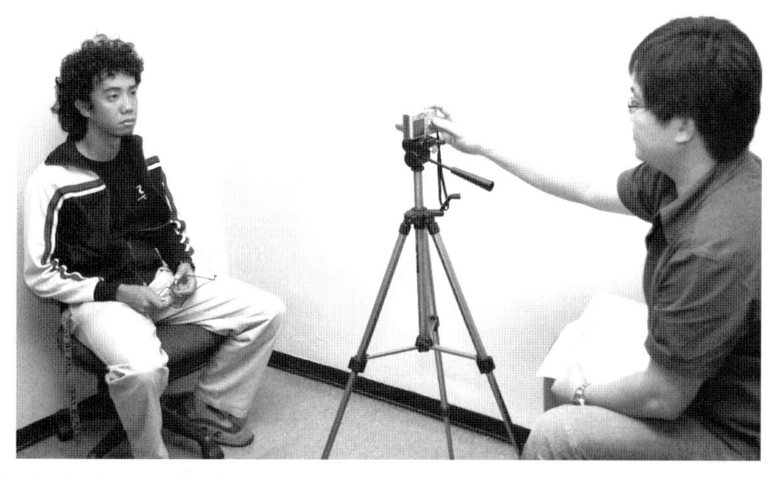

Fig. 11 During image collection process where the volunteer sitting against a white background, on the left of picture performs an expression. The camera is place 143 cm from subject. The photographer sits on the right of this picture, will give instructions to the subject for performing various expressions and poses. Subject seen here is in frontal pose position. The subject is asked to turn 45 degrees to the left and right for 2 other poses

cropped and scaled the original 2560×1920 pixels facial to around 900×1024 pixels for archival. We annotate, process, and store each image in 24-bit color Bitmap format as a ground truth. The filenames of the images were encoded in the following manner:

XXXXX_X_X_XX_X_X_X_XXX.bmp	Filename
1 2 3 4 5 6 7 8	Section

Each section of the filename stores the following information in ascending order: Subject ID, age group, gender, race, consent to publication, emo-tions, pose, and sample number. Using this filename format, the data is searchable by query. Along with the images, we included the location of the center of eyes, nose and mouth and stored in an 8-bit integer format file in a corresponding filename with "loc" as filename extension as a ground truth.

Engineering Applications of Olfactory Model from Pattern Recognition to Artificial Olfaction

Guang Li, Jin Zhang and Walter J. Freeman

Abstract Derived from biological olfactory systems, an olfactory model entitled KIII was setup. Different from the conventional artificial neural networks, the KIII model works in a chaotic way similar to biological olfactory systems. As one kind of chaotic neural network, KIII network can be used as a general classifier needing much fewer training times in comparison with other artificial neural networks. The experiments to apply the novel neural network to recognition of handwriting numerals, classification of Mandarin spoken digits, recognition of human face and classification of normal and hypoxia EEG have been carried out. Based on KIII models, an application of electronic nose on tea classification was explored. Hopefully, the K set models will make electronic noses more bionically.

1 Introduction

In recent years, the theory of chaos has been used to understand the mesoscopic neural dynamics, which is at the level of self-organization at which neural populations can create novel activity patterns [5]. From years of research in this field, it is now proposed that the chaotic attractor is an essential property of biological neural networks. Derived from the study of olfactory system, the distributed KIII model, which is a high dimensional chaotic network, in which the interactions of globally connected nodes lead to a global landscape of high-dimensional chaotic attractors. After reinforcement learning to discriminate classes of different patterns, the system forms a landscape of low-dimensional local basins, with one basin for each pattern class [13]. This formation of local basins corresponds to the memory of different patterns; the recognition of a pattern follows when the system trajectory enters into a certain basin and converges to the attractor in that basin. The covergence deletes extraneous information, which is the process of abstraction. The output of the system is controlled by the attractor, which signifies the class to which the stimulus belonged, thus exercising generalization. Abstraction and generalization are powerful attributes of biological neural networks but not easily done with artificial neural networks; however, the use of chaotic dynamics by the K set models overcomes this limitation.

L. I. Perlovsky and R. Kozma, *Neurodynamics of Cognition and Consciousness.*

KIII model was built according to the architecture of the olfactory neural system to simulate the output waveforms observed in biological experiments with EEG and unit recording. The KIII model based on deterministic chaos proved to be highly unstable. The introduction of noise modeled the biological noise sources made the KIII network stable and robust [3], which introduced 'Stochastic Chaos' and made the KIII model free from the sensitivity to variation of parameters and initial conditions, and provided a high-dimensional chaotic system capable of rapid and reliable pattern classification without gradient descent [6]. Here we present some experimental results to apply the KIII network to classify handwriting numerals, Mandarin spoken digits, human faces and normal and hypoxia EEG signals. Furthermore an electronic nose based on the KIII model is explored to classify different kinds of tea. Hopefully, the K set models will make electronic noses more bionically.

2 K Set Models

A biological olfactory neural system consists of olfactory bulb (OB), anterior nucleus (AON) and prepyriform cortex (PC). In accordance with the anatomic architecture, KIII network is a multi-layer neural network model, which is composed of several K0, KI and KII units. Figure 1 shows the topology of KIII model, in which M, G represent mitral cells and granule cells in olfactory bulb. E, I, A, B represent excitatory and inhibitory cells in anterior nucleus and prepyriform cortex respectively.

Among these models, every node is described as a second order differential equation as (1), which is derived from plenty of electro-physiological experiments:

$$\frac{1}{a \cdot b}\left[x_i^{''}(t) + (a+b)x_i^{'}(t) + a \cdot b \cdot x_i(t)\right] = \Sigma_{j \neq i}^{N}\left[W_{ij} + Q(x_j(t), q_j)\right] + I_i(t), \tag{1}$$

where $i = 1...N$, N is the number of channels, $x_i(t)$ represents the state variable of ith neural population, $x_j(t)$represents the state variable of jth neural population, which is connected to the ith, while W_{ij} indicates the connection strength between them. $I_i(t)$ is an input function which stands for the external input to the ith channel. The parameter $a = 0.220ms^{-1}$, $b = 0.720ms^{-1}$ reflect two rate constants. $Q(x_j(t), q_j)$ is a static nonlinear sigmoid function derived from Hodgkin-Huxley model and is expressed as (2).

$$Q(x_j(t), q) = \begin{cases} q(1 - e^{-\frac{e^{x}(t)-1}{q}}), & \text{if } x(t) > x_0 \\ -1, & \text{if } x(t) < x_0 \end{cases}, \tag{2}$$

$$x_0 = \ln(1 - q\ln(1 + \tfrac{1}{q})),$$

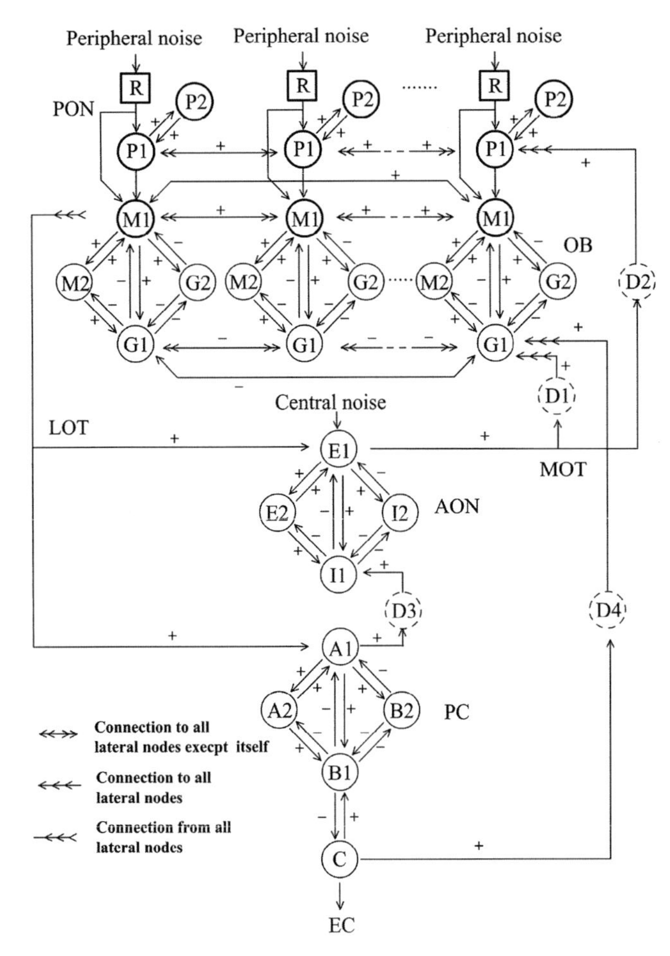

Fig. 1 Topology of the KIII model

where q represents the maximum asymptote of the sigmoid function, which is also obtained from biological experiments. Each node represents a neural population or cell ensemble. If the cell ensemble does not contain any interaction among its neurons, it is represented by a K0 model. KI model represents neural populations that are mutually connected, with excitatory or inhibitory synaptic connections. KII is a coupled nonlinear oscillator used to simulate channels in OB, AON and PC layer with both positive and negative connections. The KIII network describe the whole olfactory neural system. It populations of neurons, local synaptic connection, and long forward and distributed time-delayed feed back loops. In the topology of the KIII network, R represents the olfactory receptor, which is sensitive to the odor molecule, and offers the input to the KIII network. The periglomerular (PG) layer and the olfactory bulb (OB) layer are distributed, which in this paper contains 64 channels. The AON and PC layer are only composed of single KII network. The parameters in the KIII network, including the connection strength values between different nodes and the gain values of the lateral connections, feed forward and feed back loops, were optimized by measuring the olfactory evoked potentials and EEG, simulating their waveforms and statistical properties, and fitting the simulated functions to the data by means of nonlinear regression [5]. After the parameter optimization, the KIII network generates EEG-like waveform with $1/f$ power spectra [1, 2]. Some numerical analysis of the KIII network, using the parameter set in [2], is shown in Figs. 2, 3 and 4. When there is no stimulus, the system presents an aperiodic oscillation, which indicates that the system is in its basal state – a global chaotic attractor, as shown in Fig. 2. When there is a stimulus, corresponding to some odor molecule captured by the olfactory receptor, the trajectory of the system soon goes to specific local basin and converges to an attractor. The true series appears as a gamma range quasi-periodic burst. The system resides in the local basin for approximately the duration of the external stimulus. After the stimulus is terminated, the KIII network returns to its basal state.

Figures 3 and 4 show some phase maps of several pairs of nodes in KIII network. It is another indirect description of the basal chaotic attractor and the state transitions that take place when the stimulus begins and ends. This kind of state transition often takes less than 10 ms. During the learning process, different stimulus patterns will form different local basins and different attractors in the system trajectory space. In KIII network, the formation and consolidation of these local basins are implemented

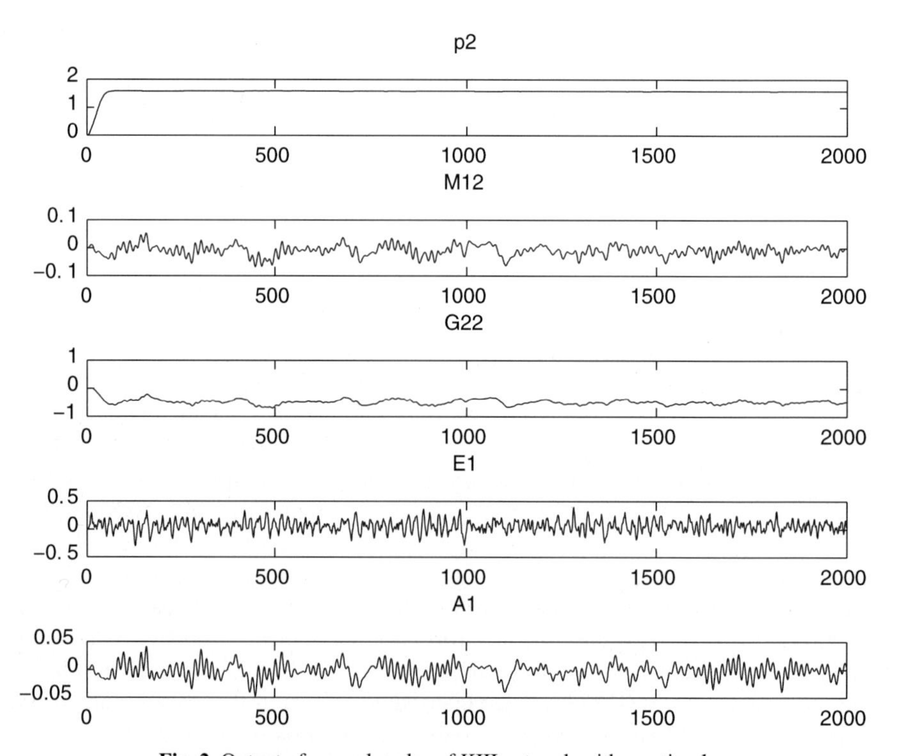

Fig. 2 Output of several nodes of KIII network with no stimulus

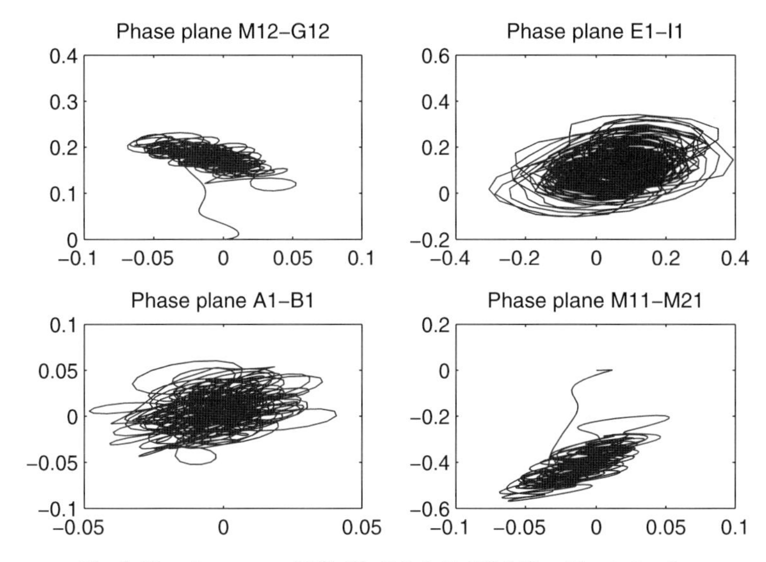

Fig. 3 The phase map of M1-G1, E-I, A-B, M1-M2, without stimulus

through changing the weights between corresponding nodes, in accordance with the biological increase and decrease of the synaptic connection strengths, which have been evaluated by curve-fitting of solutions to the equations to impulse responses of the olfactory system to electrical stimuli [5]. Compared with well-known low-dimensional deterministic chaotic systems such as the Lorenz, Rossler, and Logistics attractors, nervous systems are high-dimensional, non-autonomous and

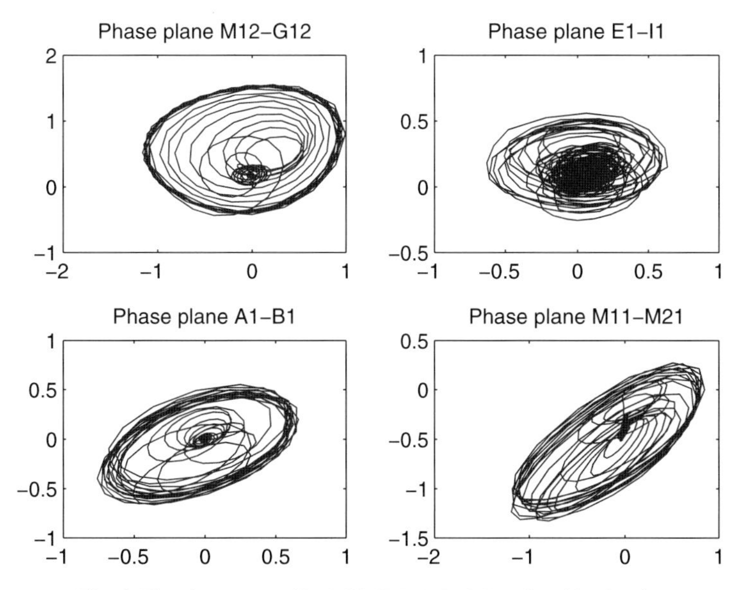

Fig. 4 The phase map of M1-G1, E-I, A-B, M1-M2, with stimulus

noisy systems [7]. In the KIII network, independent rectified Gaussian noise was introduced to every olfactory receptor to model the peripheral excitatory noise, and single channel of Gaussian noise with excitatory bias to model the central biological noise sources. The additive noise eliminated numerical instability of the KIII model, and made the system trajectory stable and robust under statistical measures, which meant that under perturbation of the initial conditions or parameters, the system trajectories were robustly stable [2]. Because of this stochastic chaos, the KIII network not only simulated the chaotic EEG waveforms, but also acquired the capability for pattern recognition, which simulated an aspect of the biological intelligence, as demonstrated by previous applications of the KIII network to recognition of one-dimensional sequences, industrial data and spatiotemporal EEG patterns [7].

3 Applications of Olfactory Models to Classification

3.1 Procedure of Classification

The procedure of the classifications is shown in Fig. 5. The features are extracted from a subject to input into the KIII model as a feature vector for further processing. The feature vector varies from one application to another and is optimized to achieve a good classification performance.

The output of the KIII network at the mitral level (M) is taken as the activity measure of the KIII model caused by the input feature vector. The activity of the ith channel is represented by $SD_{\alpha i}$, which is the mean standard deviation of the output of the ith mitral node (Mi) over the period of the presentation of input patterns, as (3). The response period with input patterns is equally divided into segments and the standard deviation of the ith segment is calculated as $SD_{\alpha ik}$, $SD_{\alpha i}$is the mean value of these s segments. $SD_{\alpha}$is an $1 * n$ activity vector containing the activity measure of all nodes in mitral level. SD^{m}_{α} is the mean activity measure over the whole OB layer with n nodes (4).

$$SD_{\alpha i} = \frac{1}{S}\Sigma^{S}_{k=1} SD_{\alpha ik} \tag{3}$$

$$SD^{m}_{\alpha} = \frac{1}{n}\Sigma^{n}_{k=1} SD_{\alpha i} \tag{4}$$

Before performance of classifications, the KIII network should be trained as well as conventional artificial neural networks. The training rules will be discussed later in Sect. 3.2.

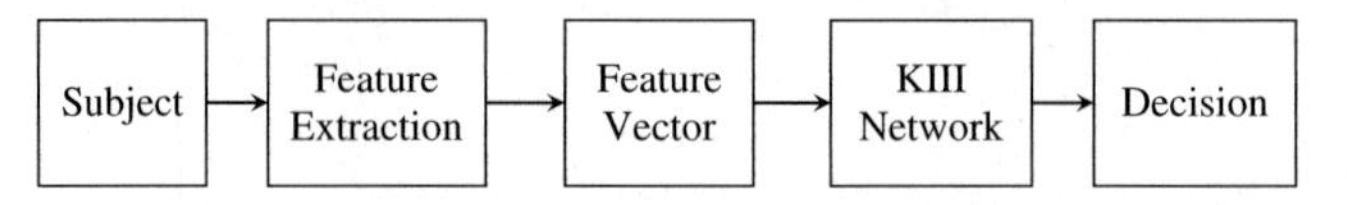

Fig. 5 Procedure of the classifications

At the end of training, the connection weights of KIII model are fixed to perform pattern classification tests. During training, several samples for each class are given. The amplitude of the N activity outputs is measured and expressed as a feature vector for every trial, as well as the mean activity of those trials that belong to the same class. The feature vector defines a point in N-space (here $N = 64$ is used); a set of training trials with the same class of stimulus forms a cluster of points in N-space; the mean is defined as the center of gravity of the cluster representing the class. Inputs of different classes that the system is trained to discriminate form multiple clusters of feature vectors, each with its center of gravity.

When a test pattern is given, its feature vector is calculated. The Euclidean distances from the corresponding point of the feature vector to those training pattern cluster centers are calculated, and the minimum distance to a center determines the classification.

To ensure correct classification, normally a threshold was introduced. When the difference between the minimum Euclidean distance and the secondary minimum distance was less than the threshold value, is the trial was regarded as a recognition failure.

3.2 Learning Rules

For KIII network there are three main learning processes: Hebbian associative learning, habituation, and normalization. Hebbian learning under reinforcement establishes the memory basins and attractors of classifying patterns, while habituation is used to reduce the impact of environment noise including non-informative background inputs to the KIII network. Global normalization is used to maintain the stability of the KIII network long-term over the changes in synaptic weights that typically increase the strengths of excitatory connections in association and reduce them during habituation. Normalization includes the modification of the Hebbian rule by imposing a maximal strength of mutual excitation to prevent runaway increases, and by requiring that Hebbian increases only occur under reinforcement. Habituation is otherwise automatic in the absence of reinforcement.

When an axon of cell A is near enough to excite a cell B and repeatedly or persistently takes part in firing it, some growth process or metabolic change takes place in one or both cells such that A's efficiency, as one of the cells firing B, is increased [15]. The principles underlying this statement have become known as Hebbian Learning. Within connectionism, Hebbian learning is an unsupervised training algorithm in which the synaptic strength (weight) is increased if both the source neuron and target neuron are active at the same time. According to our specific requirements, we made some modifications in the Hebbian learning rule: 1) we designed two methods for increasing the connection strength which is described below; 2) we introduced a bias coefficient K to the learning process.

In our experiments,the modified Hebbian rule holds that each pair of M nodes that are co-activated by the stimulus have their lateral connections, $\omega(mml)_{ij}$, strengthened. Here $\omega(mml)_{ij}$ stands for the connection weights both from Mi to Mj

and from Mj to Mi. Those nodes whose activities are larger than the mean activity of the OB layer are considered activated; those whose activity levels are less than the mean are considered not to be activated. Also, to avoid the saturation of the weight space, a bias coefficient K is defined in the modified Hebbian learning rule, as in (5)

$$\begin{aligned} &IF \quad SD_{\alpha i} > (1+K)SD_{\alpha}^{m} \ AND \ SD_{\alpha j} < (1+K)SD_{\alpha}^{m} \\ &THEN \ \omega(mml)_{ij} = w(mml)^{high}, \ Algrithm \ 1 \\ &ELSE \ \ \omega(mml)_{ij} = r \star \omega(mml)_{ij}, \ Algrithm \ 2 \end{aligned} \tag{5}$$

Two algorithms to increase the connection weight are presented; algorithm 1 is to set the value to a fixed high value $w(mml)^{high}$ as in previous references and in algorithm 2, $\omega(mml)_{ij}$ is multiplied by a coefficient $r(r > 1)$ to represent the Hebbian reinforcement.

For the habituation learning, a decrease in excitatory output synapses occurs continuously at each time of stimulus presentation (not the digitizing step) for every node that receives input without reinforcement. Unlike Hebbian learning, this reduction does not require pairwise activation, and it is reversible. For example, as for continuous habituation parameter $h_{hab} = 0.9995$ and a 400 ms simulation period, if the connection weight is not influenced by any other learning rule, it decreases at the habituation rate h_{hab} and reaches $0.9995^{400} = 0.819$ of the original value at the end of this simulation period.

At the end of a training session for all 3 types of learning, the connection weights are fixed to perform pattern classification tests.

4 Results

4.1 Classification of Handwriting Numerals

Automatic recognition of handwriting characters is a practical problem in the field of pattern recognition so that it is selected to test the classification performance of the KIII network [14].

4.1.1 Feature Extraction

The test data set contains 200 samples in 20 groups of handwritten numeric characters written by 20 different students. One group included 10 characters from zero to nine. In this application, a 64-channel KIII network was used with system parameters as [2]. Each simulation trial either for training or for classification lasted for 400 ms, while the first 100 ms was the initial period in which the KIII network entered into its basal state. The input persisted 200–300 ms. Every character in the test data was preprocessed to get the 1×64 feature vector and to place a point in a 64-dimensional feature space.

The preprocessing was in two steps. The first step included image cropping, noise removal and the 'thin' operation (a digital image processing method which shrinks an object to its minimal connected stroke). Every character is picked out and transformed to its skeleton lines [10, 17]. The second step is to extract features. The character image obtained after the first step preprocessing is first divided into $2 \times 2 = 4$ small rectangular regions according to its pixel mass center. Similar as the previous step, each small rectangular region are divided into $2 \times 2 = 4$ smaller rectangular, thus the whole character is separated into $4 \times 4 = 16$ rectangular regions, as shown in Fig. 6. In every small region the ratios of four kinds of lines in four different directions are calculated. Thus the 64 features are given as input to the KIII network as a stimulus pattern in the form of a 1×64 feature vector.

4.1.2 Experimental Results

According the method to measure the activity of the KIII network as described in Sect. 3.2, the period with stimulus patterns is divided into 5 segments to calculate the nodes' activity with each segment lasting 40 ms. Because there are 10 inter-related patterns for classification, algorithm 2 for Hebbian learning rule is chosen for this classification problem with the bias coefficient $K = 0.4$. The values of $s = 5$ and $K = 0.4$ are chosen based on the previous experiments of the application of KIII model [7]. The increasing rate r is set to 1.2. We chose 1.2 arbitrarily in the interval

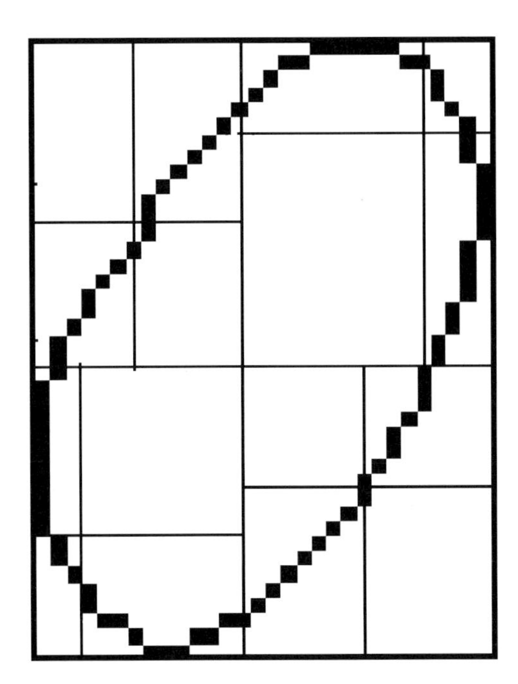

Fig. 6 An example of the segment of handwriting numerals

(1.0, 1.5). If $r > 1.5$, $\omega(mml)_{ij}$ will soon become too large and thus, the KIII model cannot work properly as a classifier. Moreover, the threshold value mentioned above was set to one twentieth of the distance between the test data cluster centroid and the training data cluster centroid. In our experiments, we have totally 200 groups of handwriting numeral characters. Each group consists of ten characters (from '0' to '9'). We arbitrarily chose 10 groups of them for training and used all the 200 groups for classification.

Furthermore, we compared our classification rates with those given by using the following types of traditional artificial neural networks (ANN): the linear filter, the two-layer perceptron and the Hopfield network respectively, to classify these same handwriting numeral characters. The selections of the training set and the testing set were both same as those of the experiments of KIII. The linear filter was very simple and has very limited classification capacity. It consisted of 64 input neurons and 10 output neurons, which corresponding to 64 features input and the 10 class output, respectively, while the Purelin transfer function was used. The Windrow-Hoff Learning Rule, which is a Least Mean Square method, was employed for training [9]. In the two-layer perceptron, we used two layers of perceptrons: an input layer of 64 neurons and an output layer of 10 neurons. The input layer used log-sigmoid transfer function and the output layer used a linear transfer function. We took each column of a 10×10 identity matrix as the target output of the two-layer perceptron , corresponding to one kind of numeric character in the training set. (i.e. the 1st column corresponded to the character '0', etc) The weight and bias values in the two-layer perceptron were updated according to Levenberg-Marquardt optimization with all the learning rates set to 0.01. The Hopfield network did not have a learning law associated with it. It was not trained, nor did it learn on its own. Instead, a design procedure based on the Lyapunov function was used to determine the weight matrix. We considered the column vector in the training set, that is, a 64×10 matrix, as the set of stable points in the network and designed a Hopfield network with a single layer using symmetric saturating linear transfer function. The test procedure of the two-layer perceptron and Hopfield network were very similar. After giving the testing input set to the network, we compared the outputs of the network to the pre-designed target vectors. The target vector that had the minimal variance with the output vector determined the category the testing set belonged to.

Table 1 shows the comparison of the classification test results.

Taking the noise level as a parameter, the influence of the noise level on the KIII network classification performance was investigated. As the stimulus pattern, which is considered as the signal, was composed of corresponding feature values, whose range is 0 to 1, the standard deviation of the Gaussian noise added at the receptor site was defined as the noise/signal rate. Repeated tests were taken as the noise level was increased. Results (Fig. 7) showed that as the noise level increased the correct classification rate of the KIII network increased to a plateau and then decreased. In previous work done by Kozma [7], an optimal noise/signal rate was found for the best classification performance of KIII, which was named 'chaotic resonance' in comparison to stochastic resonance. However, in this result, it seems difficult to find an apparent optimal noise/signal rate. The correct classification rate of the KIII network remained above 90% when the noise/signal rate was varied

Table 1 Classification result – using KIII

Pattern	Correct	Incorrect	Failure	Reliability
0	196	3	1	98.49
1	185	10	5	94.87
2	192	4	4	97.96
3	177	12	11	93.65
4	179	11	10	94.21
5	181	7	12	96.28
6	191	1	8	99.48
7	189	7	4	96.43
8	174	9	17	95.08
9	186	9	5	95.38
Total	1850	73	77	96.20
Rate	92.5	3.65	3.85	96.20

from 40% to 160%. This result indicated that the KIII network performance was insensitive to the noise level over a relatively broad range.

4.2 Face Recognition

Face recognition plays an important role in a wide range of applications, such as criminal identification, credit card verification, security system, scene surveillance, etc. In order to optimize performance, many algorithms of feature extraction are

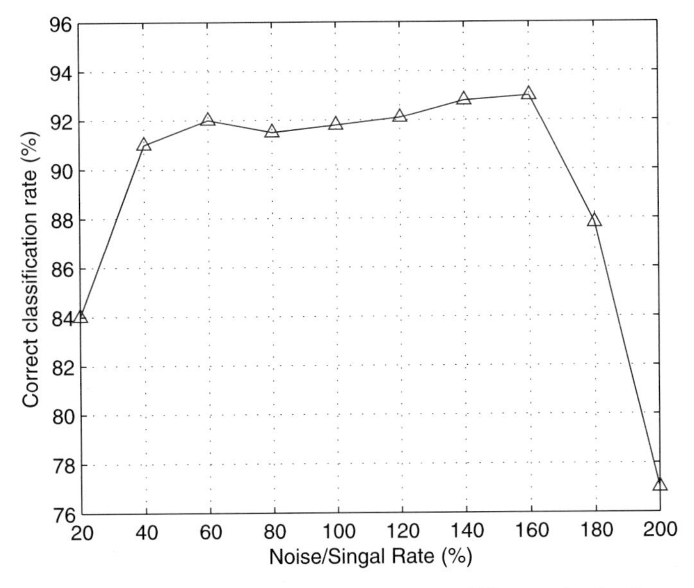

Fig. 7 Classification performance of KIII on different noise level

Table 2 Classification results – comparison

Pattern	Reliability			
	Linear filter	Perceptron	Hopfield	KIII
0	74.50	100	59.79	98.49
1	55.85	89.5	78.89	94.87
2	71.0	53.68	78.42	97.96
3	35.5	67.37	79.87	93.65
4	39.44	44.13	41.99	94.21
5	48.73	49.36	21.17	96.28
6	83.5	69.95	89.23	99.48
7	58.59	51.59	64.0	96.43
8	76.53	46.88	87.93	95.08
9	64.06	63.5	64.29	95.38
Average	60.99	64.84	66.76	96.20

applied, such as PCA, ICA and so on as well as the nearest neighbor distances, Bayesian statistics and SVM are used as classifiers.

In order to check whether the KIII are sensitive to a certain feature vector, the KIII is used to classify face images based on different feature vectors extracted [16].

4.2.1 Feature Extraction

In order to extract a feature vector from the entire image, we divide the original image into sub-images and extract the feature of each sub-image from the whole sub-image. Then, the features are combined to form the whole feature vector of the original image, Fig. 8. In our simulation, the face images are divided into 8, 16, 32, 64 and 80 sub-images individually. The discrete cosine transform (DCT), the singular value decomposition (SVD) and the wavelet packet transform (WPT) are used to extract the feature vector from face images, respectively.

2-DCT can concentrate most of the signal energy in the upper left corner. In Fig. 9 (a) is the original face image, (b) is the reconstructed image without discarding any coefficient, (c) is the reconstructed image after discarding about 90% coefficient and (d) is the reconstructed image only discarding one maximal coefficient. We select the largest coefficient as the feature of sub-image.

In SVD, the singulars compose an n-dimension feature vector and can be used to take the place of the face image. In Fig. 10, the X axis denotes 10 large singular values in each sub-image, and the Y axis denotes the number of feature $(8 \times |y_i|, namely 8, 16, 24, etc)$. The Z axis denotes the ratio of each singular value to the sum of all singulars after SVD averaged over all sub-images. In the simulation, the largest singular, which involves most energy of image, is selected as the feature.

The WPT of image results in an array of wavelet coefficients in different frequency range. Two level WPT is used to calculate the coefficients, and the norm of all the coefficients of lowest frequency range in sub-bands is selected as the feature.

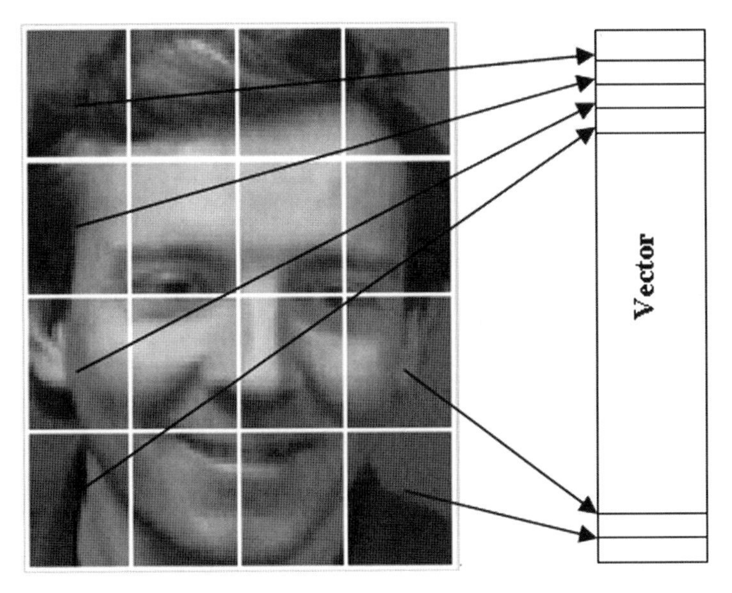

Fig. 8 Construction of a feature vector

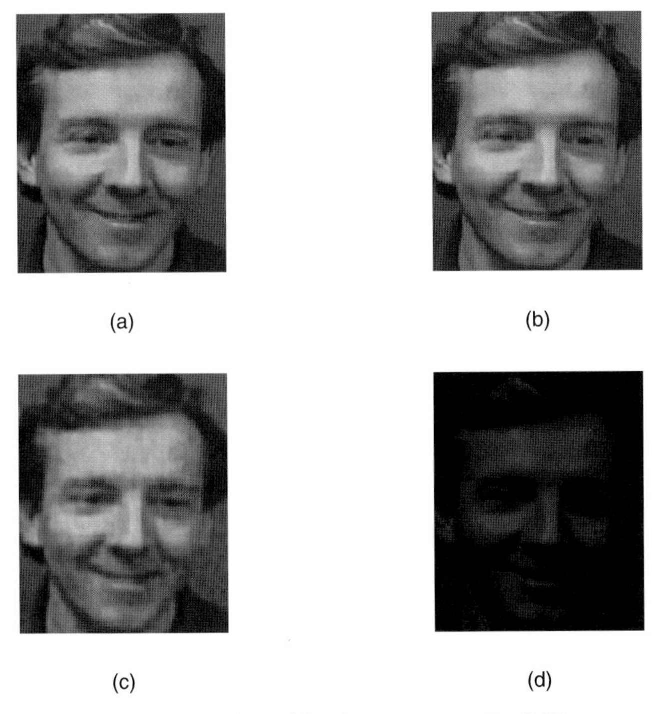

Fig. 9 Comparison of face image processed by DCT

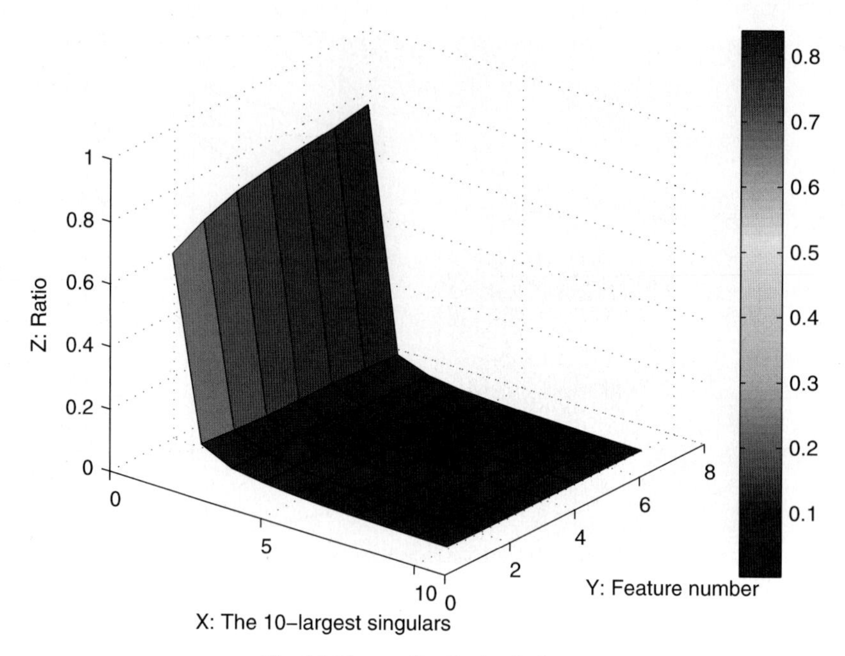

Fig. 10 Energy distribution in SVD

From Fig. 11, it is shown that most of the energy of the image concentrates in the lowest frequency range.

4.2.2 Classification Results

ORL face dataset is used to evaluate the performance. Five images of each person are selected for training and others are used to test. Compared with other ANN, each pattern is only learned 10 times in KIII. The output of OB layer is stored as cognition standard. And the nearest neighbor principle is used to classify new images. Table 3 showns that the higher is the dimension of the feature vector, the better is the performance . The DCT-based feature seems a little better, but SVD/WPT-based classifier seems more stable.

4.3 Classification of Normal and Hypoxia EEGs

Mixture of nitrogen and oxygen at normal atmosphere pressure, which simulates different altitude atmosphere by adjusting oxygen partial pressure, was provided to subjects via a pilot mask. In the first day, when the subject stayed at normal atmosphere, he carried out auditory digit span and serial addition/subtraction tests while his EEG was recorded. In the second day, after the subject stayed at environment simulating 3500m altitude for 25 minutes, they repeated the previous test procedure.

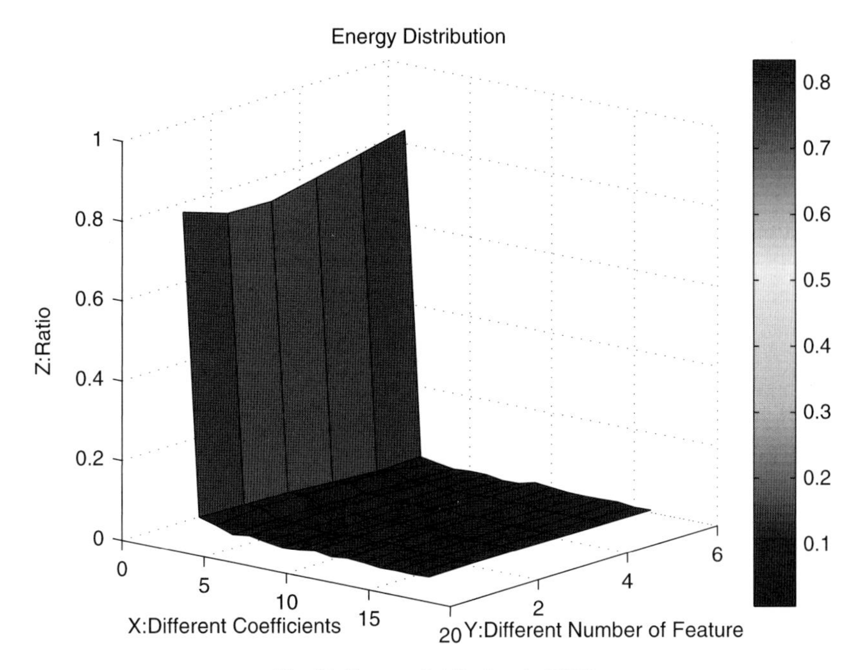

Fig. 11 Energy distribution in WPT

The experiment was carried out in the same time each day. Five healthy male volunteers around 22 years old were taken as subjects. The normal and 3500m datasets were composed of the 1.5 seconds EEG signals immediately after neurobehavioral evaluation. EEG data was taken from 30 Channels including: FP1, FP2, F7, F3, FZ, F4, F8, FT7, FC3, FCZ, FC4, FT8, T3, C3, CZ, C4, T4, TP7, CP3, CPZ, CP4, TP8, T5, P3, PZ, P4, T6, O1, OZ, O2 (10/20 system). Reference was (A1+A2)/2 (A1 = left mastoid, A2 = right mastoid). The EEG amplifier used is NuAmps Digital Amplifier (Model 7181) purchased from Neuroscan Compumedics Ltd, Texas, USA. Sampling rate was 250 $samples/sec$. All values are in μV [12].

4.3.1 Evaluation of Hypoxia

In order to evaluate the level of hypoxia, the neurobehavioral evaluation (NE), which is a sensitive and reliable tool for early detection of adverse effects of the

Table 3 Correction rate of face recognition

Feature Number						
Method	8	16	32	48	64	80
SVD	69.3%	81.5%	88.3%	90.0%	90.8%	91.0%
DCT	67.0%	82.0%	88.5%	90.0%	91.5%	91.0%
WPT	69.5%	81.8%	87.3%	89.8%	90.8%	89.8%

environmental hazards on central nervous system, was employed. In the normal and 3500m altitude simulated experiments, auditory digit span and serial addition and subtraction were utilized to evaluate the degree of hypoxia in. Result of test is shown in Table 4. T-tests were performed for two items of NE. As a result, the NE scores of normal and hypoxia were different observably ($p < 0.05$) in two items. Furthermore, the scores under the hypoxia condition were lower distinctly. According to result of NE, that subject's behavior capability under the hypoxia condition was weaker than the normal one markedly suggested pattern aroused by hypoxia come into being.

4.3.2 Feature Extraction

Wavelet representation of a neuroelectric waveform is invertible, meaning that the original waveform can be reconstructed from a set of analysis coefficients that capture all of the time (or space) and frequency information in the waveform. Greater flexibility in defining the frequency bands of a decomposition can be obtained by using a generalization of wavelets known as wavelet packets. Selection of suitable wavelet and the number of levels of decomposition is very important in analysis of signals using WPT. In our analysis, we use the COIF5 wavelet. The number of levels of de-composition is chosen as two and wavelet packet tree coefficients of a 30–60Hz sub-band are abstracted. The feature vector is a 30-dimensions vector due to 30 EEG channels. For each channel, the square of the wavelet packet tree coefficients are summed up as one dimension of the feature vector. According to the topology of the EEG channel, each feature vector can be transformed as a feature topography. A typical feature topography sample of comparing normal and hypoxia EEG collected from the same subject is illustrated in Fig. 12.

4.3.3 Classification Results

In this article, we use the KIII model to classify normal and hypoxia EEG. We train KIII network by 2-fold cross-validation. The KIII model learns the desired patterns — the normal and hypoxia EEG patterns for three times in turn. The test

Table 4 Neurobehavior evaluation scores of the subjects

	Auditory Digit Span		Addition and Subtraction	
Subject	Normal	Hypoxia	Normal	Hypoxia
No.1	30	29	28	21
No.2	28	24	31	26
No.3	25	21	20	20
No.4	32	23	24	15
No.5	19	9	24	12
Average	26.8	21.2	24.2	16.8

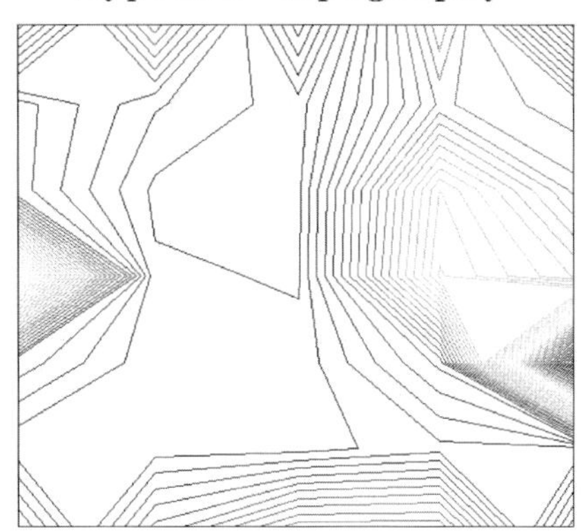

Fig. 12 A typical feature topography

data set contains 80 samples of normal and hypoxia EEG for individuals by 5 different subjects. We arbitrarily chose 40 groups samples in the odd position for training and used all the 80 groups samples for classification, and then We chose 40 samples in the even position for training and used all the 80 samples for classification. The final correction rate is from the mean of twice correction rate. In this application, a 30-channel KIII network is used with system parameters. The values of s = 5, K = 0.4 and r =1.25 are chosen based on the previous experiments of the application of KIII
indexKIIIenlargethispage*12pt model. The feature vector loaded into the R layer of the KIII model is calculated the categories of the input patterns. Only if the difference between the Euclid distances from the novel input pattern to the two kinds of stored patterns reaches the pre-defined thresh-old, the classification can be viewed as valid and persuasive. The experimental results are shown in Fig. 13. Effectively, the mean of classification rate for test data set is equal to 92%. Distinctly, hypoxia EEG can be recognized from normal EEG by KIII network. In other words, a new pattern in EEG, which is different from normal one, comes into being, induced by hypoxia. The conclusion of EEG clas-sification is consistent with NE. The minimum correction rate (85%) is from subject NO.3 by KIII network. At the same time, except subject NO.3, the scores of NE under the hypoxia condition are lower than the one under the normal condition entirely. The scores of serial addition and subtraction of NE from subject NO.3 are invariable under the normal and hypoxia conditions. The results of EEG and NE are unanimous actually. To some extent, the correction rate by KIII network represents the degree of hypoxia and probability of hypoxia pattern appearing. It provides the possibility that hypoxia can be detected quantificationally.

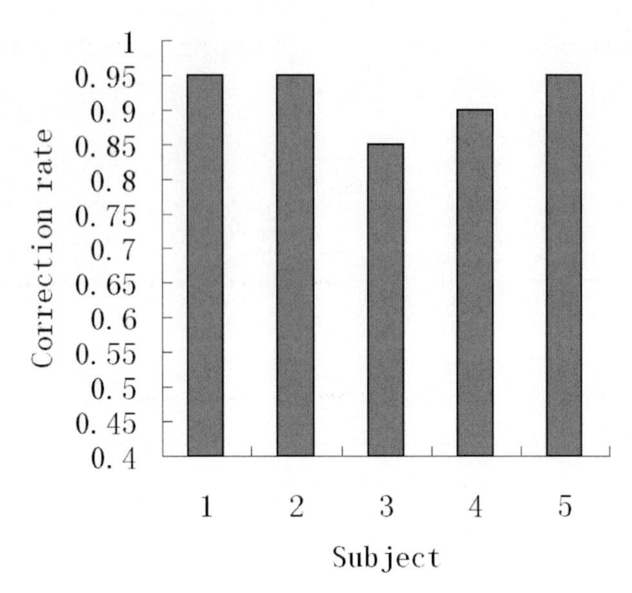

Fig. 13 EEG classification result

4.4 Tea Recognition

ANNs are generally considered as the most promising pattern recognition method to process the signals from chemical sensor array of electronic noses. Particularly, modelling biological olfaction, the KIII model should be more suitable for artificial olfaction due to its robust generalization capabilityto make the system more bionics [8]. Here the KIII model is employed by a electronic nose to classify different types of tea.

4.4.1 Data Collection

A sensor array, consisting of 7 metal oxide semiconductor (MOS) sensors (TGS2610, TGS2611, TGS800, TGS813, TGS822, TGS826, TGS880 from Figaro Co.), is constructed to mimic olfactory mucous. Each tea sample is heated on a thermostatic heater for 10 minutes before data acquirement. The volatiles emitted from tea are sensed by the sensor array [18]. Upon exposure to the volatiles, the conductivity of the MOS sensor changes causing the change in the current flowing through the MOS sensor when a constant voltage is supplied. The change in the electrical current level is a known function of the volatile concentration. A typical response recorded from the sensor array is shown in Fig. 14.

For each kind of tea, 33 tests were performed, 3 data sets for training and the other 30 data sets for testing.

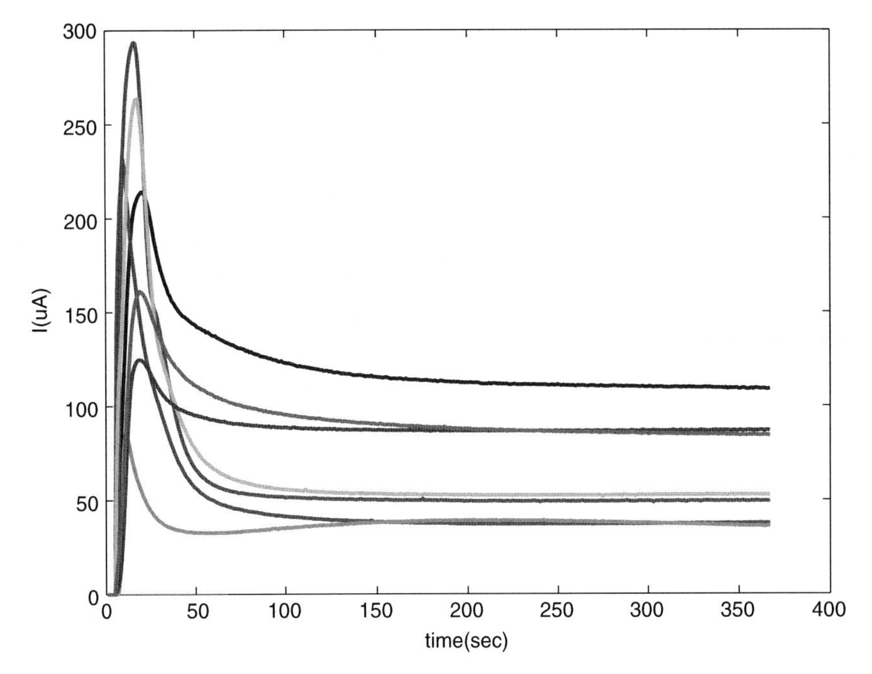

Fig. 14 A typical response of the sensor array

4.4.2 Feature Extraction

According to the constant potential provided and the current measured during tests, the resistance of the MOS sensors can be calculated. To reduce the effect of concentration fluctuations classification, a normalization approach is used. The logarithm of normalized resistance, $ln(R)$, at a steady state was calculated for each sensor. $ln(Ri)$, where $i = 1, 2, ..., 7$ denoting 7 sensor, consist a feature vector.

4.4.3 Experimental Results

At this time, a conventional artificial neural network, BP network, is carried out for comparing. The BP network consisted of an input layer of 7 neurons, a hidden layer of 6 neurons and a output layer of 4 neurons. Table 5 presents the classification

Table 5 Correction rate to classify four kinds of tea

	Chinese Green Tea	Japanese Green Tea	Indian Black Tea	Chinese Black Tea	Average
KIII	86.7%	93.3%	93.3%	80.0%	88.3%
BP	100%	80.0%	66.7%	93.3%	85.0%
Human	46.7%	80.0%	83.3%	50.0%	65.0%

result. Obviously, BP network and KIII network are both efficient. However, the classification rate of KIII model is a little more stable. The maximum classification rate of BP is 100%, but the minimum goes down to 66.7%. While to the KIII network, it varies from 80% to 93.3%.

5 Discussion and Conclusions

There are two points that should be noted in the exploration of the classification ability of the KIII network by applying it on pattern classification: the feature extraction and the learning rule. Although the feature space is used in our experiments as it is commonly used in the practical problem of pattern recognition, it is sometimes believed that feature space does not exist in real biological system. A good example is the feature-binding problem of human face recognition in which the face pattern is considered as a whole pattern, not the combination of features. The second point is about the modification of Hebbian learning rule. The modified algorithm for increasing the connection weight makes the KIII model able to memorize and classify more patterns than it used to. In previous research the pattern volume was limited to 5–8 patterns for a 64-channel KIII network to classify. Also it is more reasonable to believe that the connection weights, which represent the biological synaptic connections, change gradually in the learning process.

Another important aspect in this research concerns the role of noise in the KIII model. It is demonstrated by electrophysiological experiment and computer simulation that the additive noise in KIII network could maintain those KII components at nonzero point attractor and could stabilize those chaotic attractor landscape formed by learning [4]. However in handwriting numerals recognition an optimal classification performance was not found while adapting the noise intensity. Instead, the KIII network performs well when the noise parameter is in a broad range. This phenomenon matches the fact that the real biological neural systems can work well under a certain scope of noise level and not just at an optimal point.

In summary, as one of chaotic neural networks derived from biological neural systems, the KIII model inherits many biological characteristics. Compared with conventional artificial neural networks, the KIII network gives a more complicated and more accurate model in simulating the biological olfactory neural system, especially in respect to simulating the biological signal observed in experiments such as EEG signals.

Moreover, the KIII model has good capability for pattern recognition as a form of the biological intelligence. It can even classify some complicated patterns embedded in noisy background (such as EEG patterns) as well as relatively simple patterns (such as handwriting numerals) for human intelligence. It needs much fewer learning trials (normally only 4–8 times) than artificial neural networks when solving problems of pattern recognition. The noises play an very important role in the KIII network for pattern classification.

Although when considering the pattern volume and processing speed, the KIII network still could not replace the conventional artificial neural networks for solving practical problems due to the present digital computers implement the KIII network differs fundamentally from the analog real olfactory neural system so that both pattern volume and processing speed are limited. More work will be required in simulating the olfactory neural system with analog devices.

Acknowledgment The research is partially supported by the National Natural Science Foundation of China (No. 60421002), the National Basic Research Program of China (973 Program, No. 2004CB720302) and the Y.C. Tang Disciplinary Development Fund. All the results are based on the experiments carried out by the postgraduate students at the Center for Neuroinformatics, Zhejiang University. In particular, the authors thank Mr Xu Li, Mr Le Wang, Mr Meng Hu, Mr Jun Fu and Mr Xinling for their contributions.

References

1. Chang, H. J., Freeman, W.J.: Parameter optimization in models of the olfactory neural system, Neural Networks, **9**, 1–14(1996)
2. Chang, H. J., Freeman, W. J.: Biologically modeled noise stabilizing neurodynamics for pattern recognition, Int. J. Bifurcation and Chaos, **8**, 321–345(1998)
3. Freeman, W. J., Chang, H. J., Burke, B. C., Rose, P. A., Badler, J.: Taming chaos: Stabilization of aperiodic attractors by noise, IEEE Trans. Circuits Sys., **44**, 989–996 (1997)
4. Freeman, W. J.: Noise-induced first-order phase transitions in chaotic basin activity, Int. J. Bifurcation and Chaos **9**, 2215–2218 (1999)
5. Freeman, W. J.: Neurodynamics- An Exploration in Mesoscopic Brain Dynamics. Springer-Verlag, London (2000)
6. Freeman, W. J.:A proposed name for aperiodic brain activity: stochastic chaos, Neural Networks, **13**, 11–13 (2000)
7. Freeman, W. J., Kozma, R.:Biocomplexity: adaptive behavior in complex stochastic dynamic systems, Biosystems, **59**, 109–123 (2001)
8. Fu, J., Li, G., Qin, Y., Freeman, W. J.:A pattern recognition method for electronic noses based on an olfactory neural network, Sensors and Actuators B, (in press)
9. Hagan, M. T., Demuth, H. B., Beale, M. H.:Neural Network Design. PWS Publishing, Boston (1996)
10. Haralick, R. M., Linda, G. S.: Computer and Robot Vision, Vol. I. Addison-Wesley (1992)
11. Hebb, D. O.: The Organization of Behavior. John Wiley & Sons, New York (1949)
12. Hu, M., Li, J., Li, G., Tang, X., Freeman, W. J.: Normal and Hypoxia EEG Recognition Based on a Chaotic Olfactory Model. In: Wang, J., Yi, Z., Zurada, J. M., et al. (ed) Advances in Neural Networks. Lecture Notes in Computer Science, **3973**, 554–559, Springer-Verlag, Berlin Heidelberg New York (2006)
13. Kozma, R., Freeman, W. J.: Chaotic resonance - methods and applications for robust classification of noisy and variable patterns, Int. J. Bifurcation and Chaos, **11**, 1607–1629 (2001)
14. Li, X., Li, G., Wang, L., Freeman, W. J.: A study on a Bionic Pattern Classifier Based on Olfactory Neural System, Int. J. Bifurcation Chaos., **16**, 2425–2434(2006)
15. Li, G., Lou, Z., Wang, L., Li, X., Freeman, W. J.: Application of Chaotic Neural Model Based on Olfactory System on Pattern Recognitions. In: Wang, L., Chen, K., Ong, Y. S. (ed) Advances in Natural Computation. Lecture Notes in Computer Science, **3610**, 378–381, Springer-Verlag, Berlin Heidelberg New York (2005)
16. Li, G., Zhang, J., Wang, Y., Freeman, W. J.: Face Recognition Using a Neural Network Simulating Olfactory Systems. In: Wang, J., Yi, Z., Zurada, J. M., et al. (ed) Advances in Neural Networks. Lecture Notes in Computer Science, **3972**, 93–97, Springer-Verlag, Berlin Heidelberg New York (2006)
17. Pratt, W. K.: Digital Image Processing. John Wiley & Sons, Inc. (1991)
18. Yang, X., Fu, J., Lou, Z., Wang, L., Li, G., Freeman, W. J.: Tea Classification Based on Arti-

ficial Olfaction Using Bionic Olfactory Neural Networks. In: Wang, J., Yi, Z., Zurada, J. M., et al. (ed) Advances in Neural Networks. Lecture Notes in Computer Science, **3972**, 343–348, Springer-Verlag, Berlin Heidelberg New York (2006)

Recursive Nodes with Rich Dynamics as Modeling Tools for Cognitive Functions

Emilio Del-Moral-Hernandez

Abstract This chapter addresses artificial neural networks employing processing nodes with complex dynamics and the representation of information through spatio-temporal patterns. These architectures can be programmed to store information through cyclic collective oscillations, which can be explored for the representation of stored memories or pattern classes. The nodes that compose the network are parametric recursions that present rich dynamics, bifurcation and chaos. A blend of periodic and erratic behavior is explored for the representation of information and the search for stored patterns. Several results on these networks have been produced in recent years, some of them showing their superior performance on pattern storage and recovery when compared to traditional neural architectures. We discuss tools of analysis, design methodologies and tools for the characterization of these RPEs networks (RPEs - Recursive Processing Elements, as the nodes are named).

1 Introduction

1.1 Rich Dynamics in the Modeling of Neural Activity

In modeling real neural systems through mathematical and computational formalism, we normally have a compromise between simplicity of the model and fidelity to the modeled phenomena. This is one of the reasons why we have diverse types of models for the same modeled object: different finalities and different research communities require different levels of detail from the modeling. In addition to the diverse levels of required precision, another factor that also results in a range of alternative models, when the target is the nervous system and its functions, is the fact that the signals which are present in real neurons can be used in different forms to code information by the nervous system, according to the context and the functionality [18, 23, 28]. Therefore, in the field of artificial neural networks for neural modeling, we observed the production of several styles of models, each one emphasizing different aspects of neural behavior. We can mention models with graded activity based on frequency coding and sigmoidal transfer functions, models with binary output, and spiking models (or pulsed models), among others [19, 20].

More recently, several tools and methods from the area of complex systems

L. I. Perlovsky, R. Kozma, *Neurodynamics of Cognition and Consciousness.*

and tools of nonlinear dynamics have shown to be useful in the understanding of phenomena in brain activity and nervous system activity in general [18, 29]. As a result, a large number of different proposals on the modeling of neural activity and neural assemblies activity based on elements with diverse and complex dynamics and the representation and storage of information through spatio-temporal patterns have been presented, with interesting results [1, 2, 5, 12, 16, 18, 24, 25, 26, 27, 29, 31, 32, 33, 35, 36, 39].

We want to mention briefly some of these referred works, specifically those works that are more directly relevant to the main discussions in this chapter, since they are somehow related to model neurons in which high diversity of dynamic behavior appears, including the phenomena of bifurcation and chaos, at the level of single neural nodes.

One of the seminal works that have to be mentioned here regards the model neuron proposed by Kazuyuki Aihara and collaborators in [1]. In this type of model neuron, self-feedback represents the processes of refractory period in real neurons; this makes rich dynamics possible, with bifurcation and cascading to chaos. To a certain degree, his work extends previous models in which some elements of dynamics represented in the form of delays were introduced. In particular, we have to mention in this context the works by Caianiello [3] and Nagumo-Sato [35]. Caianiello's work already moved from the McCullock-Pitts' neuron, in which the input-to-output relationship is purely functional, with no dynamics, to a model in which the past inputs had impact on the value of the present state of the neuron. Nagumo-Sato's model added another element of dynamics, an exponential decay memory. Aihara's model includes memory elements for the inputs of the model neurons, as well as for its internal state. Additionally, it includes continuous transfer functions, an essential ingredient for the presence of rich bifurcation and cascading to chaos. Other work that can be mentioned here is that of Adachi, co-authored with Aihara. It used Aihara's chaotic neuron for associative memories [2]. From the mathematical point of view, the feedback in Aihara's model can be seen as a specific recursive node; this links Aihara's model to some discussions in this chapter.

Kunihiko Kaneko's work, primarily intended for modeling of spatio-temporal physical systems, employed the idea of coupled maps lattices [25, 26, 27]. This is a more macroscopic approach as compared to Aihara's. This work is also related to models discussed here.

Nabil Farhat and collaborators introduced the concept of the 'bifurcation neuron', a model neuron in which the phenomena of bifurcation and cascade to chaotic dynamics emerges from a simple pulsed model of type 'integrate and fire' [16]. This work and its following developments have direct relationship to ideas discussed here.

With similar motivations as in the works mentioned above, the class of models addressed in this chapter emphasizes relationships between neurocomputing and nonlinear dynamical systems that exhibit bifurcation and rich dynamic behavior such as fractality and chaotic dynamics. This chapter contributes to the application of complex neuron models and associated architectures and it contributes to move the conceptual emphasis from input-output functional mapping to a more balanced approach, in which this mapping is integrated to mathematical representations of neurodynamics, so to promote more powerful modeling environments. The discussions in this chapter show the relevance and applicability of recursive nodes for auto and hetero-association, memory, and pattern recognition, while discussing advantages of recursive nodes.

1.2 Rich Dynamics in the Modeling of Neural Activity

The neural architectures developed in this chapter are composed of recursive nodes (or 'Recursive Processing Elements' - RPEs for brevity), which can bifurcate among several dynamical modalities, and explore final cyclic states for the representation of information [4, 5, 6, 7, 12]. While in many of the traditional neuron models the node functionality is described through a nonlinear input-output mapping function, such as a sigmoid shaped function or some type of radial basis function, in RPE nodes the functionality is characterized through a parametric recursion [4, 5, 6, 7, 20]. In particular, the recursive nodes addressed in this chapter are highly nonlinear dynamical systems that are able to produce a large number of dynamical modalities, with bifurcation and cascading to chaotic dynamics. Because of that, RPEs are natural elements for the modeling of biological systems with complex temporal patterns [18, 27, 29].

An important feature of RPE nodes is local feedback. This is a straightforward way to create dynamics with controlled features. Because of this, the studied networks can generate complex spatio-temporal patterns. In other approaches to neural modeling, meaningful dynamics is usually reached only through structures of macroscopic feedback, not embedded at the level of the single node.

Interesting results have been produced employing this type of networks [6, 9, 10, 12, 14]. Resulting performance of associative neural architectures is markedly superior to similar networks based on the traditional Hopfield's associative model: Hopfield networks produce larger errors in pattern recovery than RPEs networks, by factors 1.5 to 2, as shown in Sect. 3.2 [6]. Equally important are the concepts of computation and information representation through periodic and chaotic attractors, which are related to RPEs networks [37]. Resulting models are much richer in representing dynamical phenomena than traditional model neurons and neural networks with static input to output mapping at a single node level and computation with fixed-point attractors [20].

Organization of the rest of the chapter is as follows:

Sections 2.1 to 2.5 concentrate on the features and properties of recursive nodes and coupled structures employing such nodes. The main features of recursive nodes, their functionality, and phenomenology are discussed in Sect. 2.1. This is complemented in Sect. 2.2 by the discussion of attractors, repellers and their relationship to dynamical richness. Section 2.3 discusses fully coupled structures and related collective behavior. Section 2.4 briefly addresses sensing of external information. Section 2.5 discusses coding of information through global attractors in RPEs networks.

Sections 3.1 to 3.4 concentrate on the several tools and methodologies for the study and analysis of networks of recursive nodes, as well as on their illustrative application to concrete situations. Section 3.1 briefly discusses relevant classical tools of nonlinear dynamics as well as new tools that were developed having in mind the specificities of the studied networks. Section 3.2 discusses associative memories and pattern recognition, as well as the blend of order and chaos in that context. Section 3.3 addresses quantitative characterization of associative power and robustness to input noise and to scaling of coupling strengths in RPEs networks. Section 3.4 discusses a graphical tool for the study of the landscape of attractors in RPEs networks.

Section 4 summarizes discussions and draws conclusions.

2 From Nodes to Networks

2.1 *Main Features of Recursive Nodes with Rich Dynamics*

The nodes with rich dynamics employed in the building of artificial neural networks discussed in this chapter are named 'Recursive Processing Elements', as mentioned before. This name comes from the fact that the node dynamics and the node functionality are mathematically defined through a parametric recursion of the first order of type $x_{n+1} = R_p(x_n)$. This recursion, generically named here R_p, links consecutive values of the state variable x, which evolves in discrete time n. We have thus a first order recursion, whose mapping leading the state values x_n to the state values x_{n+1} depends on a numeric parameter p. Figure 1 illustrates this concept, by showing a family of parametric recursions, which includes three instances of return maps relating x_n and x_{n+1}. Each one of these three return maps corresponds to a specific value assumed by the numeric parameter p.

A prototypical RPE node that is frequently used in the study of processes with complex collective phenomena, as well as in the implementation of neural architectures for the modeling of neural assemblies with rich dynamics, is the logistic recursion, given by (1) [21, 26, 27].

$$x_{n+1} = p.x_n.(1 - x_n). \tag{1}$$

Several studies (discussed later in this chapter) of networks using these logistic

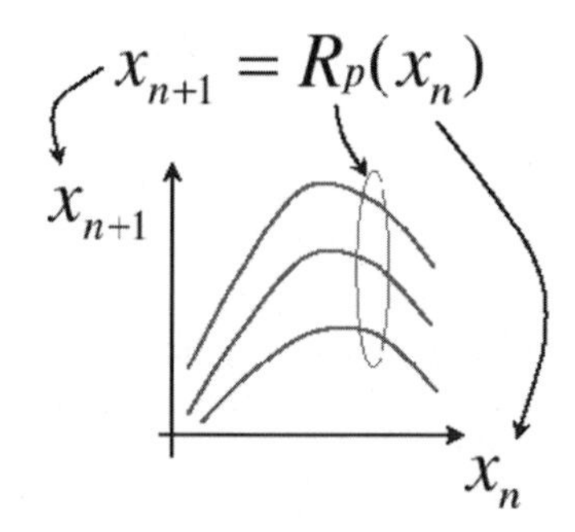

Fig. 1 Family of parametric recursions, with 3 examples of return maps relating x_n and x_{n+1}. According to the value of the parameter p, R_p maps the values of x_n and x_{n+1} in different forms [13]

nodes have demonstrated functionalities of auto-association and hetero-association. The main results and the related methods developed in these studies are valid not only for logistic RPEs, as above, but also for other recursive nodes with rich bifurcation and chaotic cascading [11, 12].

Logistic recursions are a clear demonstration that a complex and diverse phenomenology can be exhibited by recursive nodes with relatively simple mathematical definition. Figure 2 shows examples of diverse behavior exhibited by the sequence x_n produced by a logistic recursive node. In the figure, the horizontal axes represent discrete time n and the vertical axes x_n values. In many examples of RPEs evolution, we observe a transitory portion at the beginning of the sequence and gradual stabilization of the trajectory at some type of attractor. Figure 2a) illustrates stabilization, in long-term, in a fixed-point attractor. In b), there is a periodic attractor. In c), there is a chaotic sequence, with no periodic behavior; only macroscopic aspects of the sequence stabilize, such as the regions of values visited by the time series and relative frequencies of visits to each region of the state space.

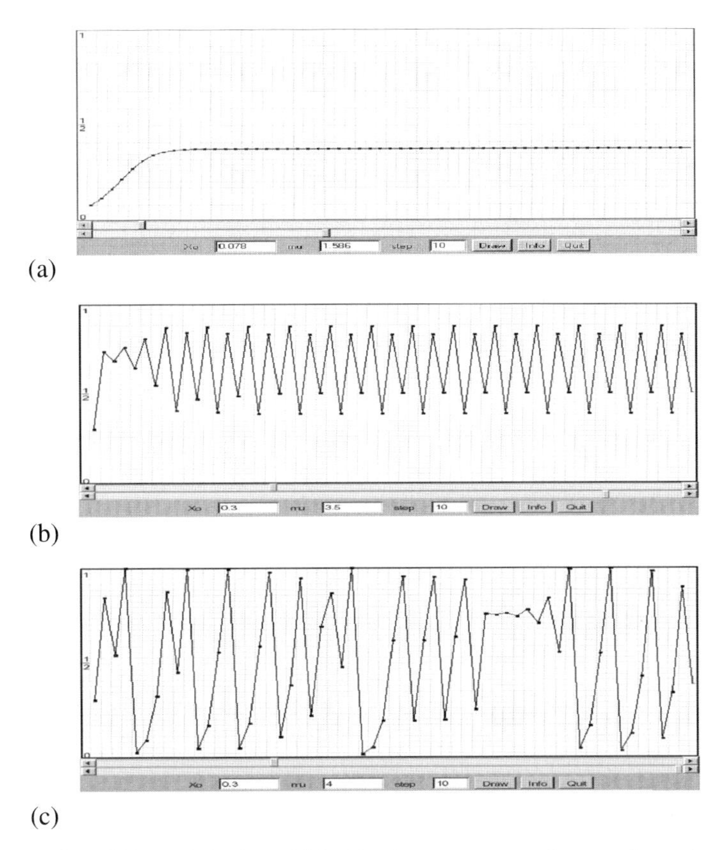

Fig. 2 Examples of diverse behavior exhibited by the sequence x_n of a recursive node with high dynamical diversity (the logistic map). The horizontal axes are discrete time n and vertical axes x_n values. (**a**) shows stabilization in a fixed-point attractor (small values of p); (**b**) a periodic attractor (medium p values); (**c**) chaotic sequence (large p values)

Central to phenomenology illustrated in Fig. 2 is the concept of cyclic trajectories consistent with a recursion R_p. For simplicity, consider a period-2 cyclic trajectory, being performed by a recursive node governed by $x_{n+1} = R_p(x_n)$. For period-2 cycles, starting from a given value x_n, recursion R_p acts twice in cascade and produces a value x_{n+2} that is equal to the original value x_n. This is expressed in (2).

$$x_{n+2} = x_n \Rightarrow x_n = R_p(R_p(x_n)). \tag{2}$$

Equation (2) determines if a trajectory x_n with period-2 can be generated by a given recursive node. In other words, it defines if such oscillation is a valid cyclic behavior for a specific recursion R_p.

In the next section, we address stability of valid cycles and the related attractors and repellers. These are important elements for building networks for pattern recognition and memory recovery with robustness to noisy and distorted stimuli.

2.2 *Attractors and Repellers as the Source of Dynamical Richness*

A central concept in RPEs, which is complementary to the concept of valid cycles discussed in the previous section, is that of stability of cyclic trajectories. A stable cycle (or stable cyclic trajectory) has the property of being robust with respect to limited perturbations, acting as an attractor in the state space [21]. For the simple case of period-2 cycles, discussed in the previous section, valid periodic trajectories (i.e., sequences x_n for which (2) holds) may not necessarily result in stable cycles, characterizing thus a repeller-type trajectory. The top part of Fig. 3 illustrates such a situation: although we have a valid period-2 trajectory involving values A and B for which $R_p(A) = B$ and $R_p(B) = A$, we observe instability in such cycles. Any small perturbation in the period-2 trajectory, even with magnitudes of 0.001 or 0.0001 (as illustrated) or even less than that, will exercise the instability of the cyclic trajectory and another trajectory with different characteristics will eventually emerge (in this case, a period-4 trajectory).

This lack of stability for valid period-2 cycles happens in the logistic map when, for example, $p = 3.50$ (this was the value used for the generation of the top part of Fig. 3).

In the bottom part of the figure, we have the opposite situation, i.e., a large stability for period-2 trajectories, which can be observed when the smaller value of $p = 3.25$ (in this example) is used. In this case, the valid period-2 trajectory, involving two values A and B linked by R_p, is not only a stable cycle, but also attracts trajectories which are originally very far from the values A and B.

The essential condition for stability of a valid period-2 trajectory is that the derivative of the double cascade of R_p recursions has to be, in magnitude, smaller than unity. This mathematical condition has to be valid for any of the x values involved in the cycle (A and B, or more generally x_{cycle}). It ensures that any small perturbation in the cycle will gradually be absorbed with time. This is mathematically expressed in (3), which in its second part rewrites this condition using the

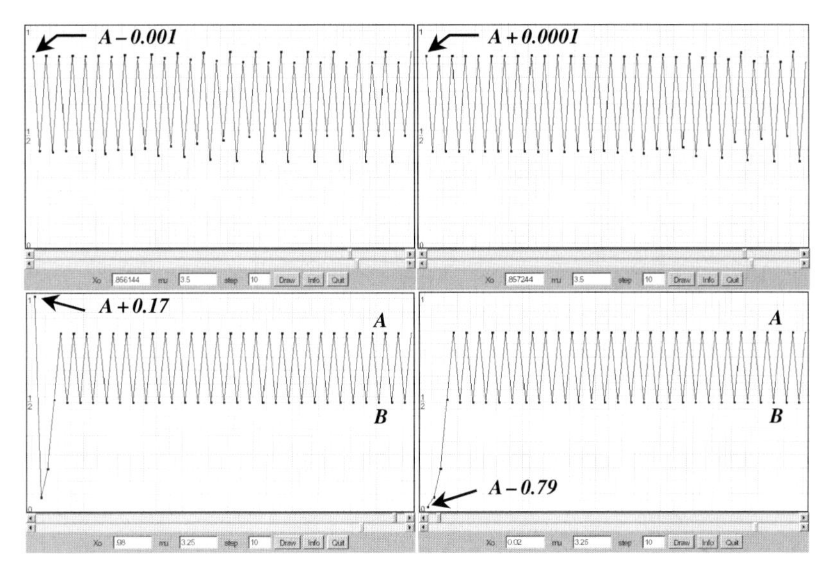

Fig. 3 Contrast between the repeller behavior (two examples at the top) and the attractor behavior (two examples at the bottom). In the two experiments at the top, the p value is 3.50, i.e., $x_{n+1} = 3.50.x_n.(1-x_n)$; even a small perturbation around the period-2 cyclic trajectory is enough to make the sequence depart from that period-2 oscillation. On the other hand, in the two experiments at the bottom, with $p = 3.25$, even large perturbations are absorbed very fast in time, and the sequence x_n tends to oscillate between values A and B

chain rule for derivatives.

$$|dRp(Rp(x))/dx| < 1, \quad for \;\; x = x_{cycle}, \quad or \;\; |Rp'(B).Rp'(A)| < 1. \qquad (3)$$

When the condition given in (3) is violated (and this happens for large p values in the logistic recursion), the period-2 trajectory becomes unstable and the RPE tends to stabilize in a trajectory of another period, as illustrated in the top of Fig. 3.

The above illustrations and discussions considered examples of a cyclic x_n trajectory with duration '$Tcycle$' equal 2. The analysis of valid cycles for more general attractors with $T_{cycle} > 2$ and the study of their stability follow similar lines as the ones taken in the simpler case of period-2 attractors discussed above. Nevertheless, we have to adequate the expressions in (2) and (3) in order to deal with a larger number of values x_{cycle} involved in these more complex cyclic trajectories.

Generation of a given cyclic trajectory with T_{cycle} duration (i.e., a trajectory involving a sequence with T_{cycle} different x_{cycle} values), by repeated R_p recursions, can only happen if each of the involved values x_{cycle} corresponds to an auto-value of the multiple cascade of recursions R_p, as in (4), where R_p appears T_{cycle} times in the chain.

$$x_{cycle} = R_p(R_p(\cdots R_p(x_{cycle})\cdots)). \qquad (4)$$

Stability condition of those valid cycles is mathematically expressed as follows [21]:

$$|dR_p(R_p(\cdots R_p(x_{cycle})\cdots))/dx| < 1 \ \ or \ \ |\prod Rp'(x)| < 1. \tag{5}$$

In the second part of (5), we have applied the chain rule, and x assumes all values of x_{cycle}.

These expressions, (4) and (5), are the general forms of the previous ones for period-2 trajectories ((2) and (3)).

The above analyses of validity of cycles and stability are important for pattern recognition, association, and decision-making based on networks of recursive nodes, for the following reasons:

- They guarantee that a given cycle can in fact be generated by the recursion R_p and therefore be used in the coding of some meaningful information (e.g., a stored memory or class label);
- They indicate that such a cycle is also an attractor, and provides some level of robustness to noisy or distorted stimuli, particularly if condition (5) is observed with large margin.

The above analysis allows us to establish an interesting relationship between richness of dynamic behavior in recursive nodes and instability of cyclic trajectories. In the genesis of each bifurcation of a recursive map [21], i.e., behind each switching between two different classes of trajectories produced by the same recursive map[1], we have that the change of the parameter p causes loss of stability of a given attractor, through the violation of the condition in (5). Once the boundary of the unity value derivative in (5) is crossed due to an increase in the p value of the logistic map, the stability of the cycle under analysis is lost, it becomes a repeller, and this gives room for a larger periodic trajectory, for which the condition of (5) returns to be observed.

The fact that the condition of stability in (5) can be repeatedly violated by changing the value of the p parameter is exactly the source of the multiple dynamical modalities observed in recursions with rich dynamics such as the logistic map and many other bifurcating maps [7, 21, 26, 27].

The rich dynamic behavior exhibited by the logistic map, with all its different periodic attractors and repeated bifurcations with the increase of p, is summarized through its bifurcation diagram, which appears at the top part of Fig. 4.

In Fig. 4, the second bifurcation diagram illustrates another recursive node, different from the logistic map. This second diagram describes the periodic attractors and bifurcations of a spiking model neuron of type ‘integrate and fire’, taking its integration slope as bifurcation parameter [14, 19]. Under certain classes of periodic stimulation, the integrate and fire neuron can be modeled by a recursive map with a large diversity of periodic and fixed-point attractors, rich dynamics, bifurcation, and cascading to chaotic dynamics [7, 14, 15, 16]. Such recursive map governs the rela-

---

[1] As a simple example of different classes of trajectories produced by the same map, we have the period-2 and period-4 trajectories illustrated in Fig. 3. Both are produced by the logistic map, given by (1), but employing two different values for the parameter p.

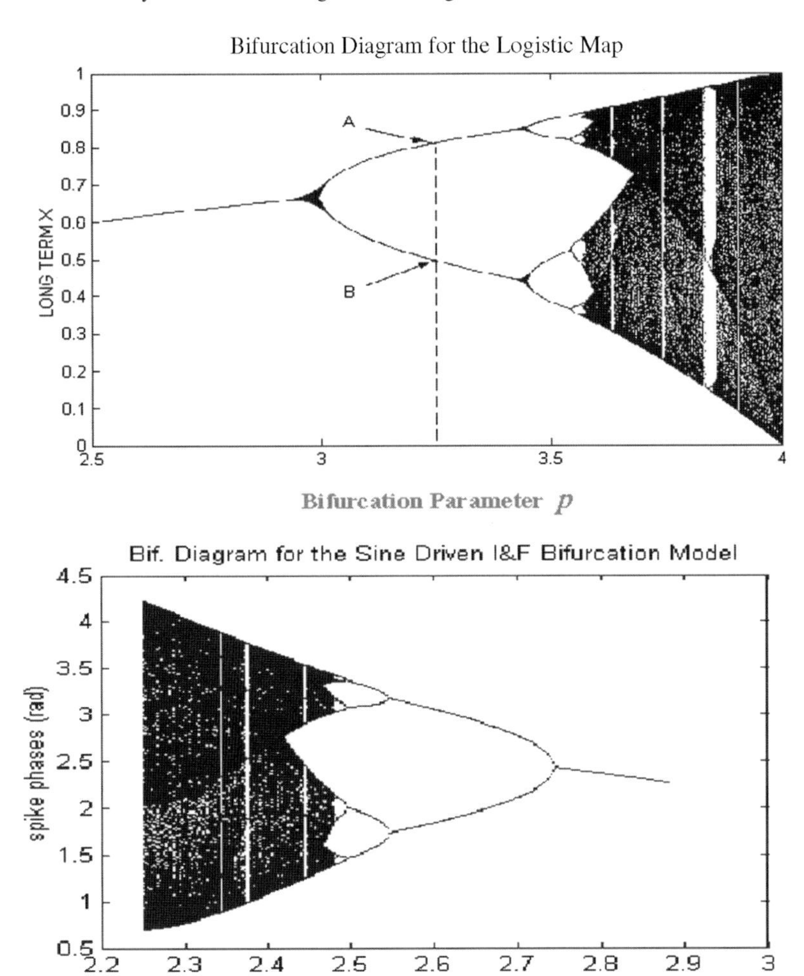

Fig. 4 The first bifurcation diagram corresponds to the logistic map, discussed in detail in the main text. The second bifurcation diagram regards another recursive node; it shows the diversity of phases of spiking in a periodically driven neural oscillator based on the integrate and fire neuron, as a function of the slope of its integrator [14]. In the first diagram, the larger the bifurcation parameter, the more chaotic the behavior of the time series produced by the recursive node. In the second diagram, the larger the bifurcation parameter, the more ordered the node behavior

tionship between the phases of consecutive spikes generated by the integrate and fire neuron. This bridges recursive nodes and spiking neurons, which is a fast growing area in artificial neural networks. In addition, this bridge between spiking phases and RPEs can involve generic recursive maps: [7] describes electronic implementations of **arbitrary recursive maps** based on the integrate and fire neurons.

2.3 Coupled Structures and the Emergence of Collective Behavior: Local Versus Global Dynamics

In networks of multiple RPEs, recursive nodes interact through coupling with neighboring neurons. The whole network evolves to collective spatio-temporal patterns. Figure 5 represents the structure of a small network of recursive nodes with full connection among them, in which the activity x_n of each node affects the recursions of all the other nodes, through changes in their p parameters. Each circle 'L_1 to L_4' represents a bifurcating recursive node, or a logistic map, in many prototypical studies [4, 6, 12].

Key features of the RPE nodes, such as the phenomenology of attractors and repellers, and the high diversity of attractors, reflect at the network level. Diverse collective phenomena emerge in networks of coupled recursions, such as the following ones:

- **Clustering of nodes' activities**, i.e., several nodes in the network evolve in a similar or identical form, in contrast with the behavior of the rest of the nodes. What exactly characterizes a cluster depends either on the goals of a given experiment / modeling or on the strategy of coding of information assumed for the network under study. The cluster can be characterized for example by the period of the state variables' cyclic trajectories, by their amplitudes of cycling, or even by their phase of cycling [17, 26].

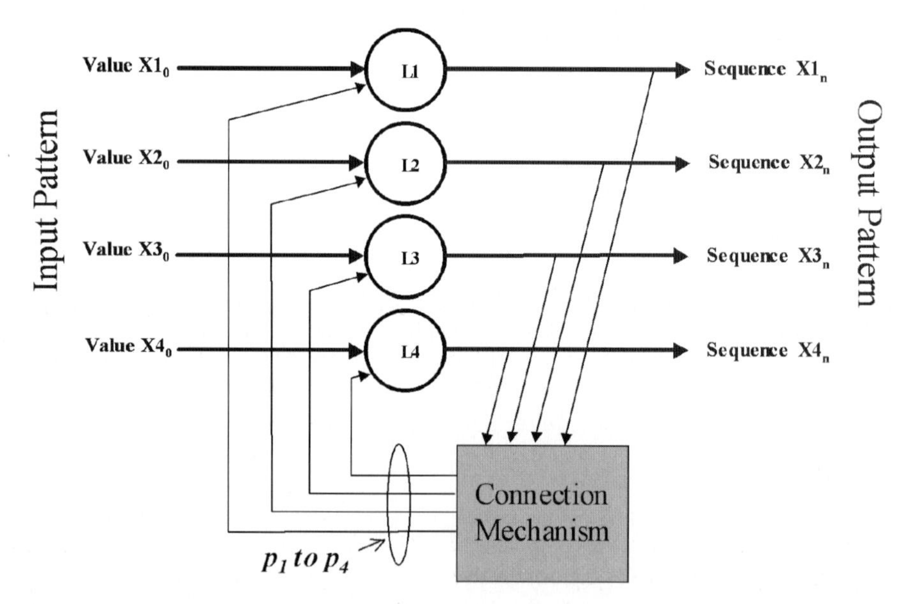

Fig. 5 A small network composed of 4 RPEs with full connection among all nodes. Each circle L_1 to L_4 represents a bifurcating recursion (e.g., a logistic map). The connection among nodes promotes changes in the bifurcation parameters of the recursive nodes (parameters p_1 to p_4). These changes are driven by the activity of the whole network [12]

- **Synchronization of the RPEs' cycling** (or 'phase locking'). In this type of collective phenomena, the cyclic activities of pairs of nodes have the same period, and they operate with constant relative phase [4, 17]. This type of behavior is important for the formation of the associative architectures discussed later [4, 6, 12].
- **Multidimensional attractors** (also named spatio-temporal attractors). These macroscopic phenomena are important for association, pattern recognition and decision-making. In the multidimensional attractor behavior, similarly to attractors and repellers at the single node level, we observe the evolution of the activity of the network towards a limited repertoire of dynamical modalities, which preferentially emerge in coupled structures. This evolution is frequently composed by initial chaotic activity followed by gradual approximation to an ordered limit cycle. This attractor represents a stored memory in association tasks, or a class label in pattern recognition tasks [6, 17].

Among other advantages, associative networks of this kind perform better in pattern recovery than traditional artificial neural networks with the same functionality (as discussed later). They also employ complex dynamics and chaos, what enhances the exploration of the state variables' universe, in searches for stored patterns and corresponding attractors [6, 14].

In the networks of coupled RPEs used for the implementation of associative memories, the strategy for the definition of synaptic weights is based on supervised Hebbian learning [20, 22]. In associative structures (addressed in Sect. 3.2), the memories to be stored are used to define the expected similarities and dissimilarities (i.e., the mathematical correlations) between activities of the neural nodes. Training samples, or stored memories, thus define a universe of collective spatio-temporal patterns modeled by the network during association. These samples of the training set are used to estimate the pair-wise similarities among nodes in the network: each synaptic weight w_{ij} linking nodes i and j is adapted to represent the correlation among the activities of these specific nodes.

The notation for subscripts in the connection weight w_{ij} is as follows: i is the target neuron (the one which receives stimulus in the synapsis ij), and j is the source neuron (the one that generates the stimulus); the synaptic weight w_{ij} regards information that 'arrives to node i, coming from node j'. A typical Hebbian expression for the connection weights can be seen in (6).

$$w_{ij} = \sum_{\mu=1}^{M} (\xi_{i,\mu} . \xi_{j,\mu}). \tag{6}$$

In (6), i and j represent the relevant nodes, and ξ_{μ} represents one of M memories (binary strings) stored in the associative network. Each term $\xi_{i,\mu}$ or $\xi_{j,\mu}$ represents a particular bit of a given memory ξ_{μ}.

Once the strengths of connection for all pairs of neurons are defined, according to (6), they can be used to define how the stimuli coming from all the nodes that compose the network are integrated and affect specifically the evolution of each node. Equation (7) ahead represents generically this integration and coupling for a given target node i: based on the activity of the network and the connection weights w_{ij}, we have the driving of the bifurcation parameter p_i of a target node i:

$$p_{i,n} = f(w_{ij}; x_{j,n}). \tag{7}$$

In (7), f represents a generic function of multiple variables (w_{ij} and $x_{j,n}$), which has to be properly chosen for the formation of the desired collective behavior for the network. The w_{ij} are the synaptic connections from neighboring neurons j, and $x_{j,n}$ their outputs at time step n.

Equation (8) bellow is a specific case of (7), for the most common modeling form of information integration used in neural networks: a model of linear composition of inputs $x_{j,n}$, followed by a function g.

$$p_{i,n} = g(\sum_j w_{ij}.x_{j,n}). \tag{8}$$

Notice that while f in the previous (7) is a function of multiple variables, g in (8) is a function of a single variable. The linear composition $\sum_j w_{ij}.x_{j,n}$ represents the synaptic-dendritic processing, and it acts as the scalar argument of function g, which defines how the total stimulation received by node i drives its bifurcation parameter.

The weighted sum of stimuli used in (8) is based on the fact that, in real neurons, the several post synaptic activities contribute jointly to the depolarization of the axon hillock membrane, through a process of linear superposition of currents generated at the several synaptic sites [28]. This linear modeling is classically done in the representation of synaptic-dendritic processing in model neurons, and it is being maintained here in the context of coupled recursive nodes.

2.4 Sensing Time-dependent Information

The structure in Fig. 5 is suitable for attractor networks. In this type of networks, the input information corresponds to the initial configuration of the state variables of the network, and the output information is given by the dynamic configuration (attractor) that emerges in the activity of the network in the long-term. The structure in Fig. 5 can be extended to keep properties of the attractor networks, while incorporating more complex modeling elements, including time-varying input patterns and hetero-associations [12, 30, 38]. This can be accomplished by a simple modification: several nodes in the network, let us say L_1 and L_2, have an exclusive function to represent input information. The activities x_1 and x_2 of these 'sensory nodes' are not affected by other nodes, but instead depend exclusively on the external information. These nodes act as input units. They are autonomous oscillators that do not monitor the network activity, neither are affected by it, but they just generate signals that feed the network. There is no need for feedback to nodes L_1 and L_2 (paths p_1 and p_2 in the figure are not present in this modified structure); the only part of the network that still receives feedback is L_3 and L_4. Nevertheless, the outputs of all the four nodes L_1 to L_4 contribute to driving p_3 and p_4.

With the described modification of architecture, we can have a blend of auto-associative functionality, which is implemented through the self connections of the coupled oscillators L_3 and L_4 (synaptic weights w_{34} and w_{43}), and hetero-associative functionality, implemented through the connections between the autonomous oscillators L_1 and L_2 and the associative nodes L_3 and L_4 (synaptic weights w_{31}, w_{32}, w_{41}, and w_{42}). Of course, this very small architecture (two sensory nodes and two associative nodes) was used here only to explain the concept of sensory nodes. Real applications of this modeling structure employ larger numbers of nodes [6, 12].

2.5 *Coding of Information Through Global Non-fixed-point Attractors*

Structures of coupled recursive nodes as the one discussed previously (Fig. 5) can exhibit preferential collective spatio-temporal patterns. This can be used to represent meaningful information. Implicit in this strategy, is coding of information through dynamic attractors. In Fig. 6, we have a more schematic illustration (as compared to Fig. 5) of a network of coupled recursions, or recursive oscillators. It emphasizes the dynamic nature of RPE networks and the relationship between limit cycles in spatio-temporal outputs and the corresponding output information. Here we have assumed that we are dealing with visual information, just to facilitate the illustration of concepts.

The coding of information in limit cycles of networks of coupled recursive elements can occur in a large variety of different forms, e.g.:

- Coding analog quantities based on the amplitude of oscillations of the state variables x (this involves the values A and B in the examples of Sect. 2.2);
- Coding analog quantities using values of the x_n time series;
- Coding discrete classes using period of cycling in closed trajectories;
- Coding discrete classes using phase of cycling;
- Mixed forms involving multiple periods, multiple amplitudes of oscillations, and multiple phases of cycling.

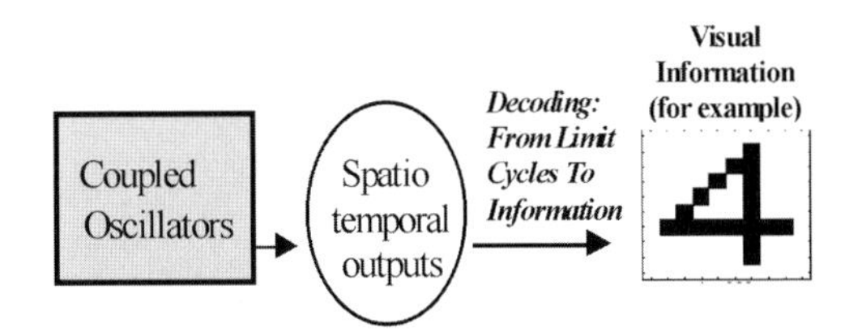

Fig. 6 Illustration of a network of coupled recursions (or oscillators) producing consistent global spatio-temporal patterns that emerge during the operation of the network in the long term, i.e., limit cycles. Thus, limit cycles represent information (adapted from [10])

Note that many of the above are closely related to the phenomena of synchronization in populations of spiking neurons and clustering of neural activity. Synchronization and phase of spiking in these neural networks are also used to represent information instead of traditional model neurons using volume of neural activity (frequency of action potentials).

In Fig. 7 we illustrate, through a tool named Limit Set Diagram, the operation of a network with 20 coupled recursive nodes. Most of them operate with the same period (period-2; only neuron number 18 does not) and with the same amplitude of cycling ($|A - B|$). Some of them oscillate with a given phase and others with the opposite phase. Here we coded binary information (or multidimensional logical information, since we have 20 nodes) using the phase of cycling.

Figure 7 represents several types of information, relevant for various coding methods listed above. The values x_n visited by the limit cycles of each recursive node are represented by the dots right above each node number. The phase of cycling of each periodic trajectory is represented through the circles around the specific x_n values visited at a given time of sampling (we observe from those circles the opposite phases being used for the representation of zeros and ones). We also have, indirectly, an indication of the level of order or chaos associated to each node, since their bifurcation parameters, the p_i, are represented through the triangles. The amplitudes of cyclic oscillations also appear in this diagram (the distance $|A - B|$), although they are not explored for coding of information in this case.

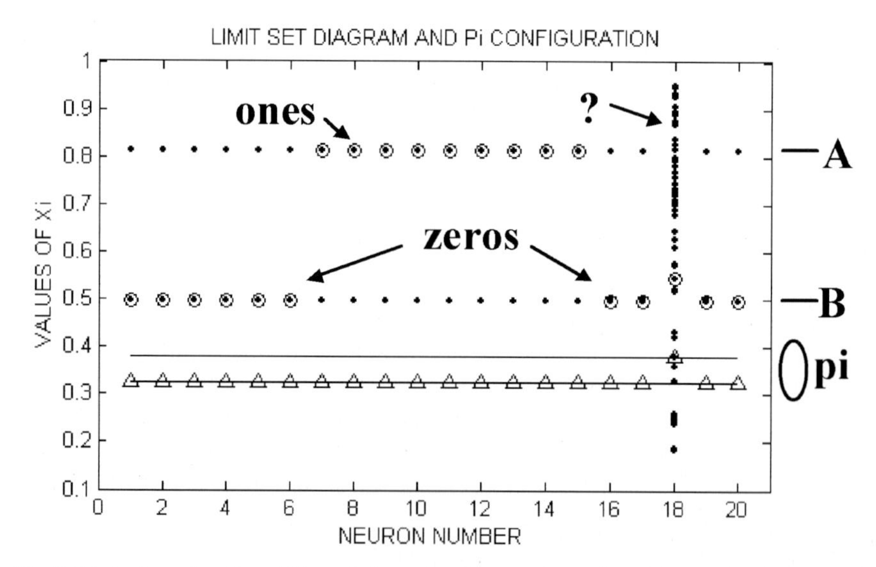

Fig. 7 Illustration of coding of binary information through phase of periodic cycling, in a network with 20 coupled recursive nodes. In this 'Limit Set Diagram', the horizontal axis represents the 20 neurons in the network (i runs from 1 to 20). The dots and the circled dots right above each value of i represent the spatio-temporal pattern of period-2 which is being exercised by the state variables x_i in the long-term. The circles around dots indicate the phase of cycling of each node. The triangles represent the values of the parameters p_i (their values divided by ten, for scale convenience) [11]. Node 18 is the only one still doing chaotic search (tagged with the question mark), and not settled in a period-2 cycle

Perhaps one of the most important differences between neural networks composed of recursive nodes and those artificial model neurons traditionally used in neural networks is the fact that the individual RPE nodes have dynamic attractors as their natural form of activity. This differs substantially from networks composed of traditional sigmoidal model neurons operating with fixed-point attractors [20]. An interesting aspect related to this difference regards the numbers of states that each class of networks can exercise. It can be shown that the number of distinct cyclic states, $CycS_{N,L}$, for a network with N nodes, and output values with L discernible levels, is given by (9), while the number of states for a similar network (i.e., the same N nodes and L discernible output levels) operating with fixed point attractors, $FixS_{N,L}$, is given by (10).

$$CycS_{N,L} = (L \sum_{m=0}^{L} (1/(L-m)))^{N}. \tag{9}$$

$$FixS_{N,L} = L^{N}. \tag{10}$$

Simple calculations using these two equations show that the value of $CycS_{N,L}$ grows much faster than the value of $FixS_{N,L}$ does, even for networks with a moderate number of nodes (N) and a moderate number of output levels (L). For example, when we consider an extremely simple network composed of only three nodes and three discernable output levels, $CycS_{N,L} = 4096$, while $FixS_{N,L} = 27$. As the number of nodes N and the number of levels L grow, such difference between the quantities $CycS_{N,L}$ and $FixS_{N,L}$ grows even more. This indicates that networks operating with cycles for the representation of information have a much larger number of distinct output patterns as compared to networks operating with fixed-point attractors.

It is important to mention that the values $CycS_{N,L}$ and $FixS_{N,L}$ have to be understood as upper limits, not as the exact indication of the number of states that can be reached by each network of these classes. Nevertheless, the marked contrast between these two upper limits ($CycS_{N,L}$ and $FixS_{N,L}$) indicates an important difference between the two classes of networks.

3 Tools of Analysis in Nonlinear Dynamics and Neural and Brain Activity Modeling

3.1 Relevant Classical and New Tools

We first briefly mention some classical tools in nonlinear dynamics (they are not detailed here; see for example [21] and [27]) and we make explicit how these classical tools are useful in architectures with recursive nodes. Then, we address some new tools developed for specific needs on RPEs networks; these new tools are explained in more detail in specific sections of the chapter.

The most basic relevant tools are probably the **Return Maps**, for the graphical description of the parametric recursions R_p, as in Fig. 1, and the **Web Diagrams** [27]. These two together are important practical tools for the understanding of the phenomenology of attractors and repellers. They are particularly useful when available through computational environments for interactive simulations, when allow for the study of the recursions' phenomenology and for observation of the emerging attractors.

Mathematical characterization of recursive dynamics can be accomplished through the **Lyapunov Exponent** and **Entropy Measure** [21, 27]. The first is useful in the design of attractors for representing information: low values of the Lyapunov exponent have direct impact on the speed of stabilization of attractors, and therefore on the speed of performing a given functionality, such as pattern recovery or pattern recognition. Large values of the Lyapunov exponent and large values of Entropy are good ingredients for the rich exploration of the state space; this is of interest for searches of stored memories and embedded patterns [6].

Bifurcation Diagrams (Sect. 2.2) summarize dynamical richness of nodes and are useful for defining its relevant features. With them, we can see which ranges of nodes' parameters are suitable for two complementary 'macro' states: 1) the wide search for patterns and 2) the stable ordered trajectories for information representation (these two states are discussed in more detail in Sect. 3.2). The **Analysis of Stability**, described in Sect. 2.2, is a classical tool instrumental in the design of periodic attractors (which are used for information representation) and for studies of bifurcation in recursive nodes.

The classical tools and methods mentioned above allow us to capture many of the important aspects of networks with nodes having rich dynamics. Still, extensions and complementary tools are needed for modeling and characterizing the RPEs neural architectures. Among important ones are:

- **Limit Set Diagrams**, as in Fig. 7, for the representation of attractors and multidimensional long-term behavior in networks of coupled recursive nodes [11, 17].
- **Error Profile Curves**, or Error Plots, for performance characterization of associative networks. These are discussed in the Sect. 3.3 [6].
- **Graphic Histograms**, for characterization of attractor landscapes. These are discussed in Sect. 3.4 [13].

More details on these three tools will appear in the rest of the chapter, where it will be clear how they can help in different aspects of the design and analysis of modeling neural architectures.

3.2 Associative Memories, Pattern Recognition, and Blending of Order and Complex Dynamics

In this section we address an important class of functionalities that can be modeled through architectures based on networks of recursive nodes: association. In particular, we explore auto-association on multidimensional logical information (or binary strings). In order to facilitate the illustration of this type of functionality, we look at the stored memories not as abstract strings of binary information or strings of logical quantities, but as binary images. This makes the description more graphical, although the same results obtained for binary images would apply to any kind of information described by binary strings, including logical information, class labeling, and decision information.

Figure 8 illustrates the functionality of associative memories of this kind. From distorted versions of binary patterns stored in the associative memory, we recover original patterns without distortion. This functionality is called auto-association, or content addressable retrieval, and it is the target of the classical neural network developed by John Hopfield in reference [22].

The successful results on the modeling of associative networks through coupled RPEs described in [4, 6, 8, 10] and [12], always employed forms of network operation with a common strategy: blend of chaos and order, and the switching between one state (chaos) and the other (order) according to the needs for either pattern search or information representation.

During the initial phases of search for a stored pattern, rich dynamics is exercised through high p values. In its evolution towards an attractor that codes the stored memory (or the class label, if we are modeling pattern recognition), the RPE nodes exercise various dynamical modalities, possibly with different periods and chaotic behavior, and their state variables x can change constantly over the state space.

During the final phases of the pattern recovery process, we have low p values and only those low period attractors representing information are exercised.

Following this general idea, (11) and (12) define a mechanism for memory search (or pattern search) with blend of chaos and order: as detailed ahead, the driving of the bifurcation parameters p_i is dynamically commanded by the network, and the switching between rich dynamics and ordered dynamics occurs according to its needs.

$$d_i = -(x_i - k1).(\sum_j w_{ij}.(x_j - k1)) + k2. \tag{11}$$

$$\Delta p_i = d_i.c - e. \tag{12}$$

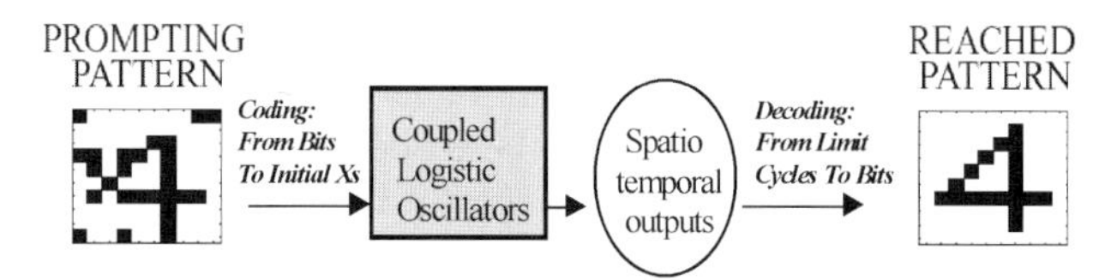

Fig. 8 Illustration of auto-association of binary patterns using a network of coupled RPEs. The input patterns are mapped to initial conditions of the attractor network and the output patterns are returned, coded through the limit cycles emerging during the operation of the network in the long-term (adapted from [10])

In (11) and (12), $k1$, $k2$, c and e are adequate constants [6, 12]. The remaining terms and variables (d_i, x_i , x_j, w_{ij}, and Δp_i) are defined together with the detailing of equations in the following two paragraphs.

Equation (11) measures the level of disagreement (d_i) / level of coordination, between the local activity x_i and the whole network (i.e., the set of all node activities x_j). According to (11), if the pair-wise activities (x_i; x_j) are mostly consistent with the expected correlations embedded in the weights w_{ij} ((6)), the disagreement measure d_i for node i will have low values (this indicates good coordination among nodes); if, on the contrary, the observed activities x_i and x_j are inconsistent with the expected correlations w_{ij}, then d_i will have high values (this indicates bad coordination among nodes). In (11), the constant $k1$ is the average value for x variables, this meaning that the terms $(x_i - k1)$ and $(x_j - k1)$ appearing in the evaluation of the disagreement d_i represent the nodes' activities, but shifted for zero average.

Under the command of the disagreement measure d_i, (12) controls the level of order and chaos in the bifurcating element i: the d_i measure (previously evaluated through (11)) drives p_i up and down, according to the increments Δp_i given by (12). In this equation, c is a scaling factor, and e helps the return of p_i to low values during network stabilization [6, 12].

From the explained above, (11) and (12) together allow for destabilization of ordered behavior whenever there is a certain level o disagreement between the network's evolution and the patterns embedded in the synaptic weights. This disagreement indicates the need for a pattern search process. After that, stabilization of the network is recovered only after the search process reaches a network configuration for which the x activities are consistent with the pair-wise correlations embedded in the w_{ij} weights.

In summary, (11) and (12) implement an autonomous self-driven process of transition between chaotic and ordered states. This is illustrated in Fig. 9, in which features of these two states are related to different aspects of the network behavior (the level of coordination among nodes, the range of values of the parameters p_i of the recursive nodes, and the level of global stimulus received by the recursive nodes) [14].

The final configuration of the coupled recursive nodes in a process of memory recovery or pattern recognition happens when a self-sustained cycle (the left part of Fig. 9) is achieved. In this configuration, ordered and coordinated activities guarantee a high level of stimulation of each node, and this, in turn, guarantees that each node operates in the ordered region of its bifurcation diagram [14]. In the Limit Set Diagram of Fig. 7 (Sect. 2.5), used in the discussion on coding of information through periodic attractors, we can find illustration for some aspects of the blend of order and chaos here discussed. In Fig. 7, all nodes except the number 18, have already found a stable mode of operation (these nodes are in the ordered state represented by the left cycle in Fig. 9). All of these ordered nodes have low values of p_i, as indicated by the triangles, and the node number 18 is the only one that is still in the search for a stable condition (this particular node is in the non-ordered state represented by the right cycle in Fig. 9).

It is interesting to present here a comparison between performances of associative RPEs networks and classical Hopfield associative networks as described in [6]. In

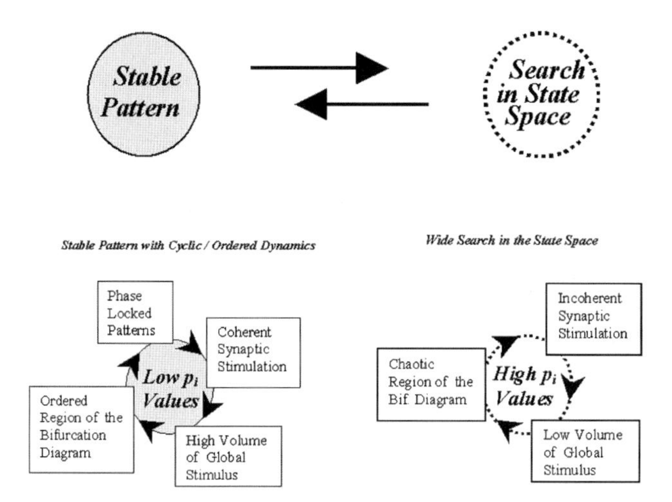

Fig. 9 Diagram showing two complementary states exercised by RPE networks: one is ordered behavior and representation of meaningful information (left part), and the other is complex dynamic behavior (needed for search of stored memories and search of patterns in recognition tasks). Each state has a certain degree of self-sustainability, as it is indicated by the two cycles represented at the bottom [14]

this study, both architectures stored binary strings (e.g., images), which were later recovered from their distorted versions. Table I presents the obtained performance measures for both architectures.

The table presents performance results with three different levels of input distortion, measured as a percentage of noisy bits in the prompting pattern received by the associative arrangements. The performance of each associative architecture is quantified based on the average Hamming error of the recovered pattern (average over 500 samples). We can see from Table I that the average recovery error is much smaller in RPEs networks, by factors in the range 1.5 to 2 [6].

Additional comments on the numbers in this table appear in the following section, together with the description of the 'Error Profile Plots'.

Before closing this section, we want to mention that the rich dynamic behavior of RPE neural networks seems suitable for implementing neural modeling fields architecture described in the chapter by Perlovsky. This neural network explores higher cognitive functions. Its dynamics is driven by the knowledge in-

Table 1 Recovery error for associative memories of hopfield and RPEs types

	Average Recovery Hamming Error in Hopfield Architecture	Average Recovery Hamming Error in RPEs Network
Input Noise 10%	0.00%	0.00%
Input Noise 20%	0.33%	0.18%
Input Noise 30%	2.12%	1.26%

stinct, mathematically represented by maximization of similarity between internal neural representations and patterns in input signals. Dynamic logic of this neural network fits internal representations to patterns in input signals with dynamic uncertainty. This uncertainty is dynamically reduced as internal representations are better adapted to input signals. RPE networks seem naturally suited for implementing dynamic logic. High values of p_i parameters may correspond to unsatisfied knowledge instinct, low similarity between internal and input patterns, and neural states in transition and adaptation. Low values of p_i parameters may correspond to satisfied knowledge instinct, high similarity between internal and input patterns, and neural states in stable dynamics. This RPE implementation of dynamic logic can be a subject for future research.

3.3 Quantifying Associative Power and Robustness to Noisy Inputs

In Table I in the previous section, the performance of associative memories in recovering patterns was quantified through two types of distances or distortions: one of them refers to the input received by the associative memory (first column), and the other refers to the output provided by the associative assembly (second and third columns). Both of these measures (input and output) indicate the distance of inputs and outputs with respect to the ideal binary patterns involved in the queries that are processed by the associative memory. In particular, we use Hamming distances, appropriate for binary strings [6]. In the first row of that table, we see that the input pattern has 10% of bits distorted at the input, but no bits distorted at the outputs (full recovery of the input pattern). In the second row, distortion is reduced from 20% at the input to 0.33% in Hopfield associative architectures, and to 0.18% in RPEs associative architectures. In both architectures, these numbers indicate considerable reduction of noise, and therefore effectiveness of the associative functionality.

To allow a more general characterization of the relationship between input and output noise (or distortion), we can use a tool named 'Error Profile Plot', or 'Error Plot'. Figure 10 shows one of these plots, for an RPEs network performing association. This plot shows the performance of recovery of stored patterns (or error in recovery, represented in the vertical axis), for arbitrary levels of input noise from 0% to 100% (represented in the horizontal axis).

The symmetry in the S shaped curve in the Error Profile Plot indicates the fact that the mirror images of the memories are attractors of the same power as the memories themselves, what also happens in the classical Hopfield network [20]. In our context, this is in fact expected for networks in which the information is coded by the phase of cycling as represented in the Limit Set Diagram of Fig. 7 (in a previous section). At the end of the recovery process, we only have a set of nodes that are 'in phase' and another set of nodes that are 'in opposite phase'. This bipartition of the network is not enough for the differentiation between memories and their mirrors

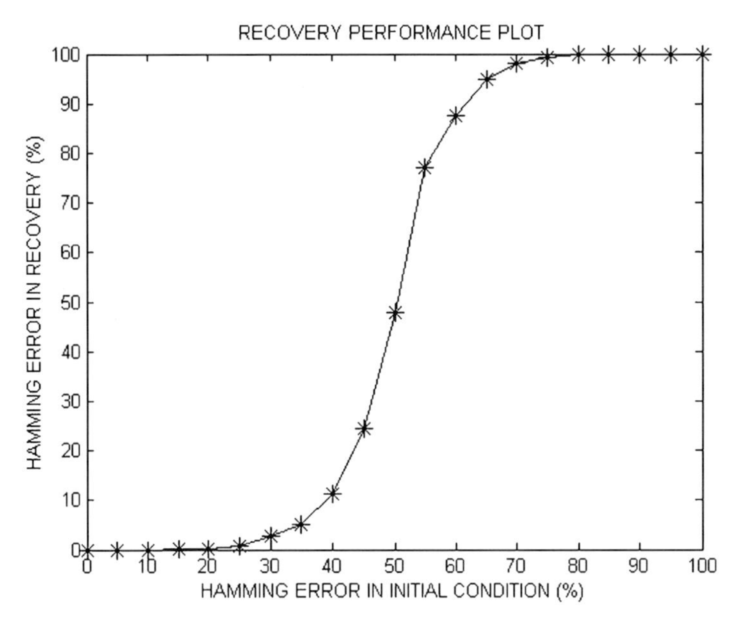

Fig. 10 Error Profile (or *'Error Plot'*) – this curve shows how the level of input noise or distortion (*horizontal axis*) affects the performance of recovery of stored patterns in the associative RPEs network. Recovery is near perfect for input errors up to 25%. Errors are represented as Hamming distances, normalized with respect to the network size (100 nodes, since the stored binary strings have 100 bits). Both axes in the plot present percentages [6]

images, given that the differentiation between ones and zeros depends on the relative phase of cycling in period-2 cycles. With that, the mirror images are represented by exactly the same bipartition of nodes in two sets with opposite phases of oscillation. Nevertheless, if desired, the ambiguity between original images and their mirror images can be solved through an external node, whose phase of cycling is taken as the reference phase for the differentiation between original images and mirror images.

In [12], we have an interesting example of application of the Error Profile Plots. They are used to show that the degradation of associative performance in RPEs networks occurs not only with the load of the stored memories [34] but also with the scaling of magnitudes in synaptic weights, as illustrated in Fig. 11.

Degradation of associative power occurs when scales of connection weights are either too high or too low. This is indicated by the increase in the recovery errors in Fig. 11, for different scaling factors applied to the synaptic weights. According to our interpretation [12], this degradation of performance relates to the loss of the optimal coupling: weak coupling leads to poor formation of collective patterns, while large strengths of coupling can lead to large increments Δp_i (according to (11) and (12)) and excessive driving to chaotic states. Both cases result in loss of performance. This result suggests that mechanisms of regulation of synaptic weight

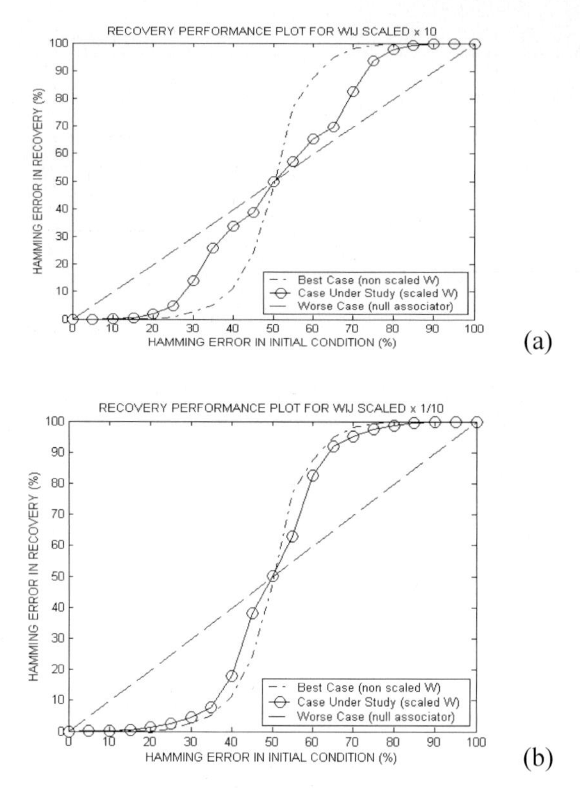

Fig. 11 Error Profiles (Hamming error in the output pattern versus Hamming error in the input noisy pattern) obtained for distinct magnitude scales of the W matrix (*composed by the* w_{ij}). With the scaling of W, we depart from the original S shaped error profile, which represents the maximal performance, and we move towards the minimal performance (*dashed straight line*) [12]. The scale factor is 10 for the top part and $1/10$ for the bottom part

are crucial to promote robustness for the coding and storage of information in RPEs networks.

3.4 Characterizing the landscape of attractors

Central aspects of the networks of coupled recursive nodes regarding their global dynamics and the representation of information include:

- Repertoire of attractors;
- Spatio-temporal features of attractors representing stored information and spurious attractors;
- Relative power of attraction of existing attractors.

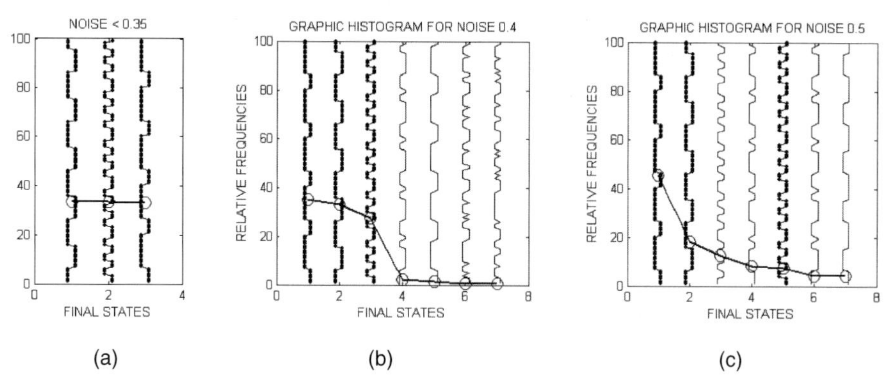

Fig. 12 Three Graphic Histograms, showing relative frequencies of patterns at the output of the associative network. The solid line linking the circles is the plot with relative frequencies. The vertical scale goes from frequency 0% to 100%. The output patterns are binary strings (100 bits), represented vertically, and ordered horizontally according to their frequencies of occurrence (1 in the horizontal axis indicates the pattern with higher frequency; larger numbers in the horizontal axis indicate patterns with lower frequencies). The three histograms show output patterns that emerge during experiments and their relative frequencies, for three different levels of noise added to the initial conditions: (**a**) represents situations with low levels of noise added to the input pattern; (**b**) medium levels of noise; and (**c**) high levels of noise. The output patterns which correspond to stored memories are presented in bold, to differentiate them from the spurious attractors [13]

Graphic Histograms, as in Fig. 12, are useful tools for understanding of the repertoire of attractors [13]. These diagrams summarize statistical scenarios in pattern recovery experiments. The name 'Graphic Histogram' comes from the fact that frequencies of attractor occurrences are accompanied by graphical representations of the respective output patterns that they represent. The histogram thus gives us a picture of the landscape of attractors in the network under study, showing not only their relative frequencies, but also similarities and differences between the involved patterns.

Graphic Histograms can be used, for example, for the analysis of the repertoire of final output configurations that emerge during experiments in RPEs networks operating with different levels of noise in the input patterns. In this way, we can evaluate relative frequencies of occurrence of each output pattern (attractor), including the planned ones and the spurious ones, compare their relative powers of attraction, and in some cases detect the emergence of undesired strong unbalance in their relative powers of attraction (this happens in Fig. 12c).

For associative memory applications, the attractors represented in the Graphic Histograms, Fig. 12, correspond to memories stored in the neural architecture. For pattern recognition and decision-making, these attractors correspond to the classes represented by the network.

The examples used here have the property that the graphical representation of attractors is relatively simple to implement. This happens because the meaningful limit cycles of individual nodes are only two, given that they represent binary information (the stored memories are binary strings). In this way, a simple and narrow vertical drawing, jumping from one level to the other according to the values of the individual bits, is sufficient to clearly represent the output binary strings (and, implicitly, it is also sufficient to represent the associated attractors that code these strings, through phase of cycling as in Fig. 7). With these output strings being graphically represented in the histogram, side by side as in Fig. 12, we can easily identify similarities and differences among the several output patterns.

4 Conclusions and Perspectives

This Chapter discussed the formalism, the features and main properties of recursive nodes and architectures built through coupled recursive nodes. We addressed roles of attractors, repellers, and blend of order and chaos; mathematical conditions for valid periodic attractors; applications to associative tasks and to pattern recognition. We analyzed dynamical richness of networks of coupled recursive nodes, and information coding using cyclic attractors. Specific tools to study and characterize networks of recursive nodes were discussed. **Limit Set Diagrams** were discussed for visualization of network activity and multi-dimensional attractors. **Error Profile Plots** were discussed for statistical characterization of associative performance. **Graphic Histograms** were discussed for summarization of data in large number of experiments, visualization of repertoire of attractors in a given network, and evaluation of relative power of each attractor.

Promising directions for future efforts include: studying asymmetric connectivity structures; bridging recursive nodes and spiking neurons; and using recursive nodes to implement cognitive dynamic architectures.

Additional developments on asymmetric connectivity structures and non-linear synaptic-dendritic processing can bring the recursive nodes to a more diverse set of neural architectures, and provide mechanisms for hetero-association and context switching directed by signals coming from higher hierarchical levels.

Recent impressive results in brain-computer interface will certainly drive the modeling of neural systems towards more detailed approaches, such as spiking neurons and cognitive architectures. We will need to develop models of the nervous functions at several levels: RPE networks modeling cognitive functions at macroscopic and symbolic levels, and also at the levels of signals and dynamical processes, where the bridge with spiking models as well as with K sets (discussed in several chapters in this book) can be important.

Among the possibilities of using recursive nodes to implement cognitive dynamic architectures in future efforts, we can mention that the chapter by Perlovsky explores neural modeling fields architecture for higher cognitive functions. RPE networks seem naturally suited for implementing its dynamic logic. Future efforts can explore such implementations and their correspondence to functioning of the brain.

Acknowledgment The author would like to thank the University of Sao Paulo, Brazil, and the funding agencies CAPES, FAPESP, FINEP and CNPq for supporting several research works discussed in this Chapter.

References

1. Aihara, K., Takabe, T., Toyoda, M.: Chaotic Neural Networks. Physica Letters A, Vol. 144(6,7 1990), 333–340
2. Adachi, M., Aihara, K.: Associative Dynamics in a Chaotic Neural Network. Neural Networks, Vol. 10(1997), 83–98
3. Caianiello, E. R.: Outline of a Theory of Thought-Processes and Thinking Machines. Journal on Theoretical Biology, Vol. 2(1961), 204–235
4. Del-Moral-Hernandez, E.: Bifurcating Pulsed Neural Networks, Chaotic Neural Networks and Parametric Recursions: Conciliating Different Frameworks in Neuro-like Computing. In: International Joint Conference on Neural Networks 2000, Vol. 6, 423–428, Como, Italy
5. Del-Moral-Hernandez, E.: A Novel Time-Based Neural Coding for Artificial Neural Networks with Bifurcating Recursive Processing Elements. In: International Joint Conference on Neural Networks 2001, Vol. 1, 44–49, Washington, USA
6. Del-Moral-Hernandez, E.: Neural Networks with Chaotic Recursive Nodes: Techniques for the Design of Associative Memories, Contrast with Hopfield Architectures, and Extensions for Time-dependent Inputs. Neural Networks, Vol. 16(2003) 675–682
7. Del-Moral-Hernandez, E., Gee-Hyuk Lee, Farhat, N.: Analog Realization of Arbitrary One Dimensional Maps. IEEE Transactions on Circuits and Systems I: Fundamental Theory and Applications, Vol. 50(Dec.2003), 1538–1547
8. Del-Moral-Hernandez, E.: Contrasting Coupling Strategies in Associative Neural Networks with Chaotic Recursive Processing Elements. In: Neural networks and Computational Intelligence IASTED Conference, NCI 2004, Grindelwald, Switzerland
9. Del-Moral-Hernandez, E., Silva, L. A.: A New Hybrid Neural Architecture (MLP+RPE) for Hetero Association: Multi Layer Perceptron and Coupled Recursive Processing Elements Neural Networks. In: International Joint Conference on Neural Networks 2004, Budapest
10. Del-Moral-Hernandez, E., Sandmann, H., Silva, L. A.: Pattern Recovery in Networks of Recursive Processing Elements with Continuous Learning. In: International Joint Conference on Neural Networks 2004, Budapest
11. Del-Moral-Hernandez, E.: Non-Homogenous Structures in Neural Networks with Chaotic Recursive Nodes: Dealing with Diverse Multi-assemblies Architectures, Connectivity and Arbitrary Bifurcating Nodes. In: International Joint Conference on Neural Networks 2005, Montreal
12. Del-Moral-Hernandez, E.: Non-homogenous Neural Networks with Chaotic Recursive Nodes: Connectivity and Multi-assemblies Structures in Recursive Processing Elements Architectures, Neural Networks, Vol. 18(n.5–6, 2005), 532–540
13. Del-Moral-Hernandez, E.: Fragmented Basins of Attraction of Recursive Processing Elements in Associative Neural Networks and its Impact on Pattern Recovery Performance. In: International Joint Conference on Neural Networks 2006, Vancouver
14. Del-Moral-Hernandez, E.: Chaotic Searches and Stable Spatio-temporal Patterns as a Naturally Emergent Mixture in Networks of Spiking Neural Oscillators with Rich Dynamics. In: International Joint Conference on Neural Networks 2006, Vancouver
15. Farfan-Pelaez, A., Del-Moral-Hernandez, E., Navarro, J. S. J., Van Noije, W.: A CMOS Implementation of the Sine-circle Map. In: 48th IEEE International Midwest Symposium on Circuits & Systems, Cincinnatti, 2005
16. Farhat, N. H., S-Y Lin, Eldelfrawy, M.: Complexity and Chaotic Dynamics in Spiking Neuron Embodiment, SPIE Critical Review, Vol. CR55 (1994), 77-88, SPIE, Bellingham, Washington
17. Farhat, N. H. Del-Moral-Hernandez, E.: Logistic Networks with DNA-Like Encoding and Interactions. Lecture Notes in Computer Science, Vol. 930 (1995), 214–222. Berlin: Springer-Verlag
18. Freeman, W. J.: Tutorial on Neurobiology: From Single Neuron to Brain Chaos. International Journal of Bifurcation and Chaos, Vol. 2(3, 1992), 451–482
19. Gerstner, W., Kistler, W. M.: Spiking Neuron Models: Single Neurons, Populations, Plasticity, Cambridge University Press, Cambridge, UK (2002)

20. Haykin, S.: Neural Networks: a Comprehensive Foundation, 2nd edn., Prentice Hall, NJ: Upper Saddle River (1999)
21. Hilborn, R.C.: Chaos and Nonlinear Dynamics: an Introduction for Scientists and Engineers, New York, Oxford University Press (1994)
22. Hopfield, J. J.: Neural Networks and Physical Systems with Emergent Collective Computational Abilities. Proceedings of the National Academy of Sciences, USA, Vol. 79(1982), 2554–2558
23. Hopfield, J. J.: Pattern Recognition Computation Using Action Potential Timing for Stimulus Representation. Nature, Vol. 376 (1995), 33–36
24. Ishii, S. et al.: Associative Memory Using Spatiotemporal Chaos. In: International Joint Conference on Neural Networks 1993, Vol. 3, 2638–2641, Nagoya
25. Kaneko, K.: Overview of Coupled Map Lattices. Chaos, Vol. 2(3, 1992), 279–282
26. Kaneko, K.: The Coupled Map Lattice: Introduction, Phenomenology, Lyapunov Analysis, Thermodynamics and Applications. In Theory and Applications of Coupled Map Lattices, 1–49. John Wiley & Sons, Chichester (1993)
27. Kaneko, K., Tsuda, I.: Complex Systems: Chaos and Beyond: a Constructive Approach with Applications in Life Sciences, Springer-Verlag (2001)
28. Kandel, E. R., Schwartz, J. H., Jessell, T. M.: Principles of Neural Science, Appleton & Lange, Norwalk, Connecticut (1991)
29. Kelso, J. A. S.: Dynamic Patterns - the Self-Organization of Brain Behavior, The MIT Press, MA: Cambridge (1995)
30. Kosko, B.: Bidirectional Associative Memories. IEEE Transactions on Systems, Man, and Cybernetics, Vol. 18 (Jan-Feb 1988), 49–60
31. Kozma, R., Freeman, W.J.: Encoding and Recall of Noisy Data as Chaotic Spatio-Temporal Memory Patterns in the Style of the Brains. In: International Joint Conference on Neural Networks 2000, Vol. 5, 33–38, Como, Italy
32. Kozma, R., Freeman, W.J.: Control of Mesoscopic/Intermediate-Range Spatio-Temporal Chaos in the Cortex. In: American Control Conference 2001, Vol. 1, 263–268
33. Lysetskiy, M., Zurada, J.M., Lozowasky, A.: Bifurcation-Based Neural Computation. In: International Joint Conference on Neural Networks 2002, Vol. 3, 2716–2720
34. McEliece, R., E., Posner, E., Venkatesh, S.: The Capacity of the Hopfield Associative Memory. IEEE Transactions on Information Theory, Vol. 33(4, Jul.1987), 461–482
35. Nagumo, J., Sato, S.: On a Response Characteristic of a Mathematical Neuron Model. Kybernetic, Vol. 10(1972), 155–164, 1972
36. Principe, J. C., Tavares, V. G., Harris J. G, Freeman, W. J.: Design and Implementation of a Biologically Realistic Olfactory Cortex in Analog VLSI. Proceedings of the IEEE, Vol. 89(2001), 1030–1051
37. Siegelmann, H. T.: Computation Beyond the Turing Limit. Science, Vol. 268(1995), 545–548
38. Wang, L.: Heteroassociations of Spatio-Temporal Sequences with the Bidirectional Associative Memory. IEEE Transactions on Neural Networks, Vol. 11(2000), 1503–1505
39. Wang, D.: A Comparison of CNN and LEGION Networks. International Joint Conference on Neural Networks 2004, 1735–1740

Giving Meaning to Cycles to Go Beyond the Limitations of Fixed Point Attractors

Colin Molter, Utku Salihoglu and Hugues Bersini

Abstract This chapter focuses on associative memories in recurrent artificial neural networks using the same kind of very simple neurons usually found in neural nets. The past 25 years have dedicated much of the research endeavor on coding the information in fixed point attractors. From a cognitive or neurophysiological point of view, this choice is rather arbitrary. This paper justifies the need to switch to another encoding mechanism exploiting limit cycles and complex dynamics in the background rather than fixed points. It is shown how these attractors encompass in many aspects the limitations of fixed points: better correspondence with neurophysiological facts, increase of the encoding capacity, improved robustness during the retrieval phase, decrease in the number of spurious attractors. However, how to exploit and learn these cycles for encoding the relevant information is still an open issue. In this paper two learning strategies are proposed, tested and compared, one rather classical, very reminiscent of the usual supervised hebbian learning, the other one, rather original since allowing the coding attractor to be chosen by the network itself. Computer experiments of these two learning strategies will be presented and explained. The second learning mechanism will be advocated both for its highly cognitive relevance and on account of its much better performance in encoding and retrieving the information. Since the kind of dynamics observed in our experiments (cyclic attractor when a stimulus is presented and a weak background chaos in the absence of such stimulation) faithfully reminds neurophysiological data and although no straightforward applications have been found so far, we limit our justification to this qualitative mapping with brain observations and the need to better explore how a physical device such as a brain can store and retrieve in a robust way a huge quantity of information.

1 Introduction

1.1 Computational Neuroscience and Artificial Intelligence

Computational neuroscience aims to model and/or explain brain functions based on neurophysiological evidence. The underlying idea is that they can provide predictions and/or theories to assist neuroscientists in their tasks. Since the neuron model

L. I. Perlovsky and R. Kozma, *Neurodynamics of Cognition and Consciousness.*

developed by Hodgin-Huxley (1952), neural networks are central in this field. The aims of Artificial Intelligence (AI) are wider, one of them being to find mechanisms to bring more power and "intelligence" to artificial devices. One obvious way is to look to the brain for some clues. Since the McCullogh and Pitts model (1942), neural networks have also become one of the grounding paradigms in AI. This common field renders the frontier between AI and computational neuroscience ambiguous (Fig. 1). The lack of a clear frontier can sometimes be misleading: results obtained from the AI field must be read with caution when applied to neuroscience.

We believe that the notion of fixed point attractor is one of these misleading notions. First discovered in the AI field, they rapidly became a paradigm of how brains store information. Although, there is no evidence for such attractors in the brain and very poor capacities were obtained, this success story strongly biased both the neuroscientist and the AI communities.

1.2 Fixed Point Attractors and Hopfield Networks

Since Amari, Grossberg and Hopfield precursor works [1, 2, 3], the privileged regime to code information has been fixed point attractors. In a landmark work, Hopfield has shown that a small set of patterns ξ^{μ} can be stored in a fully recurrent neural network by means of a Hebbian prescription rule specifying the connection weights: $w_{ij} = \frac{1}{N}\sum_{\mu=1}^{p} \xi_i^{\mu}\ \xi_j^{\mu}$ and $w_{ii} = 0$. Moreover, he demonstrated that in these networks having symmetric connections and no auto-connections, a Lyapunov function[1] can be defined on the entire state space. The learned patterns are local minima of this function and attracts noisy versions of the pattern to their steady state.

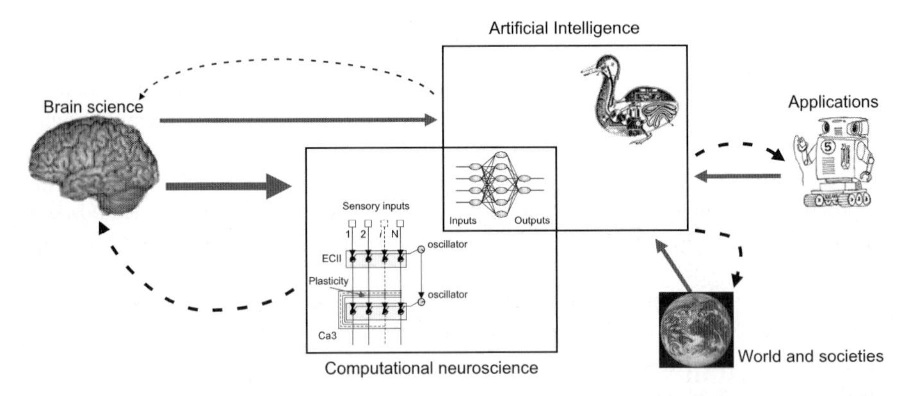

Fig. 1 Computational neuroscience and AI. Plain lines represent data flows used to ground the fields, dashed lines represent useful outcomes from the field. Flow strengths are represented by the line thickness

[1] i.e., a lower bounded function whose derivative is decreasing in time and which accordingly can be seen as an "energy function".

The distributed storage mechanism under these attractor networks makes the system very tolerant to partial damage or degradation in their connectivity. Since these two characteristics are also supposed to exist in real brains, encoding memories in fixed point attractors became the accepted viewpoint of how brains store information.

Many theoretical and experimental works have shown and discussed the limited storing capacity of these attractors network ([4, 5, 6, 7] and [8] for a review). This explains the absence of practical applications, and the loss of interest from the AI community. However, the fixed point attractors remain the image of how brains store information.

1.3 Ubiquity of Cycles

Encoding information in fixed point attractors leads to guaranteed stabilized dynamics of Hopfield type networks. Regarding the ubiquity of complex dynamics in the real world, this appears very unnatural. More specifically, in the brain, many neurophysiological reports (e.g. [9, 10, 11, 12, 13]) tend to indicate that brain dynamics is characterized by cyclic and weak chaotic regimes.

Inline with these results, in this chapter we propose to map stimuli to spatio-temporal limit cycle attractors of the network's dynamics. A learned stimulus is no longer expected to drive the network into a steady state. Instead, the stimulus is expected to drive the network into a specific spatio-temporal cyclic trajectory. This cyclic trajectory is still considered as an attractor since content-addressability is expected: before presentation of the stimulus, the network could follow another trajectory and the stimulus could be corrupted with noise.

The first problem we have to face is the difficulty to make theoretical analyses and the impossibility to provide straightforward mathematical proofs. These networks have to be analyzed mostly experimentally. The only possible way to prove that a given attractor has some content addressability will be through extensive experimental computations. The successful results obtained tend to prove that Lyapunov functions could be found around the cyclic attractors[2]. However, the presence of chaotic dynamics in various other parts of the state space would prevent one from defining these functions on the entire state space. They could be defined only in small subsets of the state space. This renders their search difficult. They have still to be found.

Bringing more complex dynamics in the network leads to the apparition of chaotic dynamics. To enable the application of mathematical tools from chaos theory, the activation function of each unit has been chosen continuous. The drawback is that attractors are continuous and difficult to compare (as an example, is 0.4 different from 0.45 or not?). To ease comparisons, the output is filtered and transformed into bit patterns (or symbols): a unit is active or inactive. After filtering, symbolic

[2] To obtain this function, the state space will have to be deployed in time such that the cycle will be equivalent to a fixed point.

analyses can be performed to characterize the dynamics. The distance between two patterns will be measured using the Hamming distance on the filtered output. It defines a natural measure obtained by counting the number of different states between them.

1.4 Hebbian Learning

Fixed point attractor networks became famous not only because they defined a way to encode associative memories, but also because the learning rule used to encode these memories was neurophysiologically plausible, strengthening again the analogy with the brain.

Synaptic plasticity is now widely accepted as a basic mechanism underlying learning and memory. There is experimental evidence that neuronal activity can affect synaptic strength, through both long-term potentiation and long-term depression [14]. Forecasting this biological fact, Hebb suggested how activity and training experience can change synaptic efficacies [15]. Since its use by Hopfield, such learning rule became essential for the construction of most models of associative memory (among others: [2, 16, 17, 18, 19, 20, 21]). In such models, the neural network maps the structure of information contained in the external and/or internal environment into embedded attractors.

However, it has been observed [22, 23] that in some conditions the time window of synaptic plasticity in pyramidal cells during tetanic stimulus conditions is more faithfully described by an asymmetric function: the connectivity between a pre-synaptic cell and a post-synaptic cell gets the most reinforcement when the post-synaptic cell fire slightly after the pre-synaptic cell. Applied to spiking neurons, this leads to the spike timing dependent plasticity learning rule (STDP). In AI, numerous kinds of time dependent Hebbian learning rules have been proposed, mostly for sequence learning applications (as an example, when applying classical Hebbian learning in time delay artificial neural network – e.g. [24]). The important point is that by applying time asymmetric learning rule, network connections become asymmetric and it is no longer possible to define an energy function which would apply for the entire state space. In fact, we have shown [25] that this kind of learning rule prevents the network from stabilizing: asymmetric learning can be seen as a road to chaotic dynamics.

2 Giving Meaning to Cycles

At the macroscopic level, brain activity is characterized by various kinds of rhythms oscillating at various frequencies [26]. During these oscillations, different populations of neurons can fire synchronously at different phases. If the synchronous neural activity is believed to play a critical role for linking multiple brain areas and solve the binding problem [27], the existence of population firing at different phases (phase locking) has been suggested to reflect the existence of temporal coding in the brain.

The analogy with our simple model is not straightforward. In fact, in the modeling, oscillations have lost their original meaning (there is no longer one frequency driving the network as a pacemaker). What remains identical is the presence of populations of units which are activated synchronously and which are phase locked to each other.

Brain processes information at many level. The important scientific question is then to know what kind of processes would require cyclic attractors. To ease comparisons with fixed point attractors, we propose here to use them as associative memories. In this mechanism, the entire cycle represents the information. However, cycles could be used in various other ways. As an example, each subcycle could contain independent information. In this case, the cycle's role would be to link several pieces of information together. This system detailed below could be used for the processing of working memories.

2.1 The Cycle is the Information

Information stored into the network consists of a set of pair-data. Each data is composed by an external stimulus and a series of patterns through which the network is expected to iterate when the stimulus feeds the network (this series corresponds to the limit cycle attractor). Traditionally, the information to be stored is either installed by means of a supervised mechanism (e.g. [2]) or discovered on the spot by an unsupervised version, revealing some statistical regularities in the data presented to the net (e.g. [28]).

In this chapter, two different forms of learning are studied and compared. In the first one, the information to be learned (the external stimulus and the limit cycle attractor) is pre-specified, and installed as a result of a classical supervised algorithm (Fig. 2B). However, such supervised learning has always raised serious problems both at a biological level, due to its top-down nature, and at a cognitive level. Who would take responsibility to look inside the brain of the learner, i.e. to decide the information to associate with the external stimulus and to exert the supervision during the learning task?

To answer at least the last question, in the second form of learning proposed here, the semantics of the attractors to be associated with the feeding stimulus is left unprescribed: the network is taught to associate external stimuli with original attractors, not specified a priori (Fig. 2C). This perspective remains in line with a very old philosophical conviction called constructivism which was modernized in neural networks terms by several authors (among others [29, 30, 31]). One operational form has achieved great popularity as a neural net implementation of statistical clustering algorithms [3, 32]. To differentiate between the two learning procedures, the first one is called "out-supervised", since the information is fully specified from outside. In contrast, the second one is called "in-supervised", since the learning maps stimuli to cyclic attractors "derived" from the ones spontaneously proposed by the network.

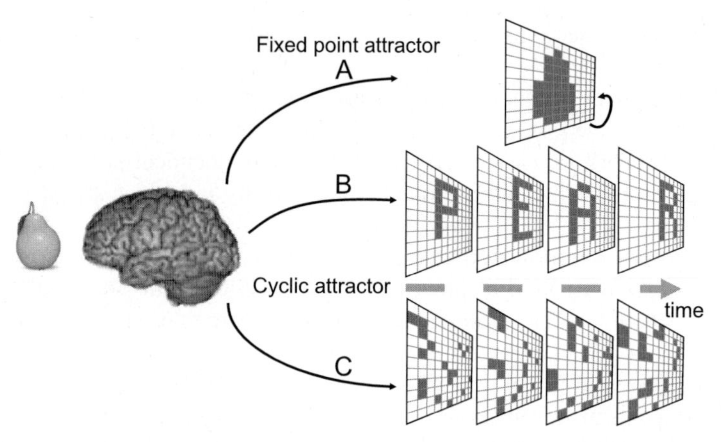

Fig. 2 How memories are represented in brain dynamics. The traditional view suggests that stable information is encoded in fixed point attractors (**A**). In this chapter, we suggest that cyclic attractors offer a better solution. The problem arising is to define the different patterns enclosed in one cycle. An obvious solution is to define them a priori, here by attributing one letter to each cycle (**B**). Another solution proposed here is to let the network choose the definition of the internal representation of the object (**C**). Layers represent the internal state of a recurrent neural network

Quite naturally, we show that this "in-supervised" learning leads to an increased storing capacity.

2.2 The Subcycle is the Information

In the 1980s, two English researchers named Baddeley and Hitch coined the term "working memory" for the ability to hold several facts or thoughts in memory temporarily while solving a problem or performing a task. How the brain manages to juggle simultaneously several pieces of information without merging them is still an open question.

According to Idiart and Lisman [33], an important clue was given by brain recordings showing that some region of the cortex have dual oscillation in which a low frequency oscillation in the theta frequency range is subdivided into subcycles by a high frequency oscillation in the gamma range [34]. They proposed that the subset of cells firing synchronously during one gamma subcycle represents one short term memory. The fact that 7 gamma oscillations can occur during one theta oscillation would explain the small capacity observed in working memories (7 ± 2 items) [35].

This working memory model appears as one possible example of where multiple bits of information are embedded in a cycle. In Fig. 3B, a cycle of size 4 is represented where each subcycle encodes one information necessary to perform a task (here cooking a dish). In this view, decoding the information kept by each cycle can be easily done through a simple feedforward network. In Fig. 3A, the information kept in each subcycle is superposed. It appears more complex (and impossible in

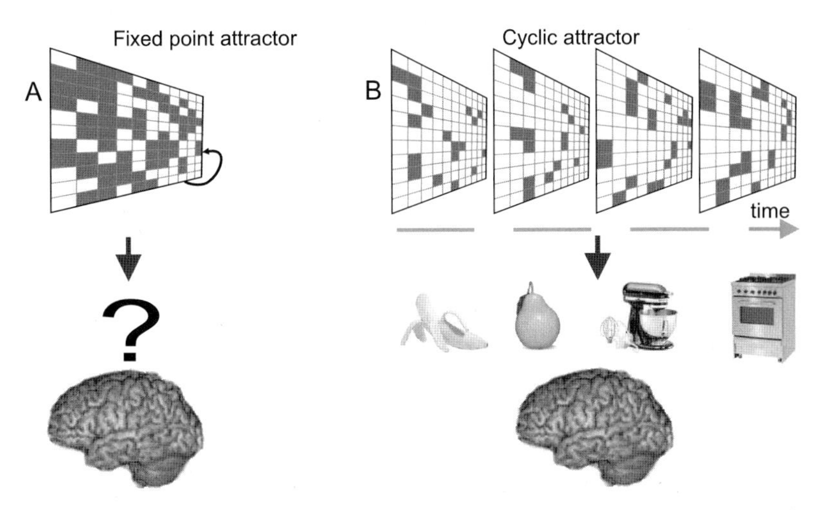

Fig. 3 In working memory tasks, multiple information have to be hold simultaneously. In (**A**), the multiple pieces of information are superposed in the same static pattern. This makes the decoding nearly impossible. In (**B**), the multiple pieces of information appear as subcycles. These patterns can them be easily decoded through a feedforward network. Layers represent the internal state of a recurrent neural network

some cases) for a feedforward network to decorrelate all the information kept in this single pattern.

3 A Simple Model for Associative Memories

3.1 The Neural Network

The network is fully connected. Each neuron's activation is a function of other neurons' impact and of an external stimulus. The neurons activation f is continuous and is updated synchronously by discrete time step. The mathematical description of such a network is the very classical one. The activation value of a neuron x_i at a discrete time step $n+1$ is:

$$\begin{aligned} x_i(n+1) &= f(g\,\mathrm{net}_i(n)) \\ \mathrm{net}_i(n) &= \sum_{j=1}^{N} w_{ij} x_j(n) + \sum_{s=1}^{M} w_{is} \iota_s \end{aligned} \tag{1}$$

where N is the number of neurons, M is the number of units composing the stimulus, g is the slope parameter, w_{ij} is the weight between the neurons j and i, w_{is} is the weight between the external stimulus' unit s and the neuron i and ι_s is the unit s of the external stimulus. The saturating activation function f is taken continuous (here tanh) to ease the study of the networks' dynamical properties.

The network's size, N, and the stimulus' size, M, have been set to 25 in this paper for legibility. Of course, this size has an impact both on the encoding capacity and on the background dynamics. This impact will not be discussed. Another impact not discussed here is the value of the slope parameter g, set to 3 in the following[3].

When storing information in fixed points, the temporal update rule can indifferently be asynchronous or synchronous. This is no longer the case when storing information in cyclic attractors, for which the updating must necessarily be synchronous: the network activity at a given time step is a function of the preceding time step.

To compare the network's continuous internal states with bit-patterns, a filter layer based on the sign function, is added. It enables us to perform symbolic investigations on the dynamical attractors.

3.2 Learning Task

Two very different learning tasks are proposed here. Both of them consists in storing a set of q external stimuli in spatio-temporal cycles of the network's internal dynamics. The data set is written:

$$\mathcal{D} = \left\{\mathcal{D}^1, \ldots, \mathcal{D}^q\right\} \tag{2}$$

where each data $\mathcal{D}^\mu$ is defined by a pair composed of:

- a pattern χ^μ corresponding to the external stimulus feeding the network;
- a sequence of patterns $\varsigma^{\mu,i}$, $i = 1, \ldots, l_\mu$ to store in a dynamical attractor.

$$\mathcal{D}^\mu = \left(\chi^\mu, (\varsigma^{\mu,1}, \ldots, \varsigma^{\mu,l_\mu})\right) \qquad \mu = 1, \ldots, q \tag{3}$$

where l_μ is the period of the pattern μ which may vary from one data to another. Each pattern μ is defined by assigning digital values to all neurons:

$$\begin{aligned} \chi^\mu &= \{\chi_i^\mu,\ i = 1, \ldots, M\} \ \texttt{with}\ \chi_i^\mu \in \{-1, 1\} \\ \varsigma^{\mu,k} &= \{\varsigma_i^{\mu,k},\ i = 1, \ldots, N\} \ \texttt{with}\ \varsigma_i^{\mu,k} \in \{-1, 1\} \end{aligned} \tag{4}$$

3.3 Learning Procedures

3.3.1 The "Out-supervised" Learning Algorithm

The learning task associated with this algorithm is very straightforward and consists in storing the well-defined data set given in (2). Each data stored in the network is

[3] The impact of this parameter has been discussed in [36]. They have demonstrated how the slope parameter can be used as a route to chaos.

fully specified a priori: given external stimuli are associated to pre-specified limit cycle attractors of the network's dynamics[4] (Fig. 2B). The algorithm, described by the authors in [37], is based on a classical iterative Hebbian algorithm described in [38]. Its principle can be described as follows: at each learning iteration, the stability of every nominal pattern ξ^{μ}, is tested. Whenever one pattern has not reached stability yet, the responsible neuron i sees its connectivity reinforced by adding a Hebbian term to all the synaptic connections impinging on it:

$$\begin{aligned} w_{ij} &\mapsto w_{ij} + \varepsilon_s \, \varsigma_i^{\mu,\nu+1} \, \varsigma_j^{\mu,\nu} \\ w_{is} &\mapsto w_{is} + \varepsilon_b \, \varsigma_i^{\mu,\nu+1} \, \chi_i^{\mu} \end{aligned} \tag{5}$$

where ε_s and ε_b respectively define the learning rate and the stimulus learning rate.

In order to not only store the patterns, but also to ensure a sufficient enough content-addressability, the basins of attraction are "excavated" by adding explicit noise during the learning phase.

3.3.2 The "In-supervised" Learning Algorithm

The encoding capacities of networks learned in the "out-supervised" way described above are fairly good. However, these results are disappointing compared to the potential capacity observed in random networks [39]. Moreover, we show that learning too many cycle attractors in an "out-supervised" way leads the network's background regime to become fully chaotic and similar to white noise.

Learning pre-specified data appears too constraining for the network. An "in-supervised" learning algorithm [40], more plausible from a biological point of view, is proposed as a radically different alternative: the network has to learn to react to an external stimulus by cycling through a sequence which is not specified a priori but which is obtained following an internal mechanism (Fig. 2C). In other words, the information is "generated" through the learning procedure assigning a "meaning" to each external stimulus: the learning procedure enforces a mapping between each stimulus of the data set and a limit cycle attractor of the network's inner dynamic, whatever it is. In fact, this is inline with a neural nets tradition of "less supervised" learning algorithms started by the seminal works of Kohonen and Grossberg. This tradition enters in resonance with writings in cognitive psychology and constructivist philosophy (among others [29, 30, 31, 41]). The algorithm to be presented now can be seen as a dynamical extension in the spirit of these preliminary works where the coding scheme relies on spatio-temporal cycles instead of single neurons.

[4] By suppressing the external stimulus and by defining all the sequences' periods l_{μ} to 1, this task is reduced to the classical learning task originally proposed by Hopfield: the storing of pattern in fixed point attractors of the underlying RNN's dynamics. The learning task described above turns out to generalize the one proposed by Hopfield.

In this algorithm, the mappings between the external stimuli and the spatio-temporal cyclic internal representation (4) are not defined before learning. As inputs, the algorithm receives only the set of stimuli to learn and a range $[\texttt{min}_{cs}, \texttt{max}_{cs}]$ which defines the bounds of the accepted periods of the limit cycle attractors coding the information. This algorithm can be broken down in three phases which are constantly iterated until a solution is found:

re-mapping stimuli into spatio-temporal cyclic attractors
During this phase, the network is presented with an external stimulus which drives it into a temporal attractor $\mathbf{output}^{\mu}$ (which can be chaotic). Since the idea is to constrain the network as little as possible, a meaning is assigned to the stimulus by associating it with a close cyclic version of the attractor $\mathbf{output}^{\mu}$, called $\mathbf{cycle}^{\mu}$, which must be an original[5] attractor respecting the periodic bounds $\texttt{min}_{cs}, \texttt{max}_{cs}$. This step is iterated for all the stimuli of the data set;
learning the information
The mappings between the stimuli and their associated proposed attractors $\mathbf{cycle}^{\mu}$ are tentatively learned by means of a out-supervised procedure. However, to avoid constraining the network too much, only a limited number of iterations (set to 10 in the following) is performed, even if no convergence has been reached;
end test
If all stimuli are successfully mapped with original cyclic attractors, the "in-supervised" learning stops, otherwise the whole process is repeated.

It has to be noted that this learning mechanism implicitly supplies the network with an important robustness to noise. First of all, the coding attractors are the ones naturally proposed by the network. Secondly, they need to have large and stable basins of attraction in order to resist the process of trials, errors and adaptations.

4 What are the Benefits or Cyclic Attractors

4.1 Capacity and Noise Tolerance

The amount of information that can be learned in an acceptable amount of time is one of the important question arising when working with memories. It is well known that the number of memories stored auto-associatively in a Hopfield networks by means of the Hebbian prescription rule is very limited (14% of the network size). By using an iterative learning rule, the capacity increases upto the network size. As shown below, this memory can be further enhanced by indexing the data using an external stimulus. However, another important measure which has to be addressed simultaneously relates to the content addressability of the obtained memories. We not only want to store data, we also want to be able to retrieve them when they are partly noised: stimuli "nearby" from a learned one have to drive the network to the same cyclic attractor (when the network's internal state is not "too far" from this attractor). Of course, everything has a price: there is a balance between the

[5] meaning "unique".

Table 1 Time required to encode specified data sets with as much content addressability as possible using the insupervised and outsupervised procedures. The learning time is measured by averaging results of 100 learnings. Time unit is one hundred processor clocks. The content addressability is measured with the normalized Hamming distance (d_H in %) between the expected cycle and the obtained cycle when both the stimulus and the internal state are initially slightly noised (noise < 10%). Each result shows the average obtained for 10 different noisy initial conditions around each learned data of each learned network (i.e. learning 20 data gives 10(noisy initial condition)*20(set size)*100(number of learning)=20000 measures). Variances (not shown here) are smaller than one percent

Algorithm	5 data		10 data		20 data		50 data		150 data	
and cycle size	time	d_H	time	d_H	time	d_H	time	d_H	time	d_H
Outsup. - size 4 cycles	4	0	26	10	–	–	–	–	–	–
Insup. - size 4 cycles	3	0	18	0	50	3	600	10	–	–
Outsup. - size 2 cycles	4	0	13	0	16	15	–	–	–	–
Insup. - size 2 cycles	3	0	18	0	51	0	81	12	650	8
Outsup. - Fixed points	1	0	12	0	40	4	–	–	–	–
Insup. - Fixed points	2	0	8	0	32	1	300	10	–	–

maximum size of the data set and the content addressability of the learned data. Table 1 compares these results for the storage of various stimuli in fixed point attractors, and in size 2 and size 4 cyclic attractors.

Different observations can be made. The first observation comes from the very different results observed between the insupervised and the outsupervised learning procedures. In the latter one, the learning of mappings to a-priori specified attractors appears too constraining. In the insupervised approach, by not constraining the network to the learning of predefined cycles, the network is free to create its desired mappings. The capacity enhancement appears valuable. As an example, if we fail to encode 50 stimuli in a priori given size 2 cycles, we successfully learn 150 mappings, with good content addressability, using the insupervised procedure. Six times the network's size! One obvious problem in the outsupervised procedure resides in the huge number of mappings required: encoding 50 stimuli in size 2 cycles requires in fact the learning of 100 mappings, i.e. the number of cycles times their size.

More generally, the possibility to "find" an important number of mappings and the difficulty to learn them a priori demonstrates how important it is to define appropriately how to represent the information.

A second observation lies in the dependence between the capacity and the size of the cyclic attractors. Regarding the insupervised mappings, similar results are obtained for the storing of fixed point attractors and the storing of size 4 cyclic attractors. Size 2 cyclic attractors give the best results. A last observation relates to the paradoxical content addressability obtained when storing size 2 cyclic attractors using the insupervised procedure: storing 150 data gives a better content addressability than storing 50 data. In the insupervised procedure the content addressability has two causes: the explicit "excavation" of the basins of attraction during the supervised routines and the inherent process of trials, errors and adaptations. Excavating explicitly the basins of attractions becomes impossible for large data set in the current implementation. Because learning 150 data implies more trials than learning 50 data, bigger content addressability is obtained.

4.2 Symbolic Analysis and Spurious Attractors

In the preceding section, we showed that size 4 cyclic attractors and fixed point attractors lead to similar results in term of encoding capacity and content addressability. This suggests that they could be used similarly. We show here that this is a true assumption.

The existence of basins of attraction around the attractors provides content addressability: learned mappings can be retrieved in presence of noise. An important question is to know what lays behind the boundaries of these basins of attraction: what happens if the network fails to "understand" the noised stimulus. A major concern in associative memories resides in the reality that "other" states besides the memory states may also be attractors of the dynamics: unexpected attractors can appear in the output. The question is to know if these attractors should be considered as spurious data or not.

The very intuitive idea we propose here is that if the attractor is chaotic or if its period is different from the learned data, it is easy to recognize it at a glance, and thus to discard it. In contrast, things become more difficult if the observed attractor has similar period as the learned ones. In that case, it is in fact impossible to know if this information is relevant without comparing it with all the learned data. This leads us to define such an attractor – having the same period as the learned data, but still different from all of them – as a spurious attractor.

Figure 4 compares the proportion of the different attractors obtained for networks mapping stimuli to fixed point attractors and to spatio-temporal attractors of size 4 by using the insupervised procedure. These proportions are measured during spontaneous activity and when the network is fed with slightly noised stimuli (noise $< 10\%$). The following categories of attractors are proposed: the chaotic attractors,

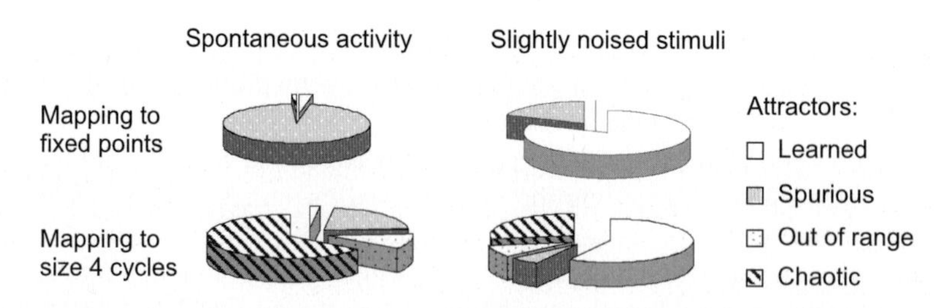

Fig. 4 Proportions of the different symbolic attractors observed in networks having learned 30 "in-supervised" mappings. Two types of results are displayed: first, the proportion of attractors obtained during spontaneous activity (i.e. for random stimuli). Second, the proportion of attractors obtained for stimuli only slightly noised (less than 10%). In both cases, networks' initial states were randomly initiated. 100 learned networks with each time 1000 configuration have been tested. Variance obtained in these measures are smaller than one percent

the attractors having a different period than the expected one (the "out of range" attractors), the learned attractors and finally the unlearned attractors having the same period as the learned one, i.e. the spurious attractors. More detailed and extensive analyses can be found in [42].

During spontaneous activity, when stimuli are mapped to fixed point attractors, it appears that the proportion of chaotic attractors and of "out of range" attractors falls rapidly to zero. In contrast, the number of spurious data increases drastically. In fact the learning procedure tends to stabilize the network by enforcing symmetric weights and positive auto-connections (which is the condition to define an energy function for the network). When stimuli are mapped to cyclic attractors the number of chaotic attractors becomes predominant[6].

When slightly noisy learned stimuli are presented to network initialized in a random state, interesting results appear. For fixed points trained networks, in approximatively 20% of the observations we are facing spurious data! By contrast, for period 4 trained networks, space left for spurious attractors becomes less than 5%! If the probability to recover the perfect cycle is only equal to 58%, the strong presence of chaos (28%) can help for recovering the expected attractor: in a procedure where the network's states are slightly modified in case a chaotic trajectory is encountered, until a cyclic attractor is found, we are nearly sure to obtain the correct mapping since the presence of spurious attractors is less than 5%. This is verified experimentally. This appears inline with the proposition that the presence of a low level of stochastic noise can result in "taming chaos" by stabilizing the itinerant trajectories [43, 44].

When learning mappings to cyclic attractors, the negligible presence of spurious data for slightly noised stimuli is contrasted by its non-negligible presence during spontaneous activity. We do believe that this presence could have a positive effect. Indeed, this reflects that the network memory is not saturated: far from the existing mappings, other potential attractors remain as good candidates for new memories.

5 Spontaneous Dynamics and Chaotic Dynamics

The preceding section has shown that chaotic dynamics becomes the network's background dynamics after learning cycles. Results obtained after learning predefined cycles using the outsupervised algorithm are similar [25]. If the presence of chaotic attractors prevents the proliferation of spurious data, it may have another important implication: by wandering across the phase space, it may help the network to converge to the originally mapped cycle. To have a better understanding of the type of chaos encountered, chaotic dynamics obtained after outsupervised learning and insupervised learning are compared here.

[6] In fact, we proposed this learning rule as a possible road to chaos: the more you learn, the more the spontaneous regime of the net tends to be chaotic [25].

5.1 Classical Analyses

Numerous techniques exist to characterize chaotic dynamics [45]. Here, chaotic dynamics are characterized by means of return maps, power spectra and by analyzing their Lyapunov spectrum. The computation of the Lyapunov spectrum is performed through Gram-Schmidt re-orthogonalization of the evolved system's Jacobian matrix (which is estimated at each time step from the system's equations), as detailed in [46].

In Fig. 5, it appears from return maps and power spectra analysis that networks using "out-supervised" learning to remember mappings show a presence of very uninformative "deep chaos" similar to white noise. This chaos can be related to "hyper-chaos" since the presence of more than one highly positive Lyapunov exponent appears [47]. In other words, having too many competing limit cycle attractors leads to unstructured dynamics.

By contrast, the learning of "in-supervised" mappings preserves more structure in the chaotic dynamics and leads to a very informative chaos we called "frustrated chaos" [48]. The return map from one single neuron indicates the high presence of saturation which is a characteristic of intermittency chaos [49]. The Lyapunov spectrum obtained shows that this chaos could be related to chaotic itinerancies [50]. In this type of chaos, the dynamics is attracted to learned memories – which is indicated by negative Lyapunov exponents – while in the same time it escapes from them – which is indicated by the presence of at least one positive Lyapunov exponent. However, this positive Lyapunov exponent is only slightly positive to preserve traces of the system's past history and of the learned memories. This regime of

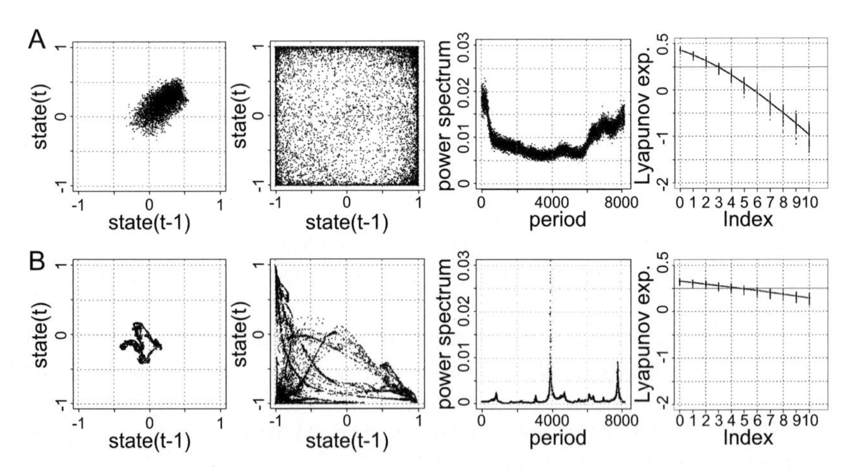

Fig. 5 Return maps of the network's mean signal (first figures from the left), return maps of a particular neuron (second figures), power spectra of the network's mean signal (third figures), and Lyapunov spectrum (last figures). (**A**) Chaotic regimes encountered in random and outsupervised learned networks is characteristics of hyperchaos. (**B**) Chaotic regimes encountered in insupervised learned networks is defined by the authors as "frustrated chaos". Modified with permission from Molter et al. (2007)

frustration is increased by some modes of neutral stability indicated by the presence of many exponents whose values are close to zero.

5.2 Frustrated Chaos

To have a better understanding of these two types of chaos, the following procedure is applied: after learning a data set, the external stimulus is slightly morphed from one learned stimulus to another one. It enables us to analyze the transition from one cyclic attractor to the other cyclic attractor. In case of fixed point attractors, one sharp transition is expected (in reality, many intermediate transitions are likely to occur due to the presence of spurious attractors). After learning cyclic attractors, chaotic dynamics is expected and effectively occurs as shown in Fig. 6. To analyze this chaotic dynamics, the probability of presence of the nearby limit cycle attractors occurring during small time windows is computed. In Fig. 6A, the network has learnt 4 stimuli to size 4 cyclic attractors by using the "out-supervised" Hebbian algorithm. This algorithm constrains strongly the network and, as a consequence, chaotic dynamics appear very uninformative: by morphing the external stimulus from one attractor to another one, a strong chaos shows up (indicated by the Lyapunov exponent) and any information concerning these two limit cycle attractors is lost.

In contrast, when mapping 4 stimuli to period 4 cycles, using the "in-supervised" algorithm (Fig. 6B), when morphing the external stimulus from one attractor to another one, the chaos encountered on the road appears much more structured: small Lyapunov exponents and strong presence of the nearby limit cycles are easy to observe.

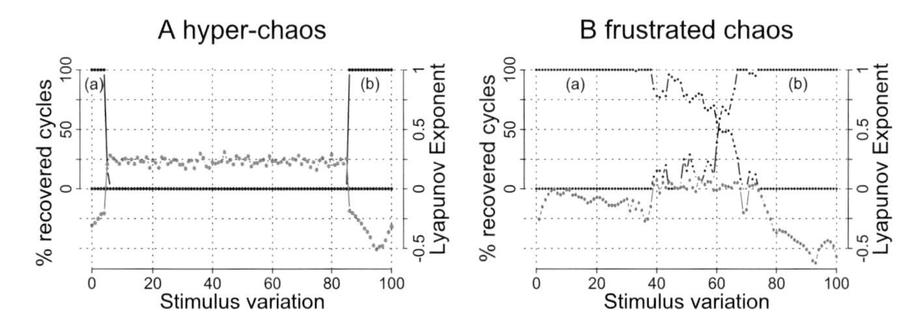

Fig. 6 Probability of presence of the nearby limit cycle attractors in a chaotic dynamics (*y-axis*). By slowly morphing the external stimulus from a stimulus previously learned (region (a)) to another stimulus learned (region(b)), the network's dynamics goes from the limit cycle attractor associated to the former stimulus to the limit cycle attractor associated to the latter stimulus. The probability of presence of cycle (*a*) and of cycle (*b*) are plotted (*black*). The Lyapunov exponent of the obtained dynamics is also plotted (*gray*). (**A**) After "out-supervised" learning, hyper-chaos is observed in-between the learned attractors. (**B**) Same conditions lead to frustrated chaos after "in-supervised" learning. Reprinted with permission from Molter et al. (2007)

The dynamical structure of the "frustrated chaos" reveals the existence of nearby competing attractors. It shows the phenomenon of frustration obtained when the network hesitates between two (or more) nearby cycles, passing from one to another.

When the network iterates through a spatio-temporal cycle, the various patterns define a sequence of activation for each neuron. For example, in a size 2 cycle, 4 possible sequences exist: $(-1, -1)$, $(-1, 1)$, $(1, -1)$ and $(1, 1)$. Accordingly, all neurons will lie in 4 different neural populations corresponding to the 4 possible activation sequences (one population is for example, all neurons having the same firing sequence $(-1, 1)$). More generally, each spatio-temporal cycle is characterized by distinct neural populations phase locked together. It is interesting to see how the transition evolves between two different combinations of neural populations when morphing the external stimulus from one learned stimulus to another one (in the same experiment as the one described above). In frustrated chaotic dynamics, we have demonstrated [51] that the neural populations obtained during short transient activity reflect simultaneously both nearby attractors. This tends to indicate that the frustration is not only temporal but also spatial.

6 Discussion

Since their discovery 25 years ago, fixed point attractors in recurrent neural networks have been suggested to be good candidates for the representation of content addressable memories. One reason lies in the possibility to establish mathematically their content addressability by showing the existence of Lyapunov functions. The other important reason lies in the neurophysiologically plausible Hebbian rule used to train these networks. Based on these considerations, it is demonstrated that limit cycle attractors are also good candidates for the representation of associative memories. First, if the content addressability of the obtained memories can not be shown mathematically, extensive computational experiments prove their existence. Then, the time asymmetric Hebbian learning rule used is not less biologically plausible.

Figure 7A shows the schematic view commonly used to represent content addressable memories in fixed point attractors. Memories are associated with the minima of an energy function and all states are attracted to one of the learned memories. However, storing information in fixed point attractors has a drawback: it tends to stabilize the entire network dynamics and it appears difficult to avoid the overwhelming presence of spurious attractors (Fig. 7B). As a consequence, the exploitation of these networks becomes very delicate: how to decide if an obtained fixed point attractor is a relevant information previously learned or if it is a spurious information. A memory of these memories has to be maintained for further comparisons! Learning data by relying on cyclic attractors brings more interesting dynamics in the network. Spurious attractors are no longer pervading the entire phase space. Instead, chaotic attractors appear as the background regime of the network (Fig. 7C). The ease to differentiate a chaotic trajectory from a cycle can prevent erroneous exploitation of these networks: when chaos appears, we directly know that the mapping has failed. Accordingly, giving meaning to cycles seems to be a better solution.

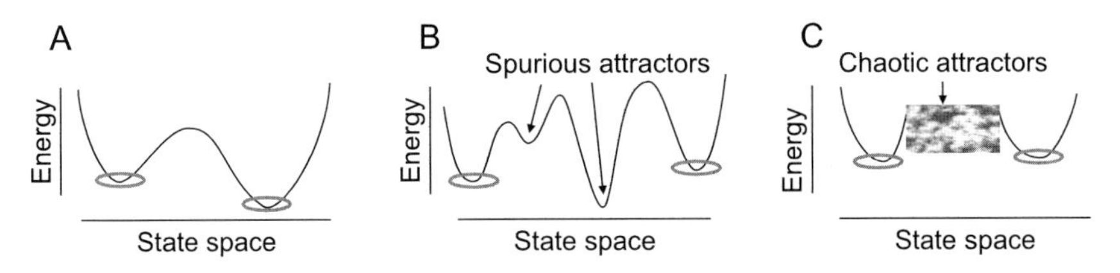

Fig. 7 Schematic picture of the learned attractors (highlighted by the oval) and their basins of attraction represented by an energy function deployed in the state space. (**A**) The ideal case: the state space is divided by the basins of attractions of the learned attractors. (**B**) The more realistic picture obtained after learning fixed point attractors: a strong presence of spurious attractors appears. (**C**) After learning cyclic attractors (still characterized by basins of attractions): a strong presence of chaotic dynamics prevails. No energy function can be defined in the chaotic regions

To further investigate the kind of chaotic dynamics appearing when learning spatio-temporal cyclic attractors in a network, two diametrically opposite versions of cyclic data are proposed. First, stimuli are mapped to predefined cycles. This defines an "out-supervised" learning algorithm. In the second version, the cyclic attractors to be associated with the external stimuli are left unprescribed before learning and are derived from the ones spontaneously proposed by the network. This defines an "in-supervised" learning algorithm. As expected, our preliminary results (Table 1) confirm the fact that by not constraining the network, "in-supervised" learning gives very good encoding performances compared to the "out-supervised" case (and compared to the learning of fixed point attractors).

When looking to the spontaneous dynamics (the dynamics observed when unlearned stimuli are presented to the network). It has been shown elsewhere [25] that the more information the network has to store in its attractors the more its spontaneous dynamical regime tends to be chaotic. Chaos is in fact the biggest pool of potential cyclic attractors. Figure 5 shows that when adopting the "out-supervised" learning chaos adopts a very unstructured shape similar to white noise. The reason lies in the constraints imposed by this learning process: the number of mappings the network must learn is equal to the number of cycles times their size. In contrast, "in-supervised" learning, by being more "respectful" of the network intrinsic dynamics, maintains much more structure in the obtained chaos. Figure 6 shows that it is still possible to observe in this chaotic regime the traces of the learned attractors.

This chaos characterized by the presence of strong cyclic components among which the dynamics randomly itinerates is very similar to a form of chaos well-known in the literature: the chaotic itinerancy [50]. However, a fundamental difference lies in the transparency and the exploitation of those cycles. Here, this chaos appears as a consequence of the learning procedure. By forcing cycles in the network (by tuning the connection parameters), in some configuration, the network hesitates between learned cycles leading to mutually competing attractors and unpredictable itinerancies among brief appearance of these attractors. This complex but still very structured and informative regime is called "frustrated chaos". This behavior can be related to experimental findings were ongoing cortical activity has been shown to encompass a set of dynamically switching cortical states which corresponds to stimulus-evoked activity [13].

From Figs. 4 and 7, we can say that the boundaries of the basin of attraction of learned cycles are mostly chaotic. Moreover, we have shown that this chaos appears very structured and representative of the nearby learned attractors after "in-supervised" learning. As a consequence, by slightly noising this chaotic dynamics, we have strong chance to recover the expected attractor. This noise enhanced retrieval property, is in line with the stochastic resonance phenomenon, well known in nonlinear dynamical systems [52]. This chaotic dynamics can be seen as a powerful mechanism for the exploration of the state space, more "intelligent" that random search since its trajectory itinerates nearby meaningful zones.

We demonstrated that dynamics are highly dependent of the information we want to learn. Fixed point attractors tend to stabilize all dynamics. Forcing predefined cycles with the "out-supervised" learning appears very difficult, not efficient and chaotic dynamics similar to white noise spreads the entire phase space. Better results are obtained with the "in-supervised" learning. This gives sense to a biological and a cognitive perspective: biological systems propose their own way to treat external impact by slightly perturbing their inner working. It emphasizes how important it is to define appropriately how to represent and to encode the information. However, the current version of the "in-supervised" algorithm appears difficult to interpret and to "engineerize". To transform this algorithm in a powerful engineering tool, we believe that the encoding procedure should be improved according to more plausible biological evidences.

References

1. S. Amari. Learning pattern sequences by self-organizing nets of threshold elements. *IEEE Transactions on computers*, 21:1197, 1972.
2. J.J. Hopfield. Neural networks and physical systems with emergent collective computational abilities. *Proceedings of the National Academy of Sciences USA*, 79:2554–2558, April 1982.
3. S. Grossberg. *Neural Networks and Natural Intelligence*. MIT Press, Cambridge, 1992.
4. D.J. Amit, G. Gutfreund, and H. Sompolinsky. Statistical mechanics of neural networks near saturation. *Ann. Phys.*, 173:30–67, 1987.
5. E. Gardner. Maximum storage capacity in neural networks. *Europhysics Letters*, 4:481–485, 1987.
6. E. Gardner and B. Derrida. Three unfinished works on the optimal storage capacity of networks. *J. Physics A: Math. Gen.*, 22:1983–1994, 1989.
7. D.J. Amit and S. Fusi. Learning in neural networks with material synapses. *Neural Computation*, 6:957–982, 1994.
8. E. Domany, J.L. van Hemmen, and K. Schulten, editors. *Models of Neural Networks*, Vol. 1. Springer, 2nd edition, 1995.
9. J. Nicolis and Tsuda. Chaotic dynamics of information processing. the 'magic number seven plus-minus two' revisited. *Bulletin of Mathematical Biology*, 47:343–365, 1985.
10. C.A. Skarda and W. Freeman. How brains make chaos in order to make sense of the world. *Behavioral and Brain Sciences*, 10:161–195, 1987.
11. A. Babloyantz and Lourenço. Computation with chaos: A paradigm for cortical activity. *Proceedings of National Academy of Sciences*, 91:9027–9031, 1994.
12. E. Rodriguez, N. George, J.P. Lachaux, B. Renault, J. Martinerie, B. Reunault, and F.J. Varela. Perception's shadow: long-distance synchronization of human brain activity. *Nature*, 397:

430–433, 1999.
13. T. Kenet, D. Bibitchkov, M. Tsodyks, A. Grinvald, and A. Arieli. Spontaneously emerging cortical representations of visual attributes. *Nature*, 425:954–956, 2003.
14. T.V. Bliss and T. Lomo. Long-lasting potentiation of synaptic transmission in the dentate area of the anaesthetized rabbit following stimulation of the perforant path. *Journal of Physiology*, 232:331–356, 1973.
15. D. Hebb. *The organization of behavior*. Wiley-Interscience, New York, 1949.
16. S. Amari. Neural theory of association and concept-formation. *Biological Cybernetics*, 26:175–185, 1977.
17. S. Amari and K. Maginu. Statistical neurodynamics of associative memory. *Neural Networks*, 1:63–73, 1988.
18. D.J. Amit. The hebbian paradigm reintegrated: local reverberations as internal representations. *Behavioral Brain Science*, 18:617–657, 1995.
19. N. Brunel, F. Carusi, and S. Fusi. Slow stochastic hebbian learning of classes of stimuli in a recurrent neural network. *Network: Computation in Neural Systems*, 9:123–152, 1997.
20. S. Fusi. Hebbian spike-driven synaptic plasticity for learning patterns of mean firing rates. *Biological Cybernetics*, 87:459–470, 2002.
21. D.J. Amit and G. Mongillo. Spike-driven synaptic dynamics generating working memory states. *Neural Computation*, 15:565–596, 2003.
22. W.B. Levy and O. Steward. Temporal contiguity requirements for long term associative potentiation/depression in the hippocampus. *Neuroscience*, 8:791–797, 1983.
23. G. Bi and M. Poo. Distributed synaptic modification in neural networks induced by patterned stimulation. *Nature*, 401:792–796, 1999.
24. A. Waibel. Modular construction of time-delay neural networks for speech recognition. *Neural Computation*, 1(1):39–46, 1989.
25. C. Molter, U. Salihoglu, and H. Bersini. The road to chaos by time asymmetric hebbian learning in recurrent neural networks. *Neural Computation*, 19(1):100, 2007.
26. G. Buzsaki. *Rhythms of the Brain*. Oxford University Press, USA, 2006.
27. W. Singer. Neuronal synchrony: A versatile code for the definition of relations? *Neuron*, 24:49–65, 1999.
28. D.J. Amit and N. Brunel. Learning internal representations in an attractor neural network with analogue neurons. *Network: Computation in Neural Systems*, 6:359–388, 1994.
29. F. Varela, E. Thompson, and E. Rosch. *The Embodied Mind: Cognitive Science and Human Experience*. MIT Press, 1991.
30. P. Erdi. The brain as a hermeneutic device. *Biosystems*, 38:179–189, 1996.
31. I. Tsuda. Towards an interpretation of dynamic neural activity in terms of chaotic dynamical systems. *Behavioral and Brain Sciences*, 24(5), 2001.
32. T. Kohonen. Self-organized formation of topologically correct feature maps. *Biological Cybernetics*, 43:59–69, 1982.
33. J.E. Lisman and M.A.P. Idiart. Storage of 7 ± 2 short-term memories in oscillatory subcycles. *Science*, 267:1512–1515, 1995.
34. A. Bragin, G. Jando, Z. Nadasdyand J. Hetke, K. Wise, and G. Buzsáki. Gamma (40-100 hz) oscillation in the hippocampus of the behaving rat. *The Journal of Neuroscience*, 15:47–60, 1995.
35. G.A. Miller. The magical number seven, plus minus two: Some limits on our capacity for processing information. *Psychol. Rev.*, 63:81–97, 1956.
36. E. Dauce, M. Quoy, B. Cessac, B. Doyon, and M. Samuelides. Self-organization and dynamics reduction in recurrent networks: stimulus presentation and learning. *Neural Networks*, 11: 521–533, 1998.
37. C. Molter, U. Salihoglu, and H. Bersini. Learning cycles brings chaos in continuous hopfield networks. *Proceedings of the conference, Montreal*, 2005.
38. B.M. Forrest and D.J. Wallace. *Models of Neural NetWorks*, Vol. 1, chapter Storage Capacity abd Learning in Ising-Spin Neural Networks, pp. 129–156. Springer, 2nd edition, 1995.
39. C. Molter and H. Bersini. How chaos in small hopfield networks makes sense of the world.

Proceedings of the IJCNN conference, Portland, 2003.
40. C. Molter, U. Salihoglu, and H. Bersini. Introduction of an hebbian unsupervised learning algorithm to boost the encoding capacity of hopfield networks. *Proceedings of the IJCNN conference, Montreal*, 2005.
41. J. Piaget. *The Psychology of Intelligence*. Routledge, New York, 1963.
42. C. Molter, U. Salihoglu, and H. Bersini. How to prevent spurious data in a chaotic brain. In IEEE Press, editor, *International Joint Conference on Neural Networks (IJCNN 2006)/. Proc WCCI*, pp. 1365–1371, 2006.
43. W. J. Freeman, H.-J. Chang, B. C. Burke, P. A. Rose, and J. Badler. Taming chaos: stabilization of aperiodic attractors by noise. *IEEE Trans. on Circuits and Systems — 1*, 44(10):989–996, 1997.
44. Robert Kozma. On the constructive role of noise in stabilizing itinerant trajectories. *Chaos, Special Issue on Chaotic Itinerancy*, 13(3):1078–1090, 2003.
45. J.P. Eckmann and D. Ruelle. Ergodic theory of chaos and strange attractors. *Reviews of modern physics*, 57(3):617–656, July 1985.
46. A. Wolf, J.B. Swift, H.Swinney, and J.A. Vastano. Determining lyapunov exponents from a time series. *Physica*, (D16):285–317, 1984.
47. O. E. Rössler. The chaotic hierarchy. *Zeitschrift für Naturforschung*, 38a:788–801, 1983.
48. H. Bersini. The frustrated and Compositionnal Nature of Chaos in Small Hopfield Networks. *Neural Networks 11*, pp. 1017–1025, 1998.
49. Y. Pomeau and P. Manneville. Intermittent transitions to turbulence in dissipative dynamical systems. *Comm. Math. Phys.*, 74:189–197, 1980.
50. K. Kaneko and I. Tsuda. Chaotic itinerancy. *Chaos: Focus Issue on Chaotic Itinerancy*, 13(3):926–936, 2003.
51. C. Molter, U. Salihoglu, and H. Bersini. Phase synchronization and chaotic dynamics in hebbian learned artificial recurrent neural networks. *CNS Workshop: Nonlinear spatio-temporal neural dynamics - Experiments and Theoretical Models*, 2005.
52. D. F. Russel, L. A. Wilkens, and F. Moss. Use of behavioural stochastic resonance by paddle fish for feeding. *Nature*, 402:291–294, 1999.

Complex Biological Memory Conceptualized as an Abstract Communication System–Human Long Term Memories Grow in Complexity during Sleep and Undergo Selection while Awake

Bruce G. Charlton and Peter Andras

Abstract Biological memory in humans and other animals with a central nervous system is often extremely complex in its organization and functioning. A description of memory from the perspective of complex systems may therefore be useful to interpret and understand existing neurobiological data and to plan future research. We define systems in terms of communications. A system does not include the communication units ('CUs') that produce and receive communications. A dense cluster of inter-referencing communications surrounded by rare set of communications constitutes a communication system. Memory systems are based on communication units that are more temporally stable than the CUs of the system which is using the memory system. We propose that the long term memory (LTM) system is a very large potential set of neurons among which self-reproducing communication networks (i.e. individual memories) may be established, propagate and grow. Long term memories consist of networks of self-reproducing communications between the neurons of the LTM. Neurons constitute the main communication units in the system, but neurons are not part of the abstract system of memory.

1 Introduction

Since neurons tend to be lost from memory systems by entropic mechanisms, there is a necessary tendency for all potentially-sustainable memories to grow by recruitment of new neuron communication units to form larger (more complex) networks of communications. Such growth of memories may occur by mechanisms such as expansion and combination of already existing memories, leading to some of the familiar distortions of memory such as confabulation and generation of standard scenarios.

Memory systems are therefore conceptualized as systems that spontaneously grow by recruitment of neurons to participate in expanding networks of communications. We suggest that growth of memories occurs mainly (although not entirely)

L. I. Perlovsky, R. Kozma, *Neurodynamics of Cognition and Consciousness.*

during sleep, and memories are subject to selection mainly while the organism is awake.

Selection of memory systems occurs by interaction with other brain systems. Memories systems that lead to further communications that are not contradicted by 'experience' of neural communications during waking behaviour may persist and continue to increase in complexity–these are provisionally assumed to be correct memories. Memories that create contradictions with other communications within the LTM system, or with communications from other neural systems, do not lead to further communications within the long term memory system. They are regarded as informational 'errors' and eliminated from the long term memory system by their failure to propagate. (Note that contradictions are qualified as such by the impossiblity of further use of them within the memory system, according to the rules of this system. There is no particular observer component or mechanism within the system that decides that some memory creates a contradiction.)

The adaptiveness of memories is constrained in the first place by the nature of the memory system, which has been shaped by the organism's evolutionary history; and in the second place by the selective pressure of the organism's continued experience interacting with the self-reproduction of memory scenarios.

The 'memory function' of a complex biological memory system represents a small proportion of the possessing of the memory system since differentially much greater amounts of internal processing are intrinsic to the existence, maintenance and growth of biological memory. In other words, most of what the biological memory system does is not related to the 'memory function' observable from outside, but it is rather internal processing that is usually inaccessible for an external observer. Internal processing in the human long term memory system probably occurs mainly during sleep. During sleep memory systems are more-or-less cut-off from communications with the rest of the organism and its environment.

The primary function of sleep is therefore to maintain and increase the complexity of the long term memory system, which includes combination and harmonization of memories. In a paradoxical sense, the LTM system exists mainly to sleep, and its memory function is merely the 'rent' that the LTM system pays to the organism in order that the organism will allow the LTM system's continued existence. This conceptualization may help explain the indirect and imprecise association between sleep and LTM function in humans, since the memory function is a secondary and subordinate attribute of the LTM system.

What follows is not intended as a contribution to the neurobiology of memory. Rather this account is an abstract reconceptualization of the nature of memory–a framework intended to lead to an improved interpretation and understanding of the neurobiology of memory and to guide future scientific investigations. The novelty of this description arises from its reversal of the usual description of memory as being formed while awake and consolidated while asleep. By contrast, we propose that memories are systems of communications which grow during sleep and are selected by interaction with other brain communications when awake. In contrast to traditional 'instructionist' ideas of learning and memory–which see the environment as instructing the mind–our theory is more akin to 'selectionist' accounts of neurobiology such as those provided by Edelmann [9] and Gazzaniga [11]. Like these authors we regard memories as a consequence of the generation of diversity and selection among variants. But we also believe that only systems can undergo selection, and not communication units such as neurons; that therefore systems are primary and selection is secondary; and that long term memories are abstract systems of communications between neurons.

First we introduce our conceptualisation of communication systems. This is followed by the interpretation of LTM in this context. Finally we discuss the role of sleep in the LTM.

2 Communications and Abstract Communication Systems

There are numerous version of systems theory, and conceptualizations of complexity. To distinguish the version deployed here we use the term 'abstract communication systems', The main source of the theory is the work of Luhmann [18] as modified by our earlier works [2, 3, 6, 7].

We take it as axiomatic that the world consists of systems and their environment. From the perspective of a specific system there is only itself (the system) and the environment–and all knowledge is knowledge within systems. The environment beyond the system is only inferred indirectly, as a system's 'explanation' of why the system is not perfectly predictive. The system only knows that it does not function perfectly, i.e. that it does not know everything about itself, and therefore infers that there is a world outside itself. A system can model what happens in this environment, it can be aware that its models of the environment have not (yet) been contradicted by experience, but the system does not know anything directly concerning the environment (i.e. all its knwoledge about its environment is derived from the problems of its predictions about itself, without being directly derived from the environment). For example, this implies that a 'memory system' is primarily a system defined by a specific processing 'logic' and only secondarily functions to provide memories. The 'memories' within a memory system should therefore be conceptualized as elements of the memory system's model of its environment. (This perspective of memory as a conjectural 'model' of its environment is in line with the concept of autopoesis described by Maturana and Varela [21].)

The critical conceptual breakthrough deriving from Luhmann is that systems are defined in terms of communications, and therefore that systems exclude the communication units ('CUs') which produce and receive communications (see a representation of an abstract communiction system in Fig. 1). Biological systems such as memory therefore do not include the physical brain communication units (such as nerve cells), and social systems such as the economy, politics and the law do not include the human beings who work in them. Nerve cells and human beings are, in this context, communication units–but are not themselves communications, hence cannot be part of the systems under consideration. Other CUs may be non-living such as books and computer disks.

Communications are sequences of symbols communicated between communication units. Abstract communication systems are made by such communications

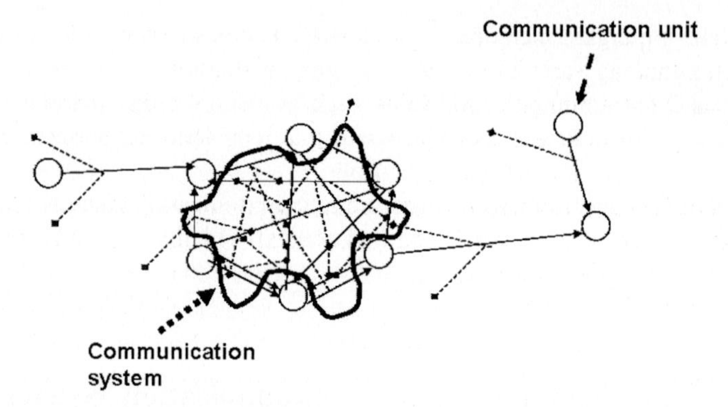

Fig. 1 Schematic views of simple behavior on a given set of possible actions

between communication units. (To count as a communication, a signal must be generated, transmitted and received.) The communication units are not part of the system, since they are not themselves communications but instead transmit and receive communications. CUs may be inert–books, computer disks and DNA molecules do not 'generate' and 'receive' communications exactly, but are structurally altered by communications in ways that enable these alterations to affect subsequent communications, and this is a more precise definition of CU.

Communications 'reference' other communications, in the sense that the sequence of symbols contained in a communication is dependent on the contents of other earlier or simultaneous communications and thereby refer to them. A dense cluster of inter-referencing communications surrounded by rare set of communications constitutes a communication system [2] (dens cluster of communications means a set of communications that are frequently referenced by other communications within the cluster, and which also reference other communications within the cluster, rare set of communications means that these communications are rarely referenced by communications within the dense cluster and they rarely reference communications from the dense cluster–see Fig. 1 for a graphical representation). In quantifiable terms it may be said that a system is a 'significantly' dense concentration of inter-referenced communications which persists over a 'significant' timescale–in which the cut-off levels of significance define the probability that there is indeed a system [6].

A communication system is defined by the regularities that specify how referenced communications determine the content of a referencing communication. In other words, each system has a specific 'logic' by which communications are evaluated, and systems have a characteristic mode of processing. All communications that follow the set of rules defining the system are included as part of the system. Other communications that do not follow the rules of the system are part of the system's environment.

A system needs to be self-reproducing in order to maintain its characterizing density of communications over time, and this self-reproduction generates surplus communications and the tendency for expansion of the system by inclusion of communications produced by more communication units contributing to the system or expansion of communications from the units already contributing–this tendency for growth of systems generates the basis of competition between systems. The self-reproduction also randomly generates variations by entropic mechanisms, which will eventually (because of the competition created by system expansion) be subject to selection pressures.

For example, the system of computer science contains all communications which reference earlier scientific communications from the domain of computer science and which follow the rules of these scientific communications (e.g., allowing the possibility of falsification, using logical reasoning, discussing admissible topics using admissible arguments etc.). A large part of these computer science communications are derived from scientific papers, which explicitly reference other scientific papers, and use the conclusions of earlier papers as premises of the logical reasoning presented in the paper. According to systems theory, the human computer scientists are not part of the system of computer science, nor are the physical objects that are scientific papers. Only the dynamic scientific communications about computer science topics are part of this system. In order that computer science continue as a dense communication cluster over time, it needs to continually generate a surplus of communications, so the system has a tendency to expand. Since communication units have a finite capacity, the expansion (sooner or later) leads to competition and selection among variant systems of communications.

3 Growth of Systems

From each system's perspective 'the world' is constituted by binary division between itself (the system) and its own environment (not-the-system)–and there are as many such 'worlds' as there are systems. The same communication will have different meanings (i.e. be a different communication) in different systems, or be included in one system but not another. The set of regularities of referencing constitutes an abstract grammar, which defines an abstract language, characteristic of the system. For example, the sciences of economics and medicine have different specialist languages, and scientific communications belong to one of these sciences according to whether they follow the rules of the specific language. (Note that a scientific communication may not be for example exclusively about economics, but it will be part of the science of economics in terms of its economics-relevant aspects. In other words, when a later economics paper reference this communication, the reference will relate to the economics aspects of the communication, and not to other aspects that are irrelavant int he context of the science of economics.)

Communication systems reproduce themselves by recruiting new communications, which follow the referencing rules of the system. This often occurs by the recruitment of new communications units to contribute to the system. For example, in the system of computer science (or any other science) many of the communications units are the scientists (human beings), and one of the ways the system grows is by increasing the numbers of computer scientists, or by increasing the proportion of time the scientists spend on communication of computer science information [14].

Also, the system grows by increasing the frequency of communications between the scientists, and this may involve the inclusion of other types of communication units such as scientific journals.

How successful is the recruitment of new communications, depends on earlier communications generated by the system and on the match between the system and its environment. We can view the system as a self-describing system made of communications, which at the same time describes its environment in a complementary sense. More complex and potentially more-adaptive descriptions of the systems environment may lead to greater success in recruiting new communications and more rapid reproduction and expansion of the system. Memories constitute (possibly compressed) descriptions of sets of earlier system communications and in a complementary sense descriptions of aspects of the system's environment. Memories potentially may increase the adaptedness of a system by increasing its complexity, and therefore the potential closeness of 'match' between the system's model of the environment and the (infinitely complex) environment itself.

The system communications are about the system itself. System communications reference other system communications in order to prove that these communications are part of the system (i.e., that they are correct according to the rules of the system). If the communications lead to continuation of further communications, the process of proving that they are correct continues. If the system is able to continue to exist, i.e. to generate/recruit new communications according to the rules of the system, then this continuation implies that the proving process of the correctness of earlier communications continues.

In general it is not possible to prove the correctness of system communications; it is possible to prove only the incorrectness of them, when there is not further continuation of communications rooted from the original communication. We term this the 'Popper Principle', i.e., that only the falsity of system communication can be proven by stopping the generation of communications rooted from the communication in question [2].

4 Selection of Systems

Systems must grow if they are to be sustained, since there is a tendency for systems to decline due to entropic loss of communications units. For example, the system of computer science would become extinct from loss of communication units as individual scientists aged and died unless new CUs were recruited (e.g. more scientists, more publications, more professional journals etc.). Therefore all viable systems have the capacity of self-reproduction–the complexity of their communications will tend to grow. Specific memories tend to disappear, for example in the case of human memories this is due to random entropic damage to neurons [16], hence memory systems must grow to ensure their own survival.

But since there are many systems all with the tendency for growth, this expansion eventually will lead to competition between systems. Systems compete for finite communications, and this may be manifested as competition relating to communication units. Communication units tend to generate communications for several systems; but communications in one system may compete with those in other systems. For example human computer scientists never expend their whole time and energy purely on computer science communications–they will also participate in many other social systems such as politics, the legal system, the mass media and the family [23]. There will be competition between social systems for participation in the system communications. The 'work versus family' dilemma is just one aspect of this kind of system competition.

Competition between systems leads to selection. There are many types of selection–such as natural selection and market economics–but all share essential formal properties [15]. One important consideration is that only systems are selected–since only systems have the property of self reproduction and growth [6]. For example, if a mountain is eroded such that soluble limestone is dissolved but resistant granite is left standing then this is not an example of selection since the granite is not capable of growth. Likewise, selection does not act upon DNA since DNA is–of itself–not capable of self-reproduction. Rather, the relevant unit of biological selection is actually the genetic system which includes DNA and all the other communication elements necessary for its reproduction–the system consisting of interactions between DNA, RNA, protein and other molecular types [1].

Selection of memories occurs by interaction with other memories within the long term memory system, and also with other brain systems. Individual LTM neurons will typically participate in 'coding' more than one memory, and some LTM neurons will also participate in other neural systems. For example, a cortical neuron may participate in several memories (i.e. networks of communication) relating to an individual person, and also in the awake processing relating to visual perception [16]. Some of the networks of communication will be compatible and may be combined and grow to generate more complex systems of communications by including more neurons into the network. Other memories will conflict such that they cannot be combined and cannot grow in complexity–these systems are more likely to become extinct.

It is plausible that memory networks will tend to combine and grow in complexity mostly during sleep, when internal processing of the LTM can proceed without interaction with perceptual information. During waking, sensory perceptual and motor communications exert a selection pressure on the long term memory system via competition for communications at the level of neurons. This is especially the case for human vision, which generates an extremely heavy computational load and involves a high proportion of cortical neurons including those used in long term memory systems [16]. The assumption is that memories which are incompatible with ongoing visual communications during waking hours will not be reinforced, and may be suppressed; for example, if visual memories conflict with current visual information then the memory will not be able to expand by recruitment of more communication units.

Memories are subject to continual selection and reshaping by the organism's ongoing waking experience, so that memories will tend to evolve over time. Most memories will become extinct, and those which are not contradicted by experience will continue to increase in complexity (mainly during sleep) until such a point that they do lead to contradiction after which the erroneous memories will be pruned-back. This process can be seen as one in which informational errors are generated, identified and eliminated.

5 Information Errors in Communication Systems

Information errors are problems that are encountered by systems which are due to the limitations of the system, even when the system is working properly [2]. Since the environment is infinitely complex, any system's modelling of the environment will be highly simplified, and contingencies may arise in which the system behaves (relatively) maladaptively. All systems necessarily have highly simplified models of the environment and the environment is more complex than the system. Therefore 'incorrect' descriptions of the system's environment are inevitable and all systems are prone to information errors.

Information errors of communication systems are therefore cases of system maladaptiveness where communications happen according to the rules of the system, but they cannot lead to continuation because of environmental constraints. From the internal perspective of the system, communication units that are expected to produce continuations of communication do not in fact do this. For instance, a 'perfectly functioning' factory may be producing fully functional drinking glasses according to proper procedure is nonetheless running at a loss and is in danger of bankruptcy. The implication is that when a system is working according to its rules and is nonetheless contracting, then there is something wrong with the system's description of its environment such that relevant aspects are not being modelled. In this case perhaps the drinking glasses are not being delivered to shops (but deliveries are not being monitored by the system) or nobody is buying the drinking glasses (but this is not known because sales are not being monitored).

System information errors are therefore signs of a mismatch between the system's description of the environment, and the actual environment. Mismatch errors imply that some of the rules defining the system are damagingly wrong (i.e., they do not fit the environment well enough to permit the continuation of the system).

We suggest that memories are selected largely in terms of whether or not they generate information errors. By the Popper Principle, memories which are leading to continued expansion of the LTM continuations are regarded as provisionally 'true', 'correct' or 'accurate'–for as long as the system continues to expand. Memories that do not lead to continued communications are regarded as 'false', 'incorrect' or 'inaccurate', and are–in effect–purged from the system. This purging of memory may occur passively simply by failure to propagate. But in addition it is likely that in a complex system such as LTM there are mechanisms for 'checking' communications, and for tracing information errors back to their originating root 'false assumption' and eliminating the branching consequences of that assumption [2].

The primary mechanism for checking memories is internal checking for consistency within the LTM. Emerging memories will grow more rapidly if they are compatible with already existing memories, because such memories can join-up to form what are sometimes termed memory 'schemata'. Presumably, at the level of communication units, the neuron network constituting one memory can increase their communications with the neurons of another memory network to expand the number of communication units in the system hence the complexity of communications in the system. By contrast, memories that are incompatible cannot join-up with existing memories, presumably because they differ in their 'semantics', and so will constitute smaller and less complex systems which are more likely to become extinct as a natural consequence of entropic events.

By such mechanisms, memories in LTM tend over time to become combined and semantically harmonized in complex, expanding, non-contradictory networks.

6 Memory Subsystems are Based on Communication Units with Longer-Lasting Communications

As described above, systems that reproduce and expand faster than other systems may drive to extinction the slower reproducing and expanding systems. The evolution of memory subsystems may play a significant role in this process–indeed some kind of memory function is probably necessary for systems to expand beyond a certain degree of complexity.

The limits of system expansion are determined by the probabilistic nature of referencing rules. A communication may reference several earlier communications indirectly through other referenced communications constituting referencing sequences of communications. The indeterminacies of referencing rules determine how long such referencing sequences of communications can be before the later communications become a random continuation.

Longer referencing sequences of communications (i.e., more detailed descriptions) allow better, more complex descriptions of the systems and its environment. In principle, the more complex the system the greater its adaptive potential. However, in practice the optimal size of the system (i.e., the number of simultaneous communications being part of the system) is also determined and constrained by the indeterminacies of referencing rules. Systems that overgrow their maximal sustainable size may split to form two or more similar but distinct systems (the maximal sustainable size depends on the system). When the system splits into two systems, each of the two systems will form part of the environment for the other, and the frequency of referencing communications from the other system will be reduced.

Communication systems may develop subsystems that are systems within the system, i.e., they constitute a denser inter-referencing cluster within the dense communication cluster of the system (the subsystem emerges if within the subsystem there are significantly more frequent references to communications of the subsystem than to communications within the system, but not within the subsystem). Communications that are part of subsystems follow system rules with additional constraints that are characteristic of the subsystem. For example there are overall rules of human brain information processing, but this is also sub-divided into specialized functional systems dealing with sensory perceptions, movement etc. and these systems have distinctive further constraints on their information 'inputs' and 'outputs' and processes. More constrained referencing rules decrease indeterminacies and allow the system to generate better complementary descriptions of the environment and expand itself faster than systems without subsystems.

Another way of extending reliable descriptions of the environment (i.e., non-random sequences of referencing communications) is by retaining records of earlier communications, i.e., by having memories of earlier communications that can be referenced by later communications. Memory systems are therefore subsystems with particular formal properties to do with their relationship with the primary system to which they are a subsystem.

Memory systems depend upon the existence of new communication units that can produce longer-lasting communications (or recruitment of existing communication units able to produce longer-lasting communications) that potentially produce for a certain period a certain communication (i.e. they produce the same communication repeatedly, with very small likelihood of errors in the reproduction of this communication) that can be referenced in place of some other communications (i.e., the ones which are represented by the memory). Having memory subsystems with CUs with longer-lasting communications reduce the indeterminacies in referencing by allowing direct referencing of earlier communications, instead of referencing early communications indirectly through a chain of references.

Traditional computer memory is based on CUs with longer-lasting communications (e.g. in magnetic changes to tapes or disks, or in binary codes etched onto CD or DVD). This is essentially a form of 'storage' rather than a true memory system, since the communication units in computer memory do not communicate significantly among themselves, so there is no 'system' of communications. The 'memory' communication units are inert except when the communication is being encoded or recalled by the primary system.

But in biological memory, the communication units with longer-lasting communications (neurons in the long term memory) communicate among themselves, and (presumably) do so to such an extent that there are more communications among and between the communication units than there are communications between the units and their environment. In other words, human LTM is a true system, defined as a dense inter-referencing cluster of communications. In evolutionary terms, our assumption is that memory systems began to evolve and differentiate from the primary system of the CNS when communication units with longer-lasting communications began to communicate among themselves with communications referenced-to (e.g. caused-by) other internal communications. A memory system can be defined as forming at that point where these communications between such communication units quantitatively exceeded those between these CUs and their CNS environment.

7 The Nature of Long Term Memory in Humans

The main requirement for LTM is among complex animals living in complex and changing environments–in which each day generates different challenges and in which animals benefit from memories of their previous experiences [16]. In such animals (including humans) LTM often has vast capacity, and therefore necessarily vast complexity. (See also the discussion in this book in the cahpter by Perlovsky [22].

Human LTM comprises a very large potential system of communications in which neurons are the main communication units. Individual memories are assumed to be communication subsystems comprising smaller numbers of neurons which are densely intercommunicating. These individual memories can be conceptualized as 'modelling' specific environmental aspects in order to enhance the adaptiveness of the LTM system in its context within the larger communication system of the brain.

Long term memory is the memory system that is used directly to 'refer to' previous states of the organism days, years or even decades ago (in systems theory, 'referring to' previous organism states carries the implication that previous states may affect present organism states by direct communication, rather than having been the remote and indirect cause of present states). LTM–like all memory systems–therefore requires communication units that are relatively more stable over these time periods than the CUs of a system which is using LTM for its memory function, retaining information relatively unchanged. Since neurons, and their synaptic connectivity, are dynamic structures over this timescale–this implies a need for mechanism for the maintenance of information [16]. A such mechanism may be represented by repetitive replay of interaction patterns between neurons participating int he LTM [17, 27].

But complex memory requires not only more-stable CUs, but also dense communications between these more-stable CUs. This implies that internal processing within LTM is relatively more complex than the exchange of information between LTM and its environment (as measured by an external observer). This primacy of internal communications reverses the usual conceptualization of memory systems. Long term memory in humans is usually conceptualized as being formed while the organism is awake, and consolidated and edited during sleep. The adaptiveness of memories (i.e. their tendency to enhance reproductive success) is assumed to arise from their being a sufficiently-accurate representation of the environment–as if the environment 'imposed' the memories on the structure of the brain. In other words, the environment 'instructs' the brain, and memory is a 'representation' of the environment [11].

By contrast, we propose that complex memories are autonomously formed by the long term memory system mainly (although not entirely) during sleep and these memories are selected by interaction with the environment (mainly) while the organism is awake. In a nutshell, the LTM system generates a superfluity of 'conjectural' memory variants during sleep, and the interaction of the LTM system with the rest of the brain culls most of these memory variants, to leave those memories that are most adaptive in terms of enabling the LTM system to survive and thrive in the context of the brain system which is its environment. During sleep the LTM system provides multiple 'guesses' concerning the environment, and only those guesses will survive and grow which are compatible with perceptual data generated by behaviour when awake. (Note that what is 'most adaptive' is determined by the context in the sense of enabling the LTM to survive, and there is no particular inside observer component of the brain or specific context-independent criterion that is applied by the brain to decide which are the most adaptive memories.)

This selection process operates because some memories continue to lead to further communications so that these memories expand in complexity, while memories that do not lead to further communications do not expand, and will tend to be eliminated from the LTM system because they contain 'information errors'.

8 The Importance of Sleep to the LTM System

To recapitulate, since human LTM is a highly complex system it follows that there must be a differentially much larger amount of internal communication between the neuron CUs in the LTM system, than between neurons in the LTM system and the rest of the brain.

The requirement for LTM to engage in substantial internal communications would presumably manifest itself to an external observer as memory activity 'autonomous' from the rest of the organism, and with little or no communication between the LTM and its environment. In other words, the memory system would need to be relatively 'cut-off' from environmental stimulation (especially visual stimulation) and likewise disengaged from initiating 'action'–not engaged in purposive movement, and most likely with the organism either temporarily inert or merely performing repetitive and stereotyped motor behaviour. This set of conditions is a closely approximated by the state of sleep [13].

Sleep may therefore be considered to be the time during which memory systems are most engaged in their primary activity of internal processing [5, 26]. There is a great deal of evidence to suggest that sleep is important for memory functions [13]–but the perspective of abstract communication systems goes considerably further than this. From the perspective of the LTM system, sleep processing is its main activity, which allows its maintenance, self-reproduction and increase in complexity–and the 'memory function' is 'merely' a subordinate activity which has evolved to enable the LTM system to emerge, survive and thrive in the context of the rest of the brain. In a metaphorical sense, the memory function is the 'rent' paid by the LTM system to the organism.

Understanding the 'function of sleep' has proved elusive [25]. While sleep very probably has to do with the consolidation and maintenance of long term memory [5, 16, 26], the specifics of this have proved hard to pin-down. The reason is that sleep does not really have 'a function' in terms of the organism as a whole. The function of sleep is specifically to do with the LTM system as a system, but only secondarily to do with the memory function that the LTM system performs for the rest of the brain. Rather, sleep is the behavioural state during which most of the internal processing of the system of LTM occurs; the primary function of sleep is therefore the maintenance and increase of complexity in the LTM.

Conversely, lack of sleep would presumably result in a reduction of complexity of communication in LTM. The consequences of this might include a reduction in potential memory capacity of LTM, less combination of individual memories to form scenarios, and a greater probability of extinction of memories–but the specific consequences of sleep deprivation may be hard to predict without knowledge of the principles (or contingencies) of internal organization of the LTM. These factors might explain the difficulties that sleep and memory researchers have experienced in precisely defining the function of sleep.

9 Interpretation of Experimental Observations

Research on memory shows that brain components (e.g. hippocampus) participating heavily in memory related activity are dominated by repetitive rhythmic activity supported by synchronous activation of large sets of neurons and expressed as theta oscillations at the macro-level (e.g. in EEG data) [4, 12]. This is in good agreement with the prediction of our interpretation of LTM, which implies the repetitive production of communication patterns between neurons that represent memories of the organism through these neural communication patterns. Such repetitive activation patterns are predicted to be more frequent during sleep, which is again confirmed by experimental data [10]. Interestingly, our interpretation, may imply that the observed large scale synchronous activation is a by-product of energetically optimal synchronisation of many different repetitive activation patterns representing various memories.

Recent experimental evidence [8] shows that fruit flies with genetically impaired ability to sleep have significantly shorter lifespan than normal fruit flies. While a genetic defect may have multiple effects, our interpretation suggests that less ability to sleep implies less ability to adapt to the environment, including the internal environment of the organism. In other words, less sleep means less ability to maintain the internal workings of the organism in a sufficiently close to the optimal state. This means that individuals with less ability to sleep are more likely to accumulate organismal functional defects, which are likely to reduce their average lifespan.

A recent study have shown that humans receiving slow wave inducement support during their deep sleep improved significantly their memories [20]. In our interpretation the support of slow wave neural activity at the right frequency (i.e. the frequency of theta and delta oscillations) is likely to help the maintenance of repetitive production of memory representing neural communication patterns. In this way such support may enhance the memory performance of humans as shown by these experiments [20].

Another recent work on memory and synchronised oscillations shows that cannabis reduces the likelihood of synchronised oscillations in the hippocampus in rats [24]. The additional effect of cannabis on rats is a reduction in their memory abilities. According to our interpretation if the repetitive reproduction of memories is impaired it is likely that the memory performance of the brain will reduce, which is in good agreement with the above described experimental finding. Furthermore, on the basis of our interpretation, it is a likely prediction that a memory system with impaired repetitive reproduction ability, will aim to extend its sleep state, in order to compensate for inconsistencies caused be potentially erroneous reproduction of memories. Consequently, our prediction is that rats under the influence of cannabis should be likely to sleep longer than rats that do not have cannabis treatment.

During sleep the LTM system provides multiple 'guesses' concerning the environment, and only those guesses will survive and grow which are compatible with perceptual data generated by behaviour when awake.

Taking another perspective, we expect that artificially prolonged sleep is likely to generate many wrong guesses about the real world, the elimination of which should take more time than int eh case of usual sleep. This implies that the error rate in the execution of non-rutine, sufficiently complex tasks shortly after awakening, should be higher in individuals who had excessive sleep compared to those who had normal sleep. This may also explain the need for re-learning of various skills in case of persons awaking from long duration coma. On another side, according to our view prolonged sleep should not reduce the memory ability of a person .

We also predict that if the barrier between lower level sensory and motor system components and the cortex is not maintained at a sufficient level during sleep, that should result in memory impairment and possibly in longer sleep periods. According to our interpretation if this barrier is not maintained than the separation of the memory from the sensory and motor systems is deficient, which implies that memory maintenance cannot proceed as normal. This should result in memory impairment, and also in a need for longer sleep periods to compensate for the insufficent functioning of the memory during sleep.

References

1. Andras P, Andras C. (2005). The origins of life; the protein interaction world hypothesis: protein interactions were the first form of self-reproducing life and nucleic acids evolved later as memory molecules. Medical Hypotheses, 64: 678–688.
2. Andras P, Charlton BG. (2005). Faults, errors and failures in communications: a systems theory perspective on organizational structure. In Jones, C, Gacek C, Bernard D (Editors). Structuring computer-based systems for dependability. Springer-Verlag: Heidelberg.
3. Andras P, Charlton BG. (2005). Self-aware software - will it become a reality? In Babaogh et al (Editors). SELF-STAR 2004, Lecture notes in computer science - LNCS. Springer-Verlag: Heidelberg.
4. Buzsaki, G (2002). Theta oscillations in the hippocampus. Neuron, 33: 325–340.
5. Buzsaki G (1998). Memory consolidation during sleep: a neurophysiological perspective. Journal of Sleep Research, 7: 17–23.
6. Charlton, B and Andras, P (2003). The Modernization Imperative, Imprint Academic: Exeter, UK.
7. Charlton, BG and Andras, P (2004). What is management and what do managers do? A systems theory account. Philosophy of Management, 3: 3–16.
8. Cirelli, C, Bushey, D, Hill, S, Huber, R, Kreber, R, Ganetzky, B, and Tononi, G (2005). Reduced sleep in Drosophila Shaker mutants. Nature, 434: 1087–1092.
9. Edelman GM (1987). Neual Darwinism: the theory of neuronal group selection. Basic Books: New York.
10. Fuller, PM, Gooley, JJ, and Saper, CB (2006). Neurobiology of the Sleep-Wake Cycle: Sleep Architecture, Circadian Regulation, and Regulatory Feedback. Journal of Biological Rhythms, 21: 482–493.
11. Gazzaniga M. (1994). Nature's mind: biological roots of thinking, emotions, sexuality and intelligence. Penguin: London.

12. Hasselmo ME, Bodelon C, Wyble BP. (2002) A proposed function for hippocampal theta rhythm: Separate phases of encoding and retrieval enhance reversal of prior learning. Neural Computation, 14: 793–817.
13. Hobson JA. (2002) The dream drugstore. MIT press: Cambridge, MA, USA.
14. Hull DL (1988) Science as a process. Chicago University Press: Chicago.
15. Hull DL (2001) Science and selection. Cambridge University Press: Cambridge, UK.
16. Kavanau JL (1996). Memory, sleep and dynamic stabilization of neural circuitry: evolutionary perspectives. Neuroscience and Biobehavioral Reviews 20: 289–311.
17. Louie K, Wilson MA (2001). Temporally structured replay of awake hippocampal ensemble activity during rapid eye movement sleep. Neuron, 29: 145–156.
18. Luhmann, N (1996). Social Systems. Stanford University Press, Palo Alto, CA.Charlton, BG and Andras, P (2003). The Modernization Imperative, Imprint Academic: Exeter, UK.
19. Nadasdy Z, Hirase H, Czurko A, Csicsvari J, Buzsaki G (1999). Replay and time compression of recurring spike sequences in the hippocampus. Journal of Neuroscience 19: 9497–9507.
20. Marshall, L, Helgadottir, H, Molle, M, Born, J (2006). Boosting slow oscillations during sleep potentiates memory. Nature, 444: 610–613.
21. Maturana HR, and Varela, FJ (1980). Autopoiesis and Cognition : the realization of the living. D. Reidel Publishing Company: Boston.
22. Perlovsky, LI (2007). Neural Dynamic Logic of Consciousness: the Knowledge Instinct. In Kozma, R, Perlovsky, LI (Editors) .Neurodynamics of Higher-Level Cognition. Springer-Verlag: Heidelberg.
23. Pokol, B. (1992) The Theory of Professional Institution Systems. Felsooktatasi Koordinacios Iroda: Budapest.
24. Robbe, D, Montgomery, SM, Thome, A, Rueda-Orozco, PE, McNaughton, BL and Buzsaki, G (2006). Cannabinoids reveal importance of spike timing coordination in hippocampal function. Nature Neuroscience, 9: 1525–1533.
25. Siegel JM. (2003) Why we sleep. Scientific American November: 92–97
26. Stickgold R (1998). Sleep: off-line memory reprocessing. Trends in Cognitive Sciences, 2: 484–492.
27. Stickgold R, Hobson JA, Fosse R, Fosse M (2001). Sleep, learning, and dreams: Off-line memory reprocessing . Science, 294: 1052–1057.

Nonlinear High-Order Model for Dynamic Synapse with Multiple Vesicle Pools

Bing Lu, Walter M. Yamada, and Theodore W. Berger

Abstract A computational framework for studying nonlinear dynamic synapses is proposed in this chapter. The framework is based on biological observation and electrophysiological measurement of synaptic function. The "pool framework" results in a model composed of four vesicle pools that are serially connected in a loop. The vesicle release event is affected by facilitation and depression in a multiple-order fashion between the presynapse and the postsynapse. The proposed high-order dynamic synapse (HODS) model, using fewer parameters, predicts the experimental data recorded from Schaffer collateral – CA1 synapses under different experimental conditions better than the basic additive dynamic synapse (DS) model and the basic multiplicative facilitation-depression (FD) model. Numerical study shows that the proposed model is stable and can efficiently explore the dynamic filtering property of the synapse. The proposed model captures biological reality with regard to neurotransmitter communication between the pre- and postsynapse, while neurotransmitter communication between neurons encodes information of cognition and consciousness throughout cortices. It is expected that the present model can be employed as basic computational units to explore neural learning functions in a dynamic neural network framework.

1 Introduction

1.1 Basic Description of Dynamic Synapse

Synapses are the basic transmission units between neurons. A typical synapse has two parts, a presynapse on the first neuron, where a signal is generated, and a postsynapse on the second neuron, where the signal is received. Synaptic transmission is a chemical process. Thousands of neurotransmitter molecules are stored in bins, called vesicles, and typically there is a pool of tens of vesicles awaiting release near the plasma membrane of the presynapse. When the presynapse is triggered by a stimulus above the action potential (AP) threshold, a sequence of events is set into motion that results in vesicles fusing with the presynaptic membrane and neurotransmitter molecules inside those vesicles being released into the space separating the pre- and postsynapses; these molecules diffuse across the synapse and bind to

L. I. Perlovsky and R. Kozma, *Neurodynamics of Cognition and Consciousness.*

receptors at the postsynapse. Binding at the postsynapse sets into motion another sequence of events that results in a potential change at the postsynaptic membrane of the second neuron – thus terminating the communication from presynaptic-neuron to postsynaptic-neuron. All chemical processes are history dependent and therefore dynamic in nature.

The complete dynamic process between presynapse and postsynapse has been described as either additive or multiplicative signal processing. The dynamic synapse (DS) model [1] additively models presynaptic effects on the postsynaptic response. The dynamic facilitation-depression (FD) model [2] is a fundamental multiplicative model in which the multiplicative product of presynaptic activities is applied to affect the postsynapse. An addition of two FD models, with one fast and one slow [3], is shown to provide a better prediction of experimental data recorded at hippocampal cells than the original FD model.

In this chapter a new model is proposed and compared to the basic additive DS model and the basic multiplicative FD model. All three models are applied to synapse data recorded from the hippocampus. The purpose of the proposed model is to introduce a description of input-output synaptic activity that is more generally and intuitively applicable to cortical synapses.

1.2 Facilitation and Depression on Synaptic Transmission

Both DS and FD models explicitly explore the role of facilitative and depressive factors underlying synaptic transmission. If a presynapse experiences a series of superthreshold input events, subsequent events in the sequence are able to elicit more vesicle release than previous events in the sequence. This enhancement of vesicle release, called facilitation, results in an accumulated postsynaptic response. The facilitation can be described as a function of the input event sequence timing [1, 2]. In balance with facilitative enhancement, the input sequences elicit a depressive effect described by a decreased ability to release a vesicle [2]. The depressive effect is also described as being induced by back-propagating postsynaptic action potentials [1].

The proposed high-order dynamic synapse model analyzes both facilitative and depressive effects as a function of the input pulses. In the absence of postsynaptic modulation, as the postsynaptic action potentials are a function of the input sequences in one single pre-and-postsynapse, depression caused by postsynaptic feedback signals can be lumped with depression caused by the input sequences. Additionally, an extension of the synaptic transmission model is illustrated in Sect. 4 that can include the postsynaptic feedback information.

1.3 Vesicle Pooling

A single type of vesicle pool at the presynapse is assumed in the DS and FD models. However, many studies of neurotransmitter recycling and refilling procedures [4, 5, 6] point to the importance of the concept that multiple vesicle pooling kinetics contribute to synaptic dynamics. Based on biophysical experiments, each active zone of presynaptic plasma membrane contains only one or at most a few vesicles that can be released immediately. That is, most vesicles are stored behind the active zones of plasma membranes, forming another two pools [6]: a prepared release pool and a reserve pool. These three kinds of pools, the immediate release pool, the prepared release pool, and the reserve pool, have different rates of transfer among themselves. The immediate release pool, which is the closest to the plasma membranes, releases vesicles into the fourth pool, the used pool [6].

Recently, using random processes to describe synaptic transmission has attracted increasing interest. A dynamic stochastic model [7] describes the vesicle release event based on the binomial distribution. A statistical analysis [8] studies vesicle mobility among multiple pools, which primarily considers short-term depression of the synaptic response to a paired-pulse stimulus.

A more general input-output synaptic activity in response to longer input pulses is investigated in this chapter, which mainly emphasizes vesicle mobilities. The postsynaptic potentials are assumed to inherently affect synaptic dynamics via the used pool and the transfer rates between vesicle pools.

2 Method

The DS model [1] considers primary, fast, and slow facilitation effects. All facilitation effects are directly triggered by a train of stimuli. The overall release probability is modeled by a first-order linear addition of fast facilitation, slow facilitation, and depression. In contrast, the FD model [2] describes only one facilitation effect and a second-order depression effect, and then multiplies both facilitation and depression effects. These models view the synapse from the perspective of how the postsynaptic response is modulated by explicit presynaptic facilitation and depression. Also, these models consider only one kind of vesicle pool inside the presynapse, resulting in limited understanding of the vesicle mobility. In contrast, the present model views the synapse from a different perspective, i.e., how able is the presynapse to elicit the postsynaptic response via vesicle dynamics between multiple pools. The general goal in the present study is to design a high-order dynamic synapse model that is both computationally manipulatable and biologically interpretable.

2.1 Motivation

Based on biological experimental findings, there are several kinds of neurotransmitter pools in the presynapse [4, 5, 6]. The abstract biophysical illustration of vesicle pooling and recycling is shown in Fig. 1(A). The names of four vesicle pools are defined based on their individual hypothesized function and relative position to the plasma membrane of the presynapse [6]. When the immediate release pool has

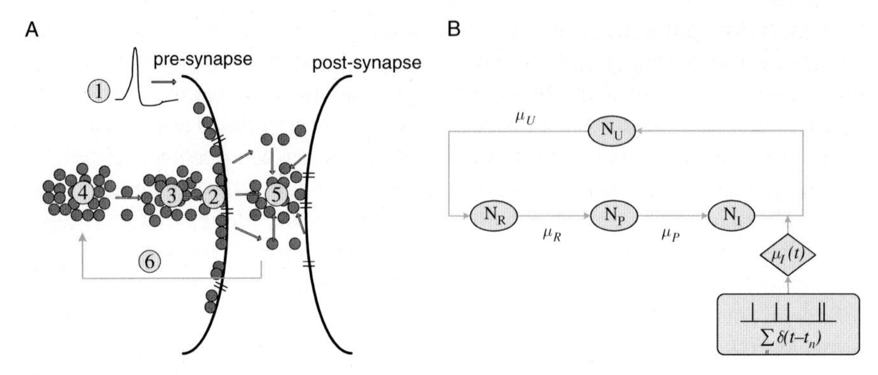

Fig. 1 (**A**) A biophysical exhibition of the vesicle pooling and cycling procedure: (1) if an input waveform above the action potential threshold arrives, an action potential pulse is generated; (2) the action potential pulse triggers the immediate release pool to release vesicle into the cleft; (3) the prepared release pool refills vesicles into the immediate release pool; (4) the reserve pool refills vesicles into the prepared release pool; (5) the used pool collects used neurotransmitters; (6) neurotransmitter vesicle recycles into the reserve pool. (**B**) A dynamic synapse model illustrates serially connected multiple vesicle pools: $\sum_n \delta(t - t_n)$ indicates arrived input spikes; N_R, N_P, N_I, and N_U denote the number of vesicles inside the reserve pool, the prepared release pool, the immediate release pool, and the used pool, respectively; μ_R, μ_P, and μ_U, represent the constant valued transfer rates between neighboring pools; $\mu_I(t)$ is the nonstationary release rate which is a function of input pulses

depleted its vesicles for previous stimuli, it cannot release vesicle for current or for future stimuli until the prepared release pool transfers its vesicles into the immediate release pool. In the meantime, the reserve pool transfers neurotransmitter vesicles into the prepared release pool. The neurotransmitters that are released from the immediate release pool, are stored in an abstract pool, the used pool. It is assumed that the process associated with the used pool includes the series activities elicited by vesicle release, such as receptor binding in the postsynapse and neurotransmitter elapsing in the synaptic cleft. Also it is assumed that the majority of neurotransmitter molecules are absorbed back into the presynapse for re-use. Therefore, neurotransmitter molecules stored in the used pool are re-formed as vesicles and then are recovered by the presynapse after cycling through the reserve pool.

2.2 A High-order Dynamic Synapse Model

Vesicle release events are modelled according to what happens between these vesicle pools. The four vesicle pools are serially connected as shown in Fig. 1(B), where N_R, N_P, N_I, and N_U denote the number of vesicles inside the reserve pool, the prepared release pool, the immediate release pool, and the used pool, respectively. μ_R, μ_P, and μ_U, represent the constant valued transfer rates between neighboring pools. When there is no stimulus to excite the immediate release pool, the immediate pool does not connect with the used pool and there is no vesicle being released. When there is a stimulus above the action potential threshold arriving at the presynapse at time t_n, an action potential $\delta(t-t_n)$ occurs (where $\delta(t)$ is the Kronecker delta function), resulting in vesicle release from the immediate release pool into the used pool. Neurotransmitters in vesicles release into the synaptic cleft, or neuro-transmission, is proportional to the transfer rate $\mu_I(t)$ between the immediate release pool and the used pool. If a series of action potentials $\sum_n \delta(t-t_n)$ excite the immediate release pool, the release rate $\mu_I(t)$ is modelled as a correspondingly varying function,

$$\tau_I \frac{\mathrm{d}\mu_I(t)}{\mathrm{d}t} = -\mu_I(t) + C_I \sum_n \delta(t-t_n), \tag{1}$$

where τ_I and C_I are the time constant and the action potential gain, respectively.

By investigating pooling kinetics, each vesicle is assumed to have the same probability to transfer and therefore each vesicle transferring event is viewed as a binomial variable with a small probability. When the number of vesicles of each pool is big enough, the total vesicle transferring event of a pool approximately follows a Poisson distribution with a mean rate [9][pp. 36–42]. In the present model, the mean rate is expressed via the transfer rates μ_R, μ_P, μ_U and the average value of $\mu_I(t)$. With the above assumptions, the dynamic system as a whole can be equivalently considered as a pure birthing process [10][p. 184], with the differentials of N_U, N_R, N_P, N_I replacing the state transition probabilities in the birthing process. In steady state the four pools can be described by the following equations, which include the sequential transfer rates between pools:

$$\frac{\mathrm{d}N_R(t)}{\mathrm{d}t} = -\mu_R N_R(t) + \mu_U N_U(t), \tag{2}$$

$$\frac{\mathrm{d}N_P(t)}{\mathrm{d}t} = -\mu_P N_P(t) + \mu_R N_R(t), \tag{3}$$

$$\frac{\mathrm{d}N_I(t)}{\mathrm{d}t} = \mu_P N_P(t) - \mu_I(t) N_I(t) \sum_n \delta(t-t_n), \tag{4}$$

$$\frac{\mathrm{d}N_U(t)}{\mathrm{d}t} = -\mu_U N_U(t) + \mu_I(t) N_I(t) \sum_n \delta(t-t_n). \tag{5}$$

Finally, the elicited post-synaptic activity (EPSC) induced by the vesicle release is given by

$$\tau_E \frac{\mathrm{d}E(t)}{\mathrm{d}t} = -E(t) + C_E \mu_I(t) N_I(t), \tag{6}$$

where τ_E represents the time course and C_E is a gain constant. From the perspective of model formulation, the designed model is a general description of synaptic function with seven parameters $\tau_I, C_I, \mu_U, \mu_R, \mu_P, \tau_E, C_E$. It is not attempted to quantify every parameter of the present model with exact biological factors.

Based on (4) and (5), the immediate pool releases vesicles at a magnitude of $\mu_I(t)N_I(t)$ into the used pool. This step can be reformulated in terms of releasable aliquots of vesicle, which is in keeping with vesicular release; graded release is chosen for numerical ease and interpretation. The multiplicative term $\mu_I(t)N_I(t)$ is nonstationary due to two-fold time-varying processes. The first time-varying process stems from the variability of the immediate release probability $\mu_I(t)$ induced by input spikes; and the second process is due to the variability of vesicle number $N_I(t)$ induced by vesicle moving kinetics between pools. The effective vesicle release event is given by a joint variation of these two processes. As proved in Appendix A, the joint variation is a third-order varying item, and the model is thereby called high-order dynamic synapse (HODS) model.

2.3 Meaning of High Order

Appendix A shows that the inverses of multiple transfer rates can express multiple timing courses of the synaptic dynamics and therefore multiple memory extents. In other words, the pools that are farther to the immediate vesicle release sites can store earlier released vesicles, which can save history information of previous input spikes. When these pools refill the immediate release pool with different timing courses, the history information can affect the current synaptic response at different levels, resulting in a multiple-order facilitative or depressive effect. These effects can be treated as a dynamic process caused by the kinetics describing vesicle transfer between multiple pools, as illustrated in Fig. 2. On the other hand, the DS model

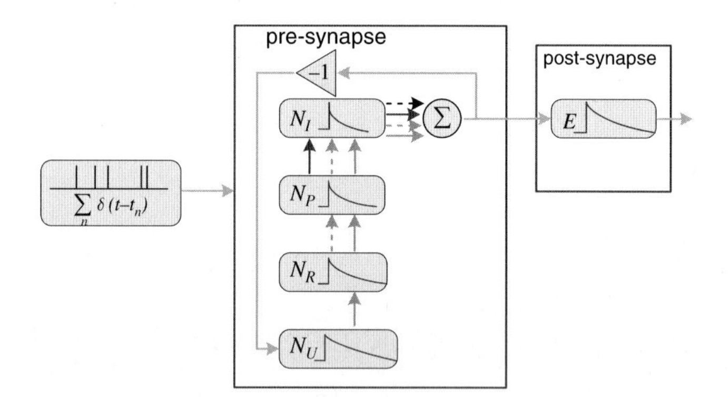

Fig. 2 An abstract functional description of synaptic transmission from a presynaptic neuron to a postsynaptic neuron. The immediate release pool has the fastest vesicle release indicated by *dashed black* arrow. Other pools indirectly influence the vesicle release event through the immediate release pool, e.g., the prepared release pool has the second fastest release indicated by *solid black* arrows, which represent vesicle transferring from the prepared release pool to the immediate release pool and then releasing to the cleft. The *dashed grey* arrows indicate vesicle kinetics in the reserve pool, and the *solid grey* arrows denote vesicle kinetics in the used pool. In contrast, released vesicles take time to recover from the used pool and then to be re-used for future release. The release event elicits a self-limiting effect (depression) proportional to the number of released vesicles. The functional combination of multiple order facilitative/depressive effects stemming from multiple pools excites the postsynaptic response

linearly adds primary and secondary facilitative effects with order one. Similarly, the FD model also considers one kind of facilitative effect. Another appealing feature of the present model is that the facilitative and depressive effects on the postsynapse are implicitly realized, while the DS and FD models explicitly model the facilitation and depression factors that describe the amplitude of postsynaptic responses. Implicit facilitation and depression induced by pooling kinetics are more biological meaningful than the explicit description, as facilitation/depression is inherently generated by bio-chemical activities in an actual synapse.

When action potentials arrive in series, the release rate $\mu_I(t)$ increases accordingly, facilitating the vesicle release if there is at least one vesicle inside the immediate release pool. When $\mu_I(t)$ is enhanced by trains of action potentials, more vesicles enter into the immediate release pool from other pools, and more vesicles can be released. Other pools impact the facilitation effect in serial. The farther the pool to the immediate pool, the slower the facilitation rate, leading to multiple order facilitative effects.

As the released vesicles increase, the used vesicles in the used pool increase. Assuming the total number of vesicles over all pools is predetermined, no vesicle is added or lost in the present model (exceptions such as neurotransmitter production or losing are discussed in Sect. 4). It takes time for used vesicles to recycle from the used pool to the immediate release pool. Hence, under the situation when vesicles inside the immediate release pool are depleted, even if the release rate $\mu_I(t)$ is high, the immediate pool is not able to release vesicle. As a result, the presynapse is not able to elicit the postsynaptic response. Since the vesicle release increases in a multiple-order fashion, the used vesicle accordingly increases in a multiple-order fashion. With depressive effects, this model never overflows with infinite facilitation effects.

3 Results

3.1 Validating Experimental Data

In this section, experimental data of synaptic input-output activity is used to validate three synapse models, DS, FD, and HODS. The collection procedure for experimental data is summarized as follows. Whole-cell recordings of EPSC were performed with an HEKA EPC-9 patch-clamp amplifier from CA1 pyramidal cells visually identified with an infrared microscope (Olympus BX50WI). Experiments were carried out under the condition that $\left[C_a^{2+}\right]_o = 2$ mM or $\left[C_a^{2+}\right]_o = 1$ mM. Two kinds of stimuli are used: one is fixed-interval trains (FIT) of electrical pulses with the frequency 10 Hz or 5 Hz; the other is Poisson random trains (RIT) of electrical pulses with mean rate 2 Hz. Both stimuli have unitary amplitudes. The smallest time unit and the duration of one stimulus is 0.1 ms.

Normalized mean square error (NMSE) is used as the cost function. The Nelder Mead simplex algorithm [11] is utilized to estimate parameters of the three models

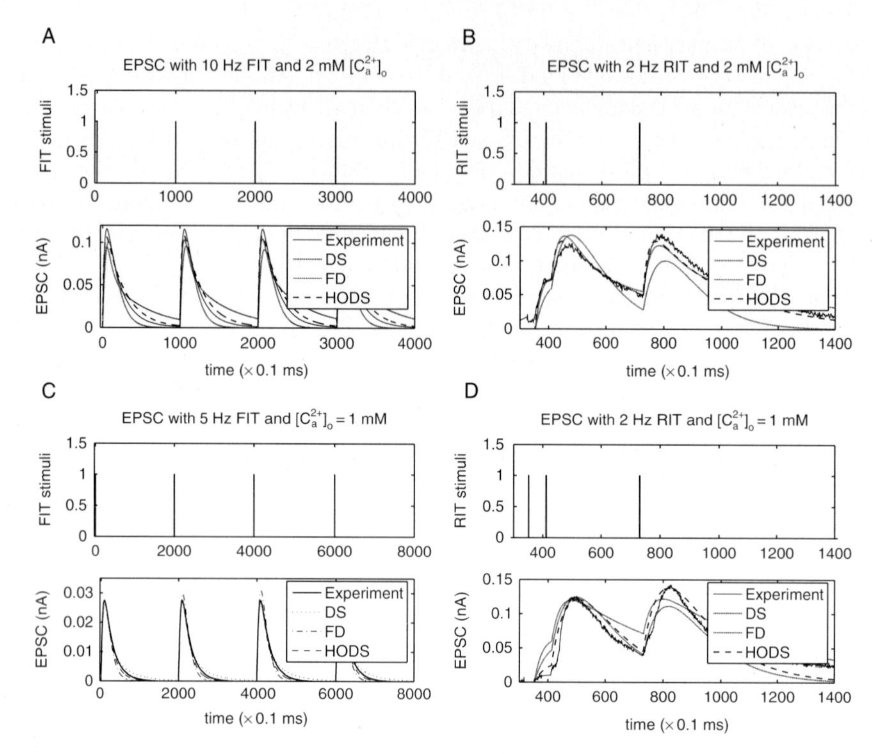

Fig. 3 Three synapse models are used to predict experimental results with respect to different inputs and experimental conditions. All upper sub-figures represent input spikes, and lower sub-figures present actual experimental results and estimated curves. The *dashed* curves estimated by the HODS model are in better agreement with the slopes of real experimental results (*solid* curves) than *dotted* curves estimated by the DS model and *dot-dashed* curves estimated by the FD model. (**A**) FIT (10 Hz) and $\left[C_a^{2+}\right]_o = 2$ mM; (**B**) RIT (2 Hz) and $\left[C_a^{2+}\right]_o = 2$ mM; (**C**) FIT (5 Hz) and $\left[C_a^{2+}\right]_o = 1$ mM; (**D**) RIT (2 Hz) and $\left[C_a^{2+}\right]_o = 1$ mM

during the training stage. At the testing stage, Fig. 3(A–D) illustrates how well the DS, FD, and HODS models predict the experimental data with either FIT or RIT stimuli. NMSE is listed in Table 1 for all three models. The estimated EPSC curves and NMSE values show that the HODS estimated results are in better agreement with the experimental results than the estimated results based on the other two models. In addition, the estimated parameters for DS, FD, and HODS models are given

Table 1 Normalized mean square errors (NMSEs) (%)

$\left[C_a^{2+}\right]_o$	2 mM		1 mM	
	FIT (10 Hz)	RIT (2 Hz)	FIT (5 Hz)	RIT (2 Hz)
DS	1.63	0.99	1.37	6.03
FD	1.12	5.23	2.84	2.79
HODS	0.28	0.72	0.78	1.74

Table 2 Estimated ten DS parameters

$[C_a^{2+}]_o$	2 mM		1 mM	
	FIT (10 Hz)	RIT (2 Hz)	FIT (5 Hz)	RIT (2 Hz)
τ_R (ms)	3.61	4.00	4.00	2.71
K_R	27.23	25.20	100.20	128.57
τ_{F_1} (ms)	4.00	4.04	4.00	4.00
K_{F_1}	25.20	24.78	560.20	100.20
τ_{F_2} (ms)	51.06	57.16	50.00	44.41
K_{F_2}	373.45	339.89	600.20	4053.34
τ_d (ms)	4.00	4.00	4.00	4.00
K_d	−25.60	−25.60	−100.60	−100.60
τ_E (ms)	4.59	5.14	150.00	57.82
K_E (%)	8.30	7.56	0.76	0.74

in Tables 2, 3, and 4, respectively. The HODS model has seven parameters only, while the DS model [1] and FS model [2] have ten and nine parameters, respectively. In a nutshell, the novel HODS model enjoys better performance as well as less controlling complexity.

Different sets of parameters of the HODS model are obtained for each experimental data set as presented in Table 4, which shows parameter independency to control this model to predict varying data. Accordingly, varying sets of parameters are expected to characterize varying experimental conditions or recording data from different kinds of neurons. In contrast, some of the optimized parameters for the DS and FD models are redundant as given in Tables 2 and 3, which can be neglected as their controlling effects on predicting the experimental data are

Table 3 Estimated nine FD parameters

$[C_a^{2+}]_o$	2 mM		1 mM	
	FIT (10 Hz)	RIT (2 Hz)	FIT (5 Hz)	RIT (2 Hz)
τ_F (ms)	89.47	95.11	161.56	161.56
F_1 (%)	9.98	3.16	2.60	2.60
ρ	0.99	1.29	1.49	1.49
τ_D (ms)	26.27	8.33	16.84	16.84
Δ_D	0.99	1.06	1.80	1.80
$K_{max}(s^{-1})$	6.83	2.65	4.46	4.46
$K_0(s^{-1})$	2.49	2.17	1.78	1.78
K_D	1.05	0.33	0.27	0.27
αN_R	1.00	2.00	1.00	2.00

Table 4 Estimated seven HODS parameters

$[C_a^{2+}]_o$	2 mM		1 mM	
	FIT (10 Hz)	RIT (2 Hz)	FIT (5 Hz)	RIT (2 Hz)
μ_U	0.45	0.51	0.02	0.01
μ_R	0.74	0.66	0.22	0.01
μ_P	0.09	0.08	0.01	0.04
τ_I (ms)	12.19	13.82	10.52	63.72
C_I	0.93	0.83	1.04	0.23
τ_E (ms)	24.29	27.53	19.54	7.41
C_E (%)	0.81	0.72	1.09	1.34

almost close to zero. The exact reason is not clear, but it is possibly due to the flat cost space of the DS and FD models defined on the recorded hippocampal experimental data.

Despite of the fact that the HODS model is not restrictively formulated to identify a biophysical structure, the parameters can still be implicitly related to underlying biophysical meanings from experimental realities. In Table 4, transferring rates μ_R, μ_P, and μ_U are larger under the condition $[C_a^{2+}]_o = 2$ mM than under $[C_a^{2+}]_o = 1$ mM. As a result, the vesicle refilling and recycling procedure is faster when the calcium concentration is bigger. For RIT stimuli, C_I is larger and τ_I is smaller under the condition $[C_a^{2+}]_o = 2$ mM than under $[C_a^{2+}]_o = 1$ mM. Hence the nonstationary release rate $\mu_I(t)$ increases when the calcium concentration increases, and the vesicle release event is facilitated due to the increasing release rate. These parameter values, to some extent, give some insight to the meaning of biophysical experimental quantities such as calcium concentration, which is consistent with the fact that calcium-dependent increase induces a facilitative effect on vesicle release probability [2, 3].

3.2 Robustness to Additive Noise

In order to test the robustness of the HODS model, an additive white Gaussian (AWGN) noise with SNR= 0 dB (signal-to-noise ratio) is added into the experimental output which has FIT input and $[C_a^{2+}]_o = 2$ mM. The estimated output of the HODS model is illustrated in Fig. 4, and compared against the noise-free experimental output. As clearly demonstrated by the figure, the HODS model is capable of removing the additive noise and recovering the clean experimental output.

Under bad experimental conditions, such as apparatus mechanical noise, or background perturbation, unexpected noise may be added into the real synaptic response. It is valued then, that the present model can retrieve the noise-free experimental data even though it is blurred by additive noise. In addition, the estimated

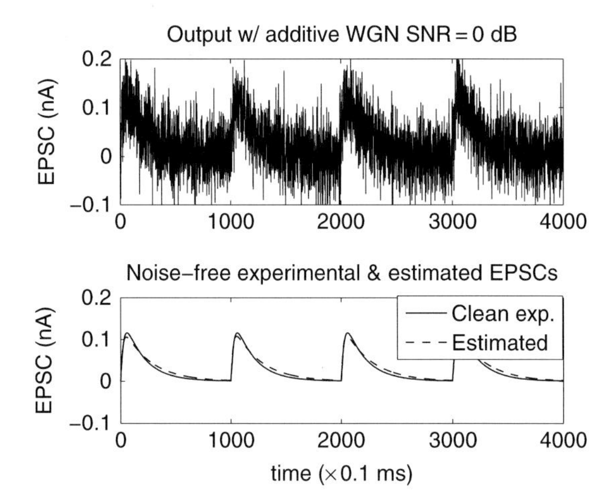

Fig. 4 The upper sub-figure displays noise-masked experimental output with AWGN (SNR = 0 dB). The lower sub-figure illustrates that the HODS model is able to recover the noise-free experimental result

parameter values are $\mu_U = 0.452$, $\mu_R = 0.741$, $\mu_P = 0.0926$, $\tau_I = 12.20$ ms, $C_I = 0.926$, $\tau_E = 24.31$ ms, $C_E = 0.806\%$, which are very close to those values estimated in column 2 in Table 4 when the output is without noise. The fact that the parameters are close to each other in the noisy case and the noise-free case indicates that the HODS model is robust and can converge from the noise-blurred point to the noise-free point in the cost space. To explore this fact further, SNR is varied from -20 dB to 10 dB, while the parameter values remain close to the values in column 2 in Table 4 with small deviation about 0.2% to 1%. For biophysical high-dimensional dynamic systems perturbed by chaotic noise, a theoretical stability analysis can be employed to describe a broad range of parameters under abruptly happened state transitions [12, 13]. As the major purpose of the present study is to establish a high-order dynamic synapse model, chaotic analysis is not covered in details in this chapter.

3.3 Response to Varying Input Pulses

Since the EPSC activity at the current time is influenced not only by the current input spike but previous spikes as well, the proposed model uses several serial pools to remember previous spiking information. To emphasize the importance of synaptic memory that is characterized by timing intervals of input pulses, the following tests are implemented.

The experimental EPSC response with RIT (rate 2 Hz) and under the condition ${C_a^{2+}}_o = 1$ mM is first given in Fig. 5(A), and the estimated EPSC output of the HODS model is also presented. The parameters are from column 5 in Table 4 and are fixed when applying the HODS model responding to varying inputs. The estimated HODS response in Fig. 5(A) is given as an original response pattern. Next, the input pulses are varied and the corresponding response is obtained. When the 5th input spike is moved closer to the 4th spike, the 5th response is enhanced with a higher amplitude than the original 5th response, as illustrated in Fig. 5(B). In Fig. 5(C), if the 7th input spike is removed the 8th response is accordingly depressed

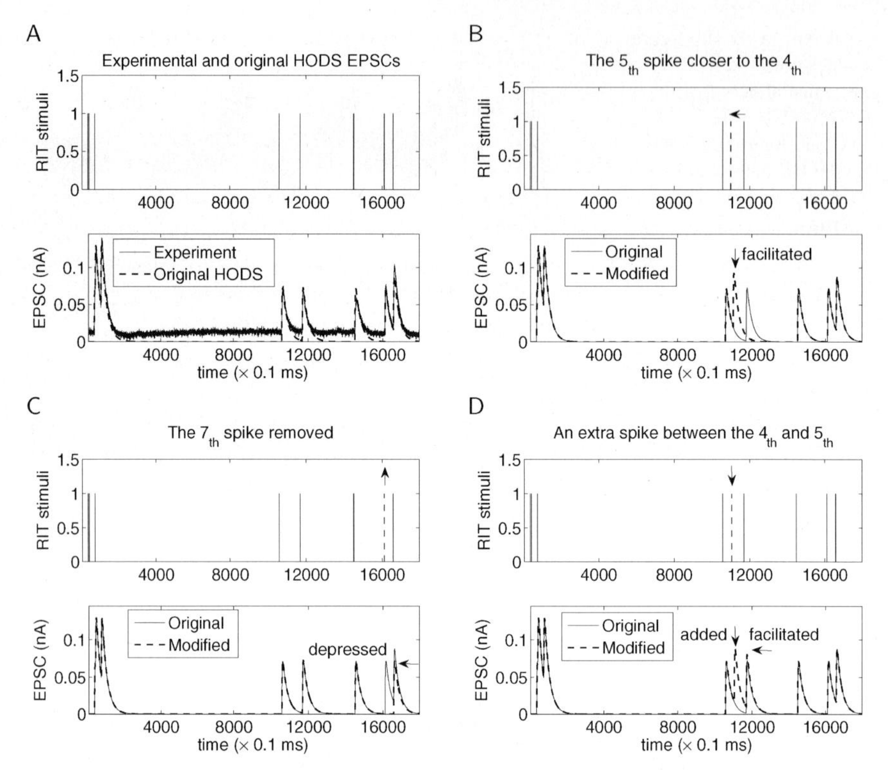

Fig. 5 (**A**) The HODS model is able to predict the experimental result and the estimated HODS EPSC curve is called the original HODS response. (**B**) When moving the 5th input spike closer to the 4th spike, the 5th HODS response is facilitated. (**C**) If the 7th spike is removed, the 8th response is depressed. (**D**) Adding an extra spike between the 4th and the 5th, the 5th response is enhanced, and the extra response is facilitated due to the closeness of the added and the 4th spikes

in comparison with the original response. Then, an extra input spike is added between the 4th and the 5th, and thereby the 5th response and the added response are enhanced as given in Fig. 5(D).

It is observed that the HODS model with a set of fixed parameters can be considered as a dynamic filter, which changes its response with respect to varying inputs, especially to temporal variations. The temporal intervals between the current input spike and previous spikes tune the HODS responses. This property indicates that the dynamic model can be applied as a filter to selectively respond to specific timing inputs. As the timing course of the HODS model is related to transfer rates, it is also hypothesized that varying transfer rates can provide varying temporal input selectivity.

3.4 Response to a Train of Input Pulses

The facilitation and depression activity in response to a repetition of presynaptic spikes [2] is studied in this section. With a short train of pulses (rate 100 Hz, duration 90 ms) exciting the synapse, the HODS response is dominated by the facilitative effect and increases accordingly when more pulses arrive, as shown in Fig. 6(A). However, if the input pulses sustain longer (duration more than 100 ms), the depressive effect starts to affect, and the response stops growing after time 100 ms in Fig. 6(B), due to balanced facilitative and depressive effects; later, for lower-rate input pulses (rate 25 Hz), both effects also maintain at a saturated state. In an extreme case, if the immediate release pool is depleted, no response happens even if there are sustained pulses to excite the synapse, such as the illustrated response curve after time 160 ms in Fig. 6(C). Then, if stopping the sustained excitation of the higher-rate impulses for a while, or using a lower-rate input as in Fig. 6(D), the immediate release pool can be gradually refilled and vesicle is again available to release to excite the postsynaptic response.

These responses induced by trains of pulses can represent synaptic facilitation and depression. The facilitation/depression that the HODS model displays is similar to biophysical experimental results [2, 6] – for repetitive input pulses the EPSC is initially enhanced by the facilitative effect, and then the response is countered by

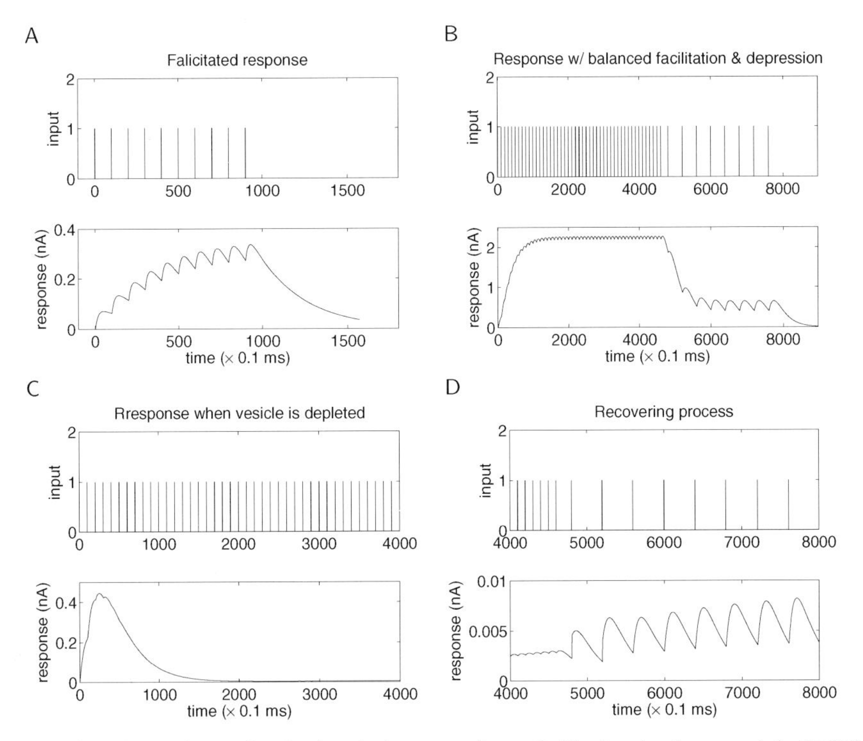

Fig. 6 (**A**) When a short train of pulses is the system input, facilitation dominates and the HODS response is enhanced. (**B**) If the train of pulses is longer, the depressive effect starts to be effective, and the HODS response reaches a plateau due to balanced facilitative and depressive effects. (**C**) In an extreme case when vesicles in the immediate release pool are depleted, the HODS model almost has no response to input pulses. (**D**) The HODS model gradually responds to a lower-rate train of pulses by refilling vesicles from other pools

depression with respect to longer pulses. The present model does not overflow even with an infinite train of input pulses, and can gradually recover by being refilled from other pools to the immediate release pool. This HODS self-limiting event is consistent with the fact that a practical biological synapse is stable.

4 Extension and Discussion

An example of the flexibility of the present dynamic paradigm is given by manipulating the model and interpreting the manipulation, i.e., other hypotheses can be speculated in the multiple-order dynamic synapse model. For example, the immediate release pool can only have a fixed number M_{N_I} of active release sites [5, 6, 8] and the pool size N_I can not exceed this number. It is hypothesized that the used pool may yield a fast recycling to the prepared release pool [6], and an arrow with rate μ_{U-P} denotes such a case in Fig. 7. Besides, the immediate release pool and the prepared release pool may have reverse transfer rates $\lambda_{I-P}, \lambda_{P-R}$ to illustrate a 'backward' vesicle transfer found in some special synapses [6], e.g., the decrease of vesicle release possibility due to the secondary messengers in a deactivated state of the NMDA (N-methyl-D-aspartic acid) receptor [14].

Before, when designing the HODS model to explore the input-output synaptic transmission, the back-propagating postsynaptic action potentials have been assumed to implicitly influence the synaptic transmission via the used pool. This assumption can also be explicitly extended such that the pre- and postsynaptic mechanisms seem to act in series, suggesting a coordinated regulation of the two sides of the synapse [15, 16]. The voltage trace of a back-propagating action potentials can

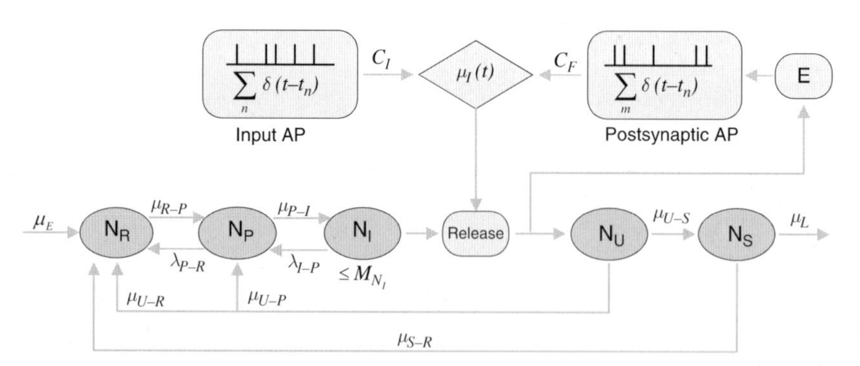

Fig. 7 Possible extensions of the HODS model: there is an upper bound of the number of vesicles in the immediate release pool, M_{N_I}; the used pool can yield a fast recycling rate μ_{U-P} to the prepared release pool; the immediate release pool and the prepared release pool can have reverse transfer rates $\lambda_{I-P}, \lambda_{P-R}$ indicating the decrease of vesicle release possibility; $\sum_m \delta(t - t_m)$ represents the back-propagating postsynaptic action potentials that affect the vesicle release rate with gain C_F; A new pool called the slow used pool with N_S neurotransmitter vesicles is defined to represent a longer reuptake process than the used pool; μ_L describes neurotransmitter losing process, while μ_E describes the biosynthetic neurotransmitter producing process

be a feedback train of pulses (*a second input*) to the presynapse, directly affecting the nonstationary release rate $\mu_I(t)$ [17]. This fact is shown in Fig. 7 with gain C_F and pulses $\sum_m \delta(t - t_m)$. Many synaptic plasticity schemes [14, 18, 19, 20] can be applied to define $\mu_I(t)$ as a function of relative pre- and postsynaptic spike timing. In addition, due to the different dynamic mechanisms of the neurotransmitter receptors at the postsynapse side [4], it is accordingly assumed that there may be another pool to infer different binding/elapsing rates, resulting in varying reuptake rates for released vesicles. In such a case, there is a new pool, called the slow used pool with N_S neurotransmitter vesicles, to represent a longer reuptake process than the used pool.

Further, supposing that some released neurotransmitters may be removed and lost outside the pre-post-synaptic pair [21], a losing rate μ_L can describe such a process. In balance with the neurotransmitter loss, an entering rate μ_E can describe the biosynthetic capabilities of neurotransmitter production in neurons, such as superior cervical ganglion neurons [22]. Furthermore, it is discovered that synaptic transmission dramatically varies with respect to temperature [23]. The transfer rates of the present model that describe synaptic activities can be defined as temperature-dependent variables.

The complicated model in Fig. 7 tries to incorporate as many synaptic activities as possible. It is expected that the possible extension may make the HODS model explore more synaptic dynamics. As a cautionary note, due to the lack of experimental data, the proposed scheme remains uncertain and needs advanced experimental validation to identify how vesicle moves among multiple pools and which moving rate it has under different biological conditions. Nevertheless, the merits of the HODS model and its possible flexibility are worthy of being studied in the field of potential synaptic dynamics.

5 Conclusions

Motivated by biological observation suggesting the existence of several types of vesicle pools in the synapse, a high-order dynamic synapse function using pooling kinetics is proposed. The proposed model describes the physiology of synaptic transmission dependent upon neurotransmitter vesicle moving between multiple vesicle pools with various transfer rates, both constant valued and nonstationary. Multiple order facilitation and depression are inherently yielded as each vesicle pool independently supports one kind of facilitative or depressive effect. When comparing the experimental data recorded at Schaffer collateral – CA1 pyramidal neurons with the HODS, DS, and FD model predictions, the HODS model achieves better performance than both the DS and the FD models with either fixed interval pulses or Poisson random interval pulses. Furthermore, the present model is capable of recovering noise-free response when the experimental data is contaminated by additive white Gaussian noise. Numerical analysis in Sects. 3.3 and 3.4 shows that the HODS model has dynamic filtering property with respect to varying input pulses.

It is observed that the proposed model embodies several characteristics that make it a valuable tool for computational description of biological synapses: independent modeling of functional sub-components, a small-size free parameter set, and flexibility. A parsimonious representation based on multiple independent pool effects is investigated. And a few reliable parameters are used to express the essential dynamics of spiking signals, which is in agreement with the goal of designing a good model – obtaining performance as well as possible while the freedom (the number of parameters) of the model being as small as possible. Besides, the present model can be flexibly extended to explore more synaptic dynamic activities. These three model characteristics result in a synaptic description that is more easily optimized than DS or FD models, for describing the dynamic nature and multiple facilitative and depressive effects expressed by chemical synapses.

Direct experimental evidence supporting the analysis has been lacking. There remains technical difficulty in obtaining recordings to verify the number of moving vesicles and to identify the transfer rates of each vesicle. Nevertheless, the formulation of the high-order dynamic synaptic model is based on basic biophysical experimental results. Hence, this model, considering the biophysical realities during its designing procedure, can exhibit some biology based learning rules in response to input timing spikes. Also, it is expected that this study will inspire more interest in experimentally validating the biophysical realities and inspire an interest in high-order synaptic modeling based on pooling kinetics.

The proposed model serves as a good starting point for the extension of dynamic neural network structure. The present high-order dynamic neural synapses have been used as basic computational units to construct a time-frequency two-dimensional dynamic neural network for speaker identification [24]. This network scheme is capable of exploring long-term temporal features of speech data. Further, the hierarchical structure that expresses brain dynamics at different levels can be composed of local-global components [25]. It is also hypothesized that neural modules at varying levels can be designed based on the present dynamic synapse. Each module can implement one special function, and different functional network modules together perform a complicated learning task. How to specify the special function for each module and how to define the connections between network modules remain future research.

Acknowledgment The authors would like to thank Dr. Dong Song for providing and explaining the experimental physiological data, the short-term synaptic responses of hippocampal Schaffer collateral – CA1 pyramidal neurons.

References

1. J.S. Liaw and T.W. Berger. Dynamic synapse: a new concept of neural representation and computation. *Hippocampus*, 6:591–600, 1996.
2. J.S. Dittman, A.C. Kreitzer, and W.G. Regehr. Interplay between facilitation, depression, and residual calcium at three presynaptic terminals. *Journal of Neuroscience*, 20:1374–1385, 2000.

3. D. Song, Z. Wang, V.Z. Marmarelis, and T.W. Berger. A modeling paradigm incorporating parametric and non-parametric methods. *IEEE EMBS*, 1:647–650, 2004.
4. C.F. Stevens and J.F. Wesseling. Identification of a novel process limiting the rate of synaptic vesicle cycling at hippocampal synapses. *Neuron*, 24:1017–1028, 1999.
5. R. Delgado, C. Maureira, C. Oliva, Y. Kidokoro, and P. Labarca. Size of vesicle pools, rates of mobilization, and recycling at neuromuscular synapses of a drosophila mutant, shibire. *Neuron*, 28:941–953, 2000.
6. R.S. Zucker and W.G. Regehr. Short-term synaptic plasticity. *Annual Review of Physiology*, 64:355–405, 2002.
7. W. Maass and A.M. Zador. Dynamic stochastic synapses as computational units. *Neural Computation*, 11:903–917, 1999.
8. V. Matveev and X.-J. Wang. Implications of all-or-none synaptic transmittion and short-term depression beyond vesicle depletion: a computational study. *Journal of Neuroscience*, 20:1575–1588, 2000.
9. H. Stark and J.W. Woods. *Probability, random processes, and estimation theory for engineers*. 2nd edition, Prentice Hall, 1994.
10. D. Bertsekas and R. Gallager. *Data networks*. 2nd edition, Prentice Hall, 1992.
11. J.A. Nelder and R. Mead. A simplex method for function minimization. *Computer Journal*, 7:308–313, 1965.
12. W.J. Freeman, R. Kozma, and P.J. Werbos. Biocomplexity: adaptive behavior in complex stochastic systems. *BioSystems*, 59:109–123, 2001.
13. R. Kozma. On the constructive role of noise in stabilizing itinerant trajectories on chaotic dynamical systems. *Chaos*, 13:1078–1089, 2003.
14. W. Senn, H. Markram, and M. Tsodyks. An algorithm for modifying neurotransmitter release probability based on pre- and postsynaptic spike timing. *Neural Computation*, 13:35–67, 2001.
15. F.-M. Lu and R.D. Hawkins. Presynaptic and postsynaptic C_a^{2+} and CamKII contribute to long-term potentiation at synapses between individual CA3 neurons. *Proceedings of the National Academy of Sciences*, 103:4264–4269, 2006.
16. X.-L. Zhang, Z.-Y. Zhou, J. Winterer, W. Müller, and P.K. Stanton. NMDA-dependent, but not group I metabotropic glutamate receptor-dependent, long-term depression at Schaffer Collateral¨CCA1 synapses is associated with long-term reduction of release from the rapidly recycling presynaptic vesicle pool. *Journal of Neuroscience*, 26:10270–10280, 2006.
17. H.Z. Shouval, M.F. Bear, and L.N. Cooper. A unified model of NMDA receptor-dependent bidirectional synaptic plasticity. *Proceedings of the National Academy of Sciences*, 99: 10831–10836, 2002.
18. P.J. Sjöström, G.G. Turrigiano, and S.B. Nelson. Rate, timing, and cooperatively jointly determine cortical synaptic plasticity. *Neuron*, 32:1149–1164, 2001.
19. H.X. Wang, R.C. Gerkin, D.W. Nauen, and G.Q. Bi. Coactivation and timing-dependent integration of synaptic potentiation and depression. *Nature Neuroscience*, 8:187–193, 2005.
20. J.-P. Pfister and W. Gerstner. Triplets of spikes in a model of spike timing-dependent plasticity. *Journal of Neuroscience*, 26:9673–9682, 2006.
21. T.S. Otis, M.P. Kavanaugh, and C.E. Jahr. Postsynaptic glutamate transport at the climbing fiber-Purkinje cell synapse. *Science*, 277:1515–1518, 1997.
22. R. Marx, R.E. Meskini, D.C. Johns, and R.E. Mains. Differences in the ways sympathetic neurons and endocrine cells process, store, and secrete exogenous neuropeptides and peptide-processing enzymes. *Journal of Neuroscience*, 19:8300–8311, 1999.
23. V.A. Klyachko and C.F. Stevens. Temperature-dependent shift of balance among the components of short-term plasticity in hippocampal synapses. *Journal of Neuroscience*, 26:6945–6957, 2006.
24. B. Lu, W.M. Yamada, and T.W. Berger. Nonlinear dynamic neural network for text-independent speaker identification using information theoretic learning technology. *IEEE EMBC*, 2006.

25. W.J. Freeman and R. Kozma. Local-global interactions and the role of mesoscopic (intermediate-range) elements in brain dynamics. *Behavioral and Brain Sciences*, 23:401, 2000.
26. W.E. Boyce and R.C. DiPrima. *Elementary differential equations and boundary value problems*. 4th edition, Wiley, New York, 1986.

Appendix A: High-order Derivation

By concatenatively replacing the variables in the (2), (3), (4) and (5) while preserving the variables $\mu_I(t)$ and $N_I(t)$, the synaptic activities can be equivalently written as a third-order differential equation

$$\begin{aligned}
&\frac{1}{\mu_U\mu_R\mu_P}\left[\frac{\mathrm{d}^3N_I(t)}{\mathrm{d}t^3}+\frac{\mathrm{d}^2(\mu_I(t)N_I(t))}{\mathrm{d}t^2}\sum_n\delta(t-t_n)\right]\\
&+\left(\frac{1}{\mu_U\mu_R}+\frac{1}{\mu_R\mu_P}+\frac{1}{\mu_P\mu_U}\right)\left[\frac{\mathrm{d}^2N_I(t)}{\mathrm{d}t^2}+\frac{\mathrm{d}(\mu_I(t)N_I(t))}{\mathrm{d}t}\sum_n\delta(t-t_n)\right]\\
&+\left(\frac{1}{\mu_U}+\frac{1}{\mu_R}+\frac{1}{\mu_P}\right)\left[\frac{\mathrm{d}N_I(t)}{\mathrm{d}t}+\mu_I(t)N_I(t)\sum_n\delta(t-t_n)\right]+N_I(t)\\
&=0,
\end{aligned}$$

where the derived coefficients for multiple-order differential items are functions of pool transfer rates. Hence it is expected that the equation root is absolutely related to addition and/or multiplication of the inverses of transfer rates. The solution that is an exponential function of equation roots is accordingly an exponential function of multiple transfer rates (for details of ordinary differential equations, please see [26]). Therefore, the inverses of transfer rates describe the timing course of a high-order vesicle kinetics.

Index

Printing: Krips bv, Meppel
Binding: Stürtz, Würzburg